A Concise Approach to Mathematical Analysis

Springer-Verlag London Ltd.

Mangatiana A. Robdera

A Concise Approach to Mathematical Analysis

With 47 Figures

 Springer

Mangatiana A. Robdera
Al Akhawayn University, School of Science and Engineering, PO Box 1828,
Avenue Hassan II, Ifrane 53000, Morocco

British Library Cataloguing in Publication Data
Robdera, Mangatiana
   A concise approach to mathematical analysis
   1. Mathematical analysis
   I. Title
   515

Library of Congress Cataloging-in-Publication Data
Robdera, Mangatiana, 1961-
   A concise approach to mathematical analysis / Mangatiana Robdera.
   p. cm.
   Includes index.

   ISBN 978-1-85233-552-6   ISBN 978-0-85729-347-3 (eBook)
   DOI 10.1007/978-0-85729-347-3
   1. Mathematical analysis. I. Title.
QA300 R56 2002
515—dc21   2001049366

http://www.springer.co.uk

© Springer-Verlag London 2003
Originally published by Springer-Verlag London Limited in 2003

Typeset by the author and Thomas Unger

12/3830-543210 Printed on acid-free paper SPIN 10849260

To Fenitra, Toky, and Mirindra

Preface

This book is intended to serve as a first course in elementary mathematical analysis or advanced calculus at the undergraduate level. Its content should be accessible to students with appropriate backgrounds from standard calculus courses but with limited or no previous experience in rigorous proofs. Although primarily written for students who desire to specialize in pure and applied mathematics, this book will prove to be useful for students taking courses in physics, engineering, computer science, and in any other applied science which employs advanced mathematical techniques. This text can also be used as a supplementary reference book for those students in graduate courses who need extra background reinforcement.

It had been noticed that although most calculus students are comfortable with notions such as limit, continuity, derivatives and integrals, a majority of them have difficulties in writing out the proof of some of the simplest calculus theorems. It is my aim in this book to help students make the transition from the problem-solving approach of standard calculus to the more rigorous task of proof-writing and the deeper understanding of mathematical analysis involving more abstract concepts. I hope this book will help students learn to think correctly, and to have a clear understanding of the idea of a mathematical proof.

This book comprises eleven chapters. Each chapter is divided into sections. To help give students a sound footing, the first part of this book (Chapters 1–7) deals with the basic foundation of analysis on the real line. Students are assumed to be more or less familiar with most of the topics covered in the first part of the book. The remaining part (Chapters 8–11) introduces the students to basic, more or less abstract, notions in mathematical analysis. The two parts of the book are of almost equal size. The six chapters of the first part and the five chapters of the second part each form a unit, most of which can be easily covered in a one-semester course (13–14 weeks). Figure 1 shows the logical interdependence of the various chapters of the book.

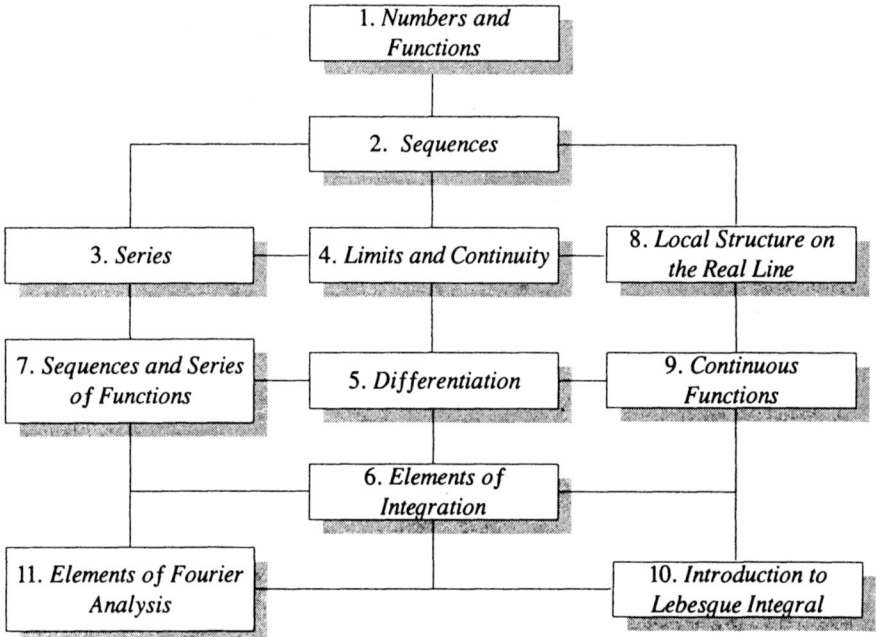

Figure 1   Logical dependence of chapters

The material in this book is reasonably self-contained. Each section provides enough examples to initiate the students into the activity of proof-writing. Students should pay particular attention to those examples. Since the ability to learn and to do mathematics can only be developed through practice and involvement, I have included at the end of each chapter a reasonable number of exercises. They range from routine to hard. They include applications as well as extensions of the main results discussed in each chapter. Some of them contain technical questions demanding only some direct computation. Others are intended to give the students the opportunity to use known proof techniques as well as to learn some new fundamental examples and ideas. A few of them may be assigned as projects. There may be some repetition in the exercises as well as in the arguments used in the proofs of examples and theorems. This is useful for the students, most of whom are trying for the first time to develop their ability to write proofs. The ends of proofs and solutions are marked by □.

I believe that the results in this book are of common knowledge. Lack of bibliographic citation should not be taken as a claim to originality. The following works were consulted during the writing of this book: [1–9]. Each of these references is recommended for further study.

It is my pleasant duty to acknowledge the debt of gratitude I owe to my family for their care, their support, and their patience during the preparation of this book. I would like to express my sincere gratitude to the many people who have helped with the daunting task of proof-reading. I would also like to thank the staff of the Springer-Verlag Publishing Company for their helpful suggestions, constructive comments and their cooperation in the realization of this project.

Contents

1
Numbers and Functions

Numbers and functions are both basic and fundamental concepts in mathematics. In this chapter, we will study the system of real numbers, emphasizing the essential properties from the axioms of a complete ordered field. We also recall the rudiments of the notion of functions without going into the details.

1.1 Real Numbers

When studying calculus we are used to thinking of real numbers as points on an infinite line on which are chosen:

- a certain point 0 called the origin;
- a positive direction: from left to right;
- a suitable unit of length.

However, in order to clarify exactly what we need to know about the real numbers, we set down a set of axioms and prove some of the related properties. We shall not concern ourselves here with the construction of the real numbers. We assume that real numbers exist and we denote by $\mathbb{R}$ the collection of all of them.

Algebraic Structure

Axiom 1.1

The system of real numbers is a set $\mathbb{R}$ equipped with two algebraic operations: addition "+" and multiplication "·" so that for every pair a, b of real numbers the **sum** $a + b$ and the **product** $a \cdot b$ represent real numbers and so that

A1 $a + (b + c) = (a + b) + c$ for all $a, b, c \in \mathbb{R}$;

A2 $a + b = b + a$ for all $a, b \in \mathbb{R}$;

A3 there exists an element 0 such that $a + 0 = a$ for all $a \in \mathbb{R}$;

A4 for all $a \in \mathbb{R}$, there exists an element $-a$ such that $a + (-a) = 0$;

A5 $a(bc) = (ab)\,c$ for all $a, b, c \in \mathbb{R}$;

A6 $ab = ba$ for all $a, b \in \mathbb{R}$;

A7 there exists a unique element 1 such that $1 \cdot a = a$ for all $a \in \mathbb{R}$;

A8 for all $a \in \mathbb{R} \setminus \{0\}$, there exists a unique element a^{-1} (also denoted by $1/a$) such that $a^{-1}a = 1$;

A9 $a\,(b + c) = ab + ac$ for all $a, b, c \in \mathbb{R}$.

The number $-a$ of **A4** is called the **opposite** of the number a, and the number a^{-1} of **A8** is called the **inverse** of the number a. Technically, we say that $\mathbb{R}$, with the two algebraic operations, is a **field**. These axioms are used to prove basic and familiar algebraic properties of $\mathbb{R}$.

Example 1.2

For $a, b, c \in \mathbb{R}$. Show that

    (1) $a + c = b + c$ implies $a = b$;

    (2) $a \cdot 0 = 0$;

    (3) $(-a)\,b = -ab$;

    (4) $a \neq 0$ and $ab = ac$ implies $b = c$;

    (5) if $ab = 0$, then $a = 0$ or $b = 0$.

Solution

Fix a, b, c in $\mathbb{R}$.

    (1) If $a + c = b + c$, then

$$(a + c) + (-c) = (b + c) + (-c),$$

and by **A1** we have

$$a + (c + (-c)) = b + (c + (-c)).$$

By **A2** we have $a + 0 = b + 0$ and so $a = b$ by **A3**.

(2) Using **A7** and **A9**, we have

$$a + a \cdot 0 = a \cdot 1 + a \cdot 0 = a(1 + 0) = a \cdot 1 = a.$$

By the previous result, we have $a \cdot 0 = 0$.

(3) Since $0 = 0 \cdot b = (a + (-a)) b = ab + (-a) b$, we have $-ab = (-a) b$. The proofs of (4) and (5) are left as exercises. $\qquad\square$

For reasons of simplicity of notation, an addition of the form "$a + (-b)$" is often written as "$a - b$".

Order Structure

Another aspect which makes $\mathbb{R}$ so interesting is its order structure. We assume the following axiom.

Axiom 1.3

$\mathbb{R}$ has a unique subset, $\mathbb{R}_+$, with the properties:

(1) $a + b \in \mathbb{R}_+$, and $ab \in \mathbb{R}_+$ whenever $a, b \in \mathbb{R}_+$;

(2) for every $a \in \mathbb{R}$, $a \neq 0$, exactly one of the following is true: $a \in \mathbb{R}_+$ or $-a \in \mathbb{R}_+$.

Members of $\mathbb{R}_+$ are called **positive** real numbers. The relation $a \in \mathbb{R}_+$ is also denoted by $a > 0$.

Definition 1.4

Let $a, b \in \mathbb{R}$. We say that a is **less than** b, denoted by $a < b$, if $b - a \in \mathbb{R}_+$. If $a < b$ or $a = b$ we say a is **less than or equal to** b, and we write this as $a \leq b$.

Note

$b > a$ (resp. $b \geq a$) means exactly the same as $a < b$ (resp. $a \leq b$).

Axiom 1.3 implies that given $a, b \in \mathbb{R}$, then either $a < b$, $a = b$, or $a > b$. We introduce the following useful notation.

$$\max\{a, b\} = a \vee b = \left\{ \begin{array}{lll} a & \text{if} & b \leq a, \\ b & \text{if} & a < b. \end{array} \right.$$

$$\min\{a, b\} = a \wedge b = \left\{ \begin{array}{lll} a & \text{if} & a \leq b, \\ b & \text{if} & b < a. \end{array} \right.$$

It is clear that $a \wedge b \leq a, b \leq a \vee b$. It turns out that the field property and the order structure of $\mathbb{R}$ closely interact with one another.

Theorem 1.5

Let a, b be in $\mathbb{R}$. Then

(1) $a < b$ implies $a + c < b + c$ for all $c \in \mathbb{R}$;

(2) $a < b$ implies $ac < bc$ for all $c > 0$.

Proof

(1) Let $a, b, c \in \mathbb{R}$ and suppose that $a < b$. First **A2** and **A4** imply

$$-a = -a + 0 = -a + (c + (-c)).$$

Then by **A2** and **A3**, $-a = c - (a + c)$. It follows that

$$b - a = b + c - (a + c) > 0,$$

and hence we have $a + c < b + c$ as desired.

(2) Let $a, b, c \in \mathbb{R}$ and suppose that $a < b$ and $c > 0$. Then $b - a$ belongs to $\mathbb{R}_+$ and so does $(b - a) c$, i.e.

$$(b - a) c = bc - ac > 0.$$

Therefore $bc > ac$. The proof is complete.   □

Other connections between the algebraic and the order properties of $\mathbb{R}$ can be derived from Theorem 1.5.

Example 1.6

Let $a, b, c \in \mathbb{R}$. Then

(1) if $a < b$, then $-b < -a$;

(2) if $a < b$ and $c < 0$, then $bc < ac$;

(3) $a^2 \geq 0$ for all a;

(4) if $a > 0$, then $a^{-1} > 0$;

(5) if $a > 0$, $b > 0$ and $a > b$, then $a^{-1} < b^{-1}$.

We leave the proof as an exercise (Exercise 1.23).

Definition 1.7

Let A be a nonempty subset of $\mathbb{R}$.

(1) An element $a_0 \in A$ is called the **greatest element** or the **maximum of** A if for every $a \in A$, $a \leq a_0$. a_0 is denoted $\max A$.

(2) An element $a_0 \in A$ is called the **least element** or the **minimum of** A if for every $a \in A$, $a_0 \leq a$. a_0 is denoted $\min A$.

For example, it is easily seen that $\max\{1, 2, 3, 4\} = 4$ and $\min\{1, 2, 3, 4\} = 1$. Also, the set $B = \{x \in \mathbb{R} : 1 < x \leq 7\}$ has a maximum equal to 7. However, a set may have no maximum and/or minimum.

Example 1.8

Let $B = \{x \in \mathbb{R} : 1 < x \leq 7\}$. Show that $\min B$ does not exist.

Solution

Assume that $m = \min\{x \in \mathbb{R} : 1 < x \leq 7\}$ exists. Then, since $m \in B$, $1 < m \leq 7$. Consider the number $(m + 1)/2$. It is clear that $1 < (m + 1)/2 \leq 7$, i.e. $(m + 1)/2$ belongs to B. On the other hand, we also have $(m + 1)/2 < m$. This contradicts the minimality of m, and therefore implies that our assumption is false. $\qquad\qquad\square$

Examples of sets which admit no maximum can also be easily given.

Example 1.9

Show that $\max\{n^2 : n = 1, 2, \ldots\}$ does not exist.

Solution

Again assume that $m = \max\{n^2 : n = 1, 2, \ldots\}$ exists. Then, there exists $n \in \mathbb{N}$ such that $m = n^2$. Now the number $(n + 1)^2$ is an element of $\{n^2 : n = 1, 2, \ldots\}$ and it is clear that

$$(n + 1)^2 = n^2 + 2n + 1 > n^2 = m.$$

This contradicts the maximality of m, and therefore implies that our assumption is false.   □

It is useful to introduce the notion of interval of numbers:

- An **interval** is the set of all real numbers x lying between two given points a and b (called the **endpoints**). An interval is said to be **closed** or **open** according to whether it does or does not include the endpoints:

  $\{x \in \mathbb{R} : a < x < b\}$ is an open interval and is denoted by (a, b);
  $\{x \in \mathbb{R} : a \leq x \leq b\}$ is a closed interval and is denoted by $[a, b]$.

- If one of the endpoints a or b belongs to the interval, while the other does not, then the interval is called **half-open** (or **half-closed**). The next two intervals are both half-open:

  $\{x \in \mathbb{R} : a < x \leq b\}$ denoted by $(a, b]$;
  $\{x \in \mathbb{R} : a \leq x < b\}$ denoted by $[a, b)$.

  It should be noticed that if $a > b$, then $(a, b) = (a, b] = [a, b) = [a, b] = \varnothing$. Here $\varnothing$ denotes the empty set.

- Finally, **infinite intervals** are sets that contain all possible values greater than (or all possible values less than) a given number.

  $\{x \in \mathbb{R} : x < b\}$ is an infinite open interval and is denoted by $(-\infty, b)$;
  $\{x \in \mathbb{R} : x \leq b\}$ is an infinite closed interval and is denoted by $(-\infty, b]$;
  $\{x \in \mathbb{R} : a \leq x\}$ is an infinite closed interval and is denoted by $[a, +\infty)$;
  $\{x \in \mathbb{R} : a < x\}$ is an infinite open interval and is denoted by $(a, +\infty)$.

  Here the symbol "∞" is read "infinity". The set $\mathbb{R}$ itself can be considered as the infinite interval containing all real numbers and is often symbolically denoted by $(-\infty, +\infty)$.

Another useful concept which connects the algebraic structure and the order structure of $\mathbb{R}$ is that of absolute value.

Definition 1.10

Let a be a real number. The absolute value of a, denoted by $|a|$ is defined by

$$|a| = \begin{cases} a & \text{if } a \geq 0; \\ -a & \text{if } a < 0. \end{cases}$$

For example, $|3| = 3, |-2| = 2, |0| = 0$.

It follows immediately from the definition that $a \leq |a|$ for all a. Let us examine some other properties of the absolute value.

Theorem 1.11

Let a and b be arbitrary real numbers. Then

(1) $0 \leq |a|$;

(2) $|ab| = |a|\,|b|$;

(3) $|a + b| \leq |a| + |b|$. **(Triangle inequality.)**

Proof

(1) is obvious from the definition.

(2) If $a \geq 0$ and $b \geq 0$, then $ab \geq 0$ and so $|a|\,|b| = ab = |ab|$. If $a \leq 0$ and $b \leq 0$, then $ab \geq 0$ and so $|a|\,|b| = (-a)(-b) = ab = |ab|$. If $a \leq 0$ and $b \geq 0$, then $ab \leq 0$ and so $|a|\,|b| = (-a)b = -ab = |ab|$. If $a \geq 0$ and $b \leq 0$, then $ab \leq 0$ and so $|a|\,|b| = a(-b) = -ab = |ab|$.

(3) If $a + b \geq 0$, then $|a + b| = a + b \leq |a| + |b|$. If $a + b \leq 0$, then $|a + b| = -(a + b) = (-a) + (-b) \leq |a| + |b|$.

This completes the proof. $\qquad\square$

The triangle inequality is very useful in many proofs in analysis. It can be used in many different but equivalent forms. For example, for the case where $a = s - t$ and $b = t - r$, the triangle inequality takes on the form

$$|s - r| \leq |s - t| + |t - r|.$$

The triangle inequality asserts that "the absolute value of the sum is no greater than the sum of the absolute values". It also implies that "the difference of absolute values is no greater than the absolute value of the difference".

Example 1.12

Show that $||a| - |b|| \leq |a - b|$, for all $a, b \in \mathbb{R}$.

Solution

Let $a, b \in \mathbb{R}$. Then $a = (a - b) + b$ and hence $|a| \leq |a - b| + |b|$, i.e. $|a| - |b| \leq |a - b|$. Similarly, $|b| - |a| \leq |b - a|$. Therefore, $||a| - |b|| \leq |a - b|$, as desired.   □

Example 1.13

Show that for every $M > 0$, $|a| < M$ if and only if $-M < a < M$.

Solution

Suppose that $|a| < M$. Then $-M < -|a| \leq a \leq |a| < M$. Conversely suppose $-M < a < M$. If $a \geq 0$, then $|a| = a < M$. If $a < 0$, then $|a| = -a < M$.   □

Example 1.14

Show that if $a \leq b + \varepsilon$ for every $\varepsilon > 0$, then $a \leq b$.

Solution

Suppose that we have $a \leq b + \varepsilon$ for every $\varepsilon > 0$. Then we shall prove that we cannot have $a > b$. If we had $a > b$, then letting $\varepsilon = (a - b)/2$ which is a positive number, we would have

$$a \leq b + \varepsilon = b + (a - b)/2$$
$$= (a + b)/2 < (a + a)/2 = a.$$

Since we cannot have $a < a$, we have reached a contradiction and so we must have $a \leq b$.   □

Suprema and Infinima

One of the fundamental concepts of the order structure of $\mathbb{R}$ is the notion of boundedness. First we lay down some definitions.

Definition 1.15

Let A be a nonempty subset of $\mathbb{R}$.

(1) A number $M \in \mathbb{R}$ is an **upper bound** for A if $a \leq M$ for every $a \in A$. The set A is said to be **bounded above** if it has an upper bound.

(2) A number $m \in \mathbb{R}$ is a **lower bound** for A if $a \geq m$ for every $a \in A$. The set A is said to be **bounded below** if it has a lower bound.

(3) The set A is said to be **bounded** if it is bounded above and bounded below.

For example, the numbers 5, 12, and 1000 are all upper bounds for $A = \{1, 2, 3, 4, 5\}$, while $0, -1, -10, -1000$ are examples of lower bounds for A. Thus A is bounded. The set $\{1, 2, 3, \ldots\}$ is bounded below (the real number 1 is a lower bound) and A is not bounded above.

It follows immediately from the definitions that the maximum (resp. minimum) of a set A, if it exists, is an upper bound (resp. a lower bound.)

On the basis of Example 1.13, we also notice:

Remark 1.16

A set A is bounded if and only if there exists $M > 0$ such that $|a| \leq M$ for all $a \in A$.

Definition 1.17

Let A be a nonempty subset of $\mathbb{R}$.

(1) Suppose that $M \in \mathbb{R}$ is an upper bound for A such that $M \leq M'$ for any upper bound M' of A. Then M is called a **supremum** or **least upper bound** of A and is denoted by $M = \sup A$.

(2) Suppose that $m \in \mathbb{R}$ is a lower bound for A such that $m \geq m'$ for any lower bound m' of A. Then m is called an **infimum** or **greatest lower bound** of A and is denoted by $m = \inf A$.

Notation

If a set A is not bounded above (resp. below) we write $\sup A = \infty$ (read infinity) (resp. $\inf A = -\infty$).

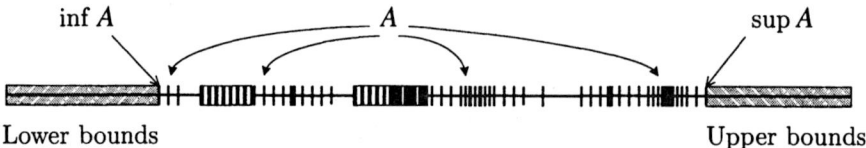

Figure 1.1   Extrema

The next example gives a characterization that could help in better understanding the definition of a supremum.

Example 1.18

Suppose that a set A is bounded above. Show that $M = \sup A$ if and only if it satisfies the following properties:

(1) There are no element a of A with $M < a$.

(2) If $a < M$, then there is an element $x \in A$ such that $a < x$.

Solution

Suppose that $M = \sup A$. Since M is an upper bound of A, property (1) is satisfied. If $a < M$, then a is not an upper bound of A. Thus there exists $x \in A$ such that $a < x$. This proves property (2).

Conversely, property (1) implies that M is an upper bound. If $a < M$, then property (2) shows that a cannot be an upper bound of A. Therefore $M = \sup A$.   □

The readers should convince themselves about the validity of the following examples, which show in particular that the supremum and the infimum of a set A **may** or **may not** belong to the set A.

A	$\sup A$	$\inf A$	$\max A$	$\min A$
$\mathbb{N}$	$+\infty$	1	DNE	1
$(a, b]$	b	a	b	DNE
$\{1, \frac{1}{2}, \frac{1}{3}, \ldots\}$	1	0	1	DNE
$\mathbb{R}$	$+\infty$	$-\infty$	DNE	DNE

Here "DNE" stands for "does not exist".

Example 1.19

Let A and B be nonempty bounded subsets of $\mathbb{R}$. Show that if $A \subset B$, then $\inf B \leq \inf A \leq \sup A \leq \sup B$.

Solution

It is clear that $\inf A \leq \sup A$. Let $a \in A$. Since $A \subset B$, $a \in B$. Therefore $a \leq \sup B$ holds for every $a \in A$. Hence $\sup A \leq \sup B$.

Similarly, let $a \in A$. Since $A \subset B$, a is in B and therefore $\inf B \leq a$. This being true for every element a in A, we conclude that $\inf B \leq \inf A$. The proof is complete. $\qquad\square$

Example 1.20

Show that if A and B are subsets of $\mathbb{R}$ which are both bounded above, then the set $A + B = \{a + b : a \in A, \ b \in B\}$ is also bounded above and

$$\sup (A + B) = \sup A + \sup B.$$

Solution

First we prove that the set $A + B$ is bounded above. Since $a \leq \sup A$ and $b \leq \sup A$ for every $a \in A$ and every $b \in B$, we have

$$a + b \leq \sup A + \sup B,$$

for every $a + b \in A + B$. It then follows that the set $A + B$ is bounded above and $\sup A + \sup B$ is an upper bound for $A + B$. Hence

$$\sup (A + B) \leq \sup A + \sup B.$$

To prove the reverse inequality, let $\varepsilon > 0$. Then $\sup A - \varepsilon/2$ is not an upper bound for the set A. Therefore there exists an element a in A such that $\sup A - \varepsilon/2 < a$. Similarly, there exists an element b in B such that $\sup B - \varepsilon/2 < b$. It follows that

$$\left(\sup A - \frac{\varepsilon}{2}\right) + \left(\sup B - \frac{\varepsilon}{2}\right) < a + b \leq \sup (A + B),$$

or equivalently

$$\sup A + \sup B \leq \sup (A + B) + \varepsilon.$$

Since $\varepsilon > 0$ is arbitrary, we have (see Example 1.14)

$$\sup A + \sup B \leq \sup (A + B)$$

and so our proof is complete. $\qquad\square$

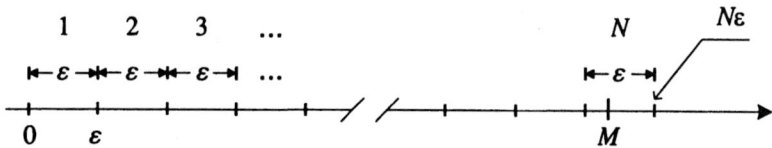

Figure 1.2   Archimedean property

Example 1.21

Let A and B be subsets of $\mathbb{R}$. Suppose that

(1) the set B is bounded above;

(2) for every $a \in A$, there exists $b \in B$ such that $a \le b$.

Show that A is bounded and $\sup A \le \sup B$.

Solution

Property (1) ensures that $\sup B$ exists as a real number. Property (2) implies that every $a \in A$ is less than $\sup B$. Hence, it is immediately seen that A is bounded and $\sup A \le \sup B$. $\square$

This last example sets up the following remarkable fact: *no real number is infinitely small or infinitely large.* Such a statement is made more precise in the next theorem and is called the **Archimedean property** of $\mathbb{R}$.

Theorem 1.22

Let $\varepsilon > 0$. For every $M \in \mathbb{R}$, there exists $N \in \mathbb{N}$ such that $M < N\varepsilon$.

Proof

Since $\mathbb{N}$ is not bounded above, the real number M/ε is not an upper bound for $\mathbb{N}$. Therefore, there must exist an $N \in \mathbb{N}$ such that $M/\varepsilon < N$. Since $\varepsilon > 0$, we have $M < N\varepsilon$ as desired. $\square$

If M is a supremum of a set A, then $M - \varepsilon$ for any positive number ε is no longer an upper bound for A. The following theorem is a useful characterization of the supremum $\sup A$ of a set A.

Theorem 1.23

Let A be a subset of $\mathbb{R}$. Then $M = \sup A$ if and only if

$$\text{for every } \varepsilon > 0, \text{ there exists } a \in A \text{ such that } M - \varepsilon < a \leq M. \qquad (1.1)$$

Proof

First suppose that $M = \sup A$. It is clear that $a \leq M$ for each $a \in A$. Let $\varepsilon > 0$. Suppose that one cannot find $a \in A$ such that $M - \varepsilon < a$. This means that for every $a \in A$, $M - \varepsilon \geq a$. Hence $M - \varepsilon$ is an upper bound for A. This contradicts our assumption that $M = \sup A$ is the least upper bound. Thus (1.1) holds.

Conversely, suppose that (1.1) holds but $M \neq \sup A$. Then there exists an upper bound M' of A such that $M' < M$. Let $\varepsilon = M - M' > 0$. Then there exists $a \in A$ such that

$$M' = M - \varepsilon = M - (M - M') < a \leq M.$$

This contradicts the fact that M' is an upper bound. Hence we must have $M = \sup A$. The proof is complete. $\qquad\square$

In a similar fashion, our next theorem gives a characterization of the infimum of a set.

Theorem 1.24

Let A be a subset of $\mathbb{R}$. Then $m = \inf A$ if and only if

$$\text{for every } \varepsilon > 0, \text{ there exists } a \in A \text{ such that } m \leq a < m + \varepsilon.$$

We leave the proof Theorem 1.24 as an exercise.

Dedekind Cuts

Definition 1.25

An ordered pair (A, B) of nonempty subsets of $\mathbb{R}$ is said to be a **cut**[1] if $A \cap B = \varnothing$, $A \cup B = \mathbb{R}$ and $a < b$ for all $a \in A$ and $b \in B$.

---

[1] R. Dedekind (1831–1916) used cuts in the rational number system in order to give a beautiful construction of the real number system.

If $b \in \mathbb{R}$, then it is clear that the pair (A, B) with $A = (-\infty, b)$ and $B = [b, +\infty)$ is an example of a cut. So every real number defines a cut. Conversely, we have

Theorem 1.26

For every cut (A, B), there corresponds a unique real number x such that $a \leq x$ for all $a \in A$ and $x \leq b$ for all $b \in B$.

Proof

Any element of B is an upper bound of A. Thus $x = \sup A$ exists as a real number. It is clear that $a \leq x \leq b$ for all $a \in A$ and all $b \in B$.

For the proof of the uniqueness, suppose that y satisfies $a \leq y$ for all $a \in A$ and $y \leq b$ for all $b \in B$. Since y is an upper bound for A, necessarily $x \leq y$. Suppose that $x < y$. Let $z = (x + y)/2$. Then $x < z < y$. Hence $z \notin A$ and $z \notin B$. This contradicts the fact that $A \cup B = \mathbb{R}$. $\square$

Extended Real Number System

It is frequently convenient to adjoin the two symbols $-\infty$ and $+\infty$ to the system of real numbers (∞ is read "infinity"). In fact, we have already used these symbols before (see pages 6 and 9). We stress the fact that $-\infty$ and $+\infty$ are not real numbers. We agree that $-\infty < x < +\infty$ for all real numbers x. The extended real number system is defined to be the collection $\mathbb{R}^* = \mathbb{R} \cup \{-\infty, +\infty\}$ sometimes simply denoted by $[-\infty, \infty]$.

The algebraic operations between the two symbols $-\infty$ and $+\infty$ and elements $x \in \mathbb{R}$ are defined as follows:

$$(\pm\infty) + (\pm\infty) = x + (\pm\infty) = (\pm\infty) + x = (\pm\infty);$$
$$(\pm\infty)(\pm\infty) = +\infty;$$
$$(\pm\infty)(\mp\infty) = -\infty;$$
$$(\pm\infty)x = x(\pm\infty) = (\pm\infty) \quad \text{if } x > 0;$$
$$(\pm\infty)x = x(\pm\infty) = 0 \quad \text{if } x = 0;$$
$$(\pm\infty)x = x(\pm\infty) = (\mp\infty) \quad \text{if } x < 0.$$

Notice that expressions such as $\infty - \infty$ and $-\infty + \infty$ are not defined.

1.2 Subsets of $\mathbb{R}$

Natural Numbers

Definition 1.27

A subset S of $\mathbb{R}$ is said to be **inductive** if it satisfies the following two conditions

(1) $1 \in S$;

(2) $s \in S$ implies $s + 1 \in S$.

$\mathbb{R}$ and $\mathbb{R}_+$ are examples of inductive sets. It is also very natural to see that the set $\{1, 1 + 1, (1 + 1) + 1, \ldots\}$ is an inductive set.

Definition 1.28

The inductive subset $\{1, 2, 3, \ldots\}$ of $\mathbb{R}$ is called the set of **natural numbers** and is denoted by $\mathbb{N}$.

It follows from this definition that

- $\mathbb{N} \subset \mathbb{R}_+$: every natural number is a positive real number;

- 1 is the smallest element of $\mathbb{N}$: if $n \in \mathbb{N}$, then $n \geq 1$.

It turns out that $\mathbb{N}$ is the smallest inductive subset of $\mathbb{R}$ in the following sense: if S is a subset of $\mathbb{R}$ with the property that

(1) $1 \in S$;

(2) $s \in S$ implies $s + 1 \in S$,

then $\mathbb{N} \subset S$.

This fact sets up a very important and useful property of $\mathbb{N}$ known as the **principle of mathematical induction**:

Axiom 1.29

Let A be a subset of $\mathbb{N}$ with the properties that

(1) $1 \in A$;

(2) $k \in A$ implies $k + 1 \in A$.

Then $A = \mathbb{N}$.

This axiom is essential in proving properties depending on $n \in \mathbb{N}$. Next we give a few examples of the use of the principle of mathematical induction.

Example 1.30

Show that $1 + 2 + \cdots + n = \frac{1}{2}n(n+1)$ for all $n \in \mathbb{N}$.

Solution

Let $A = \left\{ n \in \mathbb{N} : 1 + 2 + \cdots + n = \frac{1}{2}n(n+1) \right\}$. We wish to show that $A = \mathbb{N}$.

If $n = 1$, we have $1 = \frac{1}{2}(1)(1+1)$. Thus $1 \in A$. Next suppose that $k \in A$, that is $1 + 2 + \cdots + k = \frac{1}{2}k(k+1)$. Then

$$1 + 2 + \cdots + k + (k+1) = \frac{1}{2}k(k+1) + (k+1)$$
$$= (k+1)\left(\frac{1}{2}k + 1\right)$$
$$= \frac{1}{2}(k+1)(k+2).$$

So $k + 1 \in A$. By the principle of mathematical induction $A = \mathbb{N}$ and the proof is concluded. $\qquad \square$

Example 1.31

Show that $2^n > n^2$ for all $n \geq 5$.

Solution

First we notice that the statement "$2^n > n^2$ for all $n \geq 5$" is equivalent to "$2^{m+4} > (m+4)^2$ for all $m \geq 1$". So we are done if we prove that $2^{m+4} > (m+4)^2$ for all $m \geq 1$.

Let $A = \left\{ m \in \mathbb{N} : 2^{m+4} > (m+4)^2 \right\}$. We wish to show that $A = \mathbb{N}$.

Clearly $2^5 = 32 > 5^2 = 25$. Thus $1 \in A$. Suppose that $k \in A$. Then

$$((k+1)+4)^2 = ((k+4)+1)^2$$
$$= (k+4)^2 + 2(k+4) + 1$$
$$< (k+4)^2 + 2(k+4) + (k+4)$$
$$= (k+4)^2 + 3(k+4)$$
$$< (k+4)^2 + (k+4)(k+4)$$
$$= 2(k+4)^2 < 2 \cdot 2^{k+4} = 2^{(k+1)+4}.$$

Thus $k + 1 \in A$ and so $A = \mathbb{N}$. $\square$

Example 1.32

Show that $\frac{d}{dx} x^n = n x^{n-1}$ for all $n \in \mathbb{N}$.

Solution

Let $A = \{n \in \mathbb{N} : \frac{d}{dx} x^n = n x^{n-1}\}$. Again we wish to show that $A = \mathbb{N}$.

Since $\frac{d}{dx} x^1 = 1 = 1 x^{1-1}$, $1 \in A$. Suppose that $k \in A$, i.e. $\frac{d}{dx} x^k = k x^{k-1}$. Then using the product rule for derivatives, we have

$$\frac{d}{dx} x^{k+1} = \frac{d}{dx} \left(x x^k \right) = x^k + x k x^{k-1} = (k + 1) x^k.$$

Thus $k + 1 \in A$ and so $A = \mathbb{N}$. $\square$

We can also appeal to the principle of mathematical induction to prove the following properties of $\mathbb{N}$.

Proposition 1.33

If $m, n \in \mathbb{N}$, then $m + n \in \mathbb{N}$, and $m \cdot n \in \mathbb{N}$.

Proof

First, consider the set $A = \{m \in \mathbb{N} : m + n \in \mathbb{N} \text{ for all } n \in \mathbb{N}\}$. We wish to prove that $A = \mathbb{N}$.

It is obvious that $1 \in A$. If $m \in A$, then for all $n \in \mathbb{N}$, we have $m + (n + 1) \in \mathbb{N}$ and so

$$(m + 1) + n = m + (n + 1) \in \mathbb{N} \text{ for all } n \in \mathbb{N}.$$

Thus $m + 1 \in A$, and hence by the principle of mathematical induction $A = \mathbb{N}$, as desired.

Similarly, let $B = \{m \in \mathbb{N} : mn \in \mathbb{N} \text{ for all } n \in \mathbb{N}\}$. Again it is obvious that $1 \in B$. If $m \in B$, then for all $n \in \mathbb{N}$, since $mn \in \mathbb{N}$, we have (by the first part of the proof) $mn + n \in \mathbb{N}$. Thus

$$(m + 1) n = mn + n \in \mathbb{N} \text{ for all } n \in \mathbb{N},$$

i.e. $m + 1 \in B$. Hence $B = \mathbb{N}$. This concludes the proof. $\square$

We finish this section with another important feature of $\mathbb{N}$ which directly follows from the principle of mathematical induction: the so-called **well-ordering principle.**

Theorem 1.34

Every nonempty subset of $\mathbb{N}$ has a least element.

Proof

Let A be a nonempty subset of $\mathbb{N}$. Suppose to the contrary that A has no least element. Let

$$B = \{n \in \mathbb{N} : k \notin A \text{ for all } k \leq n\}.$$

It is clear that if n is in B, then $1, 2, \cdots, n \in B$. We will show that $B = \mathbb{N}$.

First, we claim that $1 \in B$. Indeed, if $1 \notin B$, then $1 \in A$. Since 1 is the least natural number, 1 would be a least element in A, contradicting our assumption and proving our claim.

Now suppose that $k \in B$. Then $1, 2, \cdots, k \notin A$. It follows that if $k + 1$ were in A, then it would be a least element for A. Thus $k + 1 \notin A$, and so $k + 1 \in B$.

Hence by the principle of mathematical induction $B = \mathbb{N}$ and thus A is empty, a contradiction. Therefore A must have a minimum. $\qquad\square$

Example 1.35

Let $\mathbb{Z}$ denote the subset of real numbers $\{\cdots, -3, -2, -1, 0, 1, 2, 3, \cdots\}$. The elements of $\mathbb{Z}$ are called the **integers.** Given any $a \in \mathbb{R}$, show that there exists a unique $n \in \{\ldots, -2, -1, 0, 1, 2, \ldots\}$ such that $n \leq a < n + 1$. n is called the **integer part** of a and is denoted int a.

Solution

Suppose first that $a > 0$. Then applying Theorem 1.22 for $\varepsilon = 1$, we see that the set $\{k \in \mathbb{N} : k > a\}$ is not empty. It is clearly bounded above. Thus by the well-ordering principle $\inf \{k \in \mathbb{N} : k > a\} = n + 1$ exists as a natural number (Theorem 1.34). It is then clear that int $a = n$. Similarly, if $a < 0$, the set $\{k \in \mathbb{N} : k > -a\}$ is not empty. Thus $\inf \{k \in \mathbb{N} : k > -a\} = m$ exists as a natural number, and int $a = -m$. $\qquad\square$

The following facts are immediate and the proofs are left as exercises.

Remark 1.36

For any $a, b \in \mathbb{R}$,

(1) $\operatorname{int} a = \sup\{n \in \{\ldots, -2, -1, 0, 1, 2, \ldots\} : n \leq a\}$;

(2) $\operatorname{int}(\operatorname{int} a) = \operatorname{int} a$;

(3) $\operatorname{int} a \leq \operatorname{int} b$, whenever $a \leq b$.

Example 1.37

Show that for any real number a, $(\operatorname{int} a + 1) = \operatorname{int}(a + 1)$.

Solution

Fix a in $\mathbb{R}$. Since $(\operatorname{int} a) + 1$ is an integer and $(\operatorname{int} a) + 1 \leq a + 1$, we have $(\operatorname{int} a) + 1 \leq \operatorname{int}(a + 1)$. Conversely, if $n \in \mathbb{Z}$ with $n \leq a + 1$, then $n - 1 \leq a$ and thus $n - 1 \leq \operatorname{int} a$ or equivalently $n \leq (\operatorname{int} a) + 1$. Thus $\operatorname{int}(a + 1) \leq \operatorname{int} a + 1$. We have finished the proof. $\qquad \square$

Since $\operatorname{int} a \leq a < \operatorname{int} a + 1$, it is clear that $0 \leq a - \operatorname{int} a < 1$. The real number $\operatorname{fra} a = a - \operatorname{int} a$ is called the **fractional part** of a.

Rational Numbers

An arbitrary set which has the same algebraic properties (Axiom 1.1) as $\mathbb{R}$ is called a field. Neither 0 nor the opposite of any real number in $\mathbb{R}$ is in $\mathbb{N}$. We denote by $\mathbb{N}_0 = \mathbb{N} \cup \{0\}$ and by $\mathbb{N}_-$ the set of all the opposites of the natural numbers, i.e. $\mathbb{N}_- = \{-n \in \mathbb{R} : n \in \mathbb{N}\}$. Then $\mathbb{N}_0 \cup \mathbb{N}_- = \mathbb{Z}$. It is easily checked that $\mathbb{Z}$ satisfies all of Axiom 1.1 with the exception of **A8**. If $a \in \mathbb{Z}$, $a \neq 0$, then the real number $a^{-1} = \frac{1}{a}$ does not belong to $\mathbb{Z}$. We need a larger system to include the inverses of all integers. Consider then the set

$$\mathbb{Q} = \{\frac{p}{q} \in \mathbb{R} : p \in \mathbb{Z}, q \in \mathbb{N}\}.$$

$\mathbb{Q}$ is called the set of **rational numbers.** In the representation p/q of a rational number, we always assume the lowest simplified form, that is to say, p and q have no common divisor. For example, the rational number 3/6 will be represented by 1/2.

It is a good exercise to show that $\mathbb{Q}$ satisfies all of Axiom 1.1, i.e. to show that $\mathbb{Q}$ is a field. It is plain from the definitions that $\mathbb{N} \subset \mathbb{Z} \subset \mathbb{Q} \subset \mathbb{R}$. All of these subsets inherit the order structure of $\mathbb{R}$.

Completeness axiom   It looks as if we have found in $\mathbb{Q}$ a "satisfactory" system of numbers which suits all the required properties of a field and respects the order structure. We say that $\mathbb{Q}$ is an **ordered field**. However, we will see that $\mathbb{Q}$ is not perfect in a certain sense. First we will prove that one cannot find any rational number a such that $a^2 = 2$. We say that the equation $a^2 = 2$ has no solution in $\mathbb{Q}$.

Claim 1.38

There exists no rational number a satisfying $a^2 = 2$.

Proof

Suppose to the contrary that there exists $a = \frac{m}{n} \in \mathbb{Q}$ where m and n have no common factors, such that $a^2 = 2$. Then we have $m^2 = 2n^2$. Thus m^2 is an even integer and so is m (why?). Thus $m = 2k$ for some integer k. It follows that $4k^2 = 2n^2$ or $2k^2 = n^2$. Therefore n is also an even integer. This contradicts our assumption that m and n have no common factors. Hence we have proved that $a^2 = 2$ has no solution in $\mathbb{Q}$. $\square$

Consider the following two sets

$$A = \left\{ r \in \mathbb{Q} : r > 0 \text{ and } r^2 < 2 \right\}, \quad B = \left\{ q \in \mathbb{Q} : q > 0 \text{ and } q^2 \geq 2 \right\}.$$

It is clear that $A \cap B = \varnothing$.

Claim 1.39

Every element of B is an upper bound for A.

Proof

Let $r \in A$ and $q \in B$. Then $0 < r^2 < 2 \leq q^2$. It follows that

$$q^2 - r^2 = (q - r)(q + r) > 0,$$

and since $q + r > 0$, we infer that $q > r$. This proves our claim. $\square$

Claim 1.40

Every rational upper bound for A belongs to B.

Proof

Let $m \in \mathbb{Q}$ be an upper bound for A. Suppose that $m^2 < 2$. Then $2 - m^2 > 0$. By the Archimedean property, we can choose $n \in \mathbb{N}$ so that $n > (2m + 1) / (2 - m^2)$, i.e. so that $m^2 + 2m/n + 1/n < 2$. Then for such n, we have

$$\left(m + \frac{1}{n}\right)^2 = m^2 + \frac{2m}{n} + \frac{1}{n^2} \leq m^2 + \frac{2m}{n} + \frac{1}{n} < 2.$$

Thus $m + 1/n \in A$. This contradicts our assumption that m is an upper bound for A. Thus $m^2 \geq 2$, i.e. $m \in B$. $\qquad\square$

Thus if $M = \sup A$ exists as a rational number, then $M^2 \geq 2$ must hold.

Claim 1.41

If $M = \sup A$ exists as a rational number, then $M^2 \leq 2$.

Proof

Suppose that $M = \sup A \notin A$, i.e. $M^2 > 2$. Then $M^2 - 2 > 0$. Choose n large enough, so that $\frac{1}{n} < \frac{M^2 - 2}{2M}$, i.e. so that $M^2 - \frac{2M}{n} > 2$. Then

$$\left(M - \frac{1}{n}\right)^2 = M^2 - \frac{2M}{n} + \frac{1}{n^2} > M^2 - \frac{2M}{n} > 2.$$

Thus $M - 1/n \in B$ and therefore it is a rational upper bound of A. Thus M could not be the supremum of A. This contradiction proves our claim. $\qquad\square$

Thus if $M = \sup A$ exists as a rational number, then we must have $M^2 = 2$. According to Claim 1.38, such a number could not exist. The set A has upper bounds but no supremum in $\mathbb{Q}$. If we only consider the system of rational numbers, then we come to the conclusion that there is a gap between the set A and its upper bounds B. We say that the ordered field $\mathbb{Q}$ is not **complete**. This observation leads to a fundamental property, known as the **completeness axiom,** which distinguishes $\mathbb{R}$ from $\mathbb{Q}$.

Axiom 1.42

If A is a nonempty subset of $\mathbb{R}$ which is bounded above, then A has a least upper bound, that is $\sup A$ exists as a real number.

For example, the set $A = \{r \in \mathbb{Q} : r > 0 \text{ and } r^2 < 2\}$ is a nonempty subset of $\mathbb{R}$, and we saw that it is bounded. Thus $\sup A = M$ exists as a real number and cannot be rational: $M \in \mathbb{R} \setminus \mathbb{Q}$. An element of the set $\mathbb{R} \setminus \mathbb{Q}$ is called an **irrational** number. We also saw that $M^2 = 2$. M is denoted by $\sqrt{2}$ and called the **square root** of 2. More generally, given a positive number a, the square root of a is defined to be a number x such that $x^2 = a$. Such a number x is denoted by $\sqrt{a}$. Slight modifications to the arguments in the proofs of Claim 1.40 and Claim 1.41 show that if $a \geq 0$, then $\sqrt{a}$ exists (see Exercise 1.20).

The following is an immediate consequence (and is in fact considered as a part) of the completeness axiom. We leave the proof as an exercise.

Theorem 1.43

Every nonempty subset A of $\mathbb{R}$ that is bounded below has a greatest lower bound.

Density theorem   Another important property of $\mathbb{R}$ is the fact that the rational numbers come arbitrarily close to any real number. Technically, the theorem says that the set of rational numbers $\mathbb{Q}$ is **dense** in $\mathbb{R}$.

Theorem 1.44

If a and b are real numbers such that $a < b$, then there exists a rational number $r \in \mathbb{Q}$ such that $a < r < b$.

Proof

Since $b - a > 0$, there exists $n \in \mathbb{N}$ such that $n(b - a) > 1$, i.e. such that $1/(b - a) < n$. Let $m = \text{int } na$. Thus $m \leq na \leq m + 1$. Set $r = (m + 1)/n$. Then $a < r$, and $r - a < 1/n < b - a$. Hence $r < b$ and thus $a < r < b$. $\square$

It follows from this theorem that given a real number x, no matter how we choose $\varepsilon > 0$, we will be able to find a rational number q such that $x - \varepsilon < q < x$. Thus real numbers are as close as one wants to rational numbers. The next example proves exactly the same idea.

Example 1.45

Given a real number x, show that for each $n \in \mathbb{N}$, there exists a rational number q such that $q \leq x < q + 1/10^n$.

Solution

Fix x in $\mathbb{R}$. Consider the set

$$A = \{n \in \mathbb{N} : q_n \leq x < q_n + 1/10^n \text{ for some } q_n \in \mathbb{Q}\}.$$

We are done if we show that $A = \mathbb{N}$. First we notice that from the inequalities

$$\text{int}(10\,\text{fra}\,x) \leq 10\,\text{fra}\,x < \text{int}(10\,\text{fra}\,x) + 1,$$

it follows that

$$\text{int}\,x + \frac{\text{int}(10\,\text{fra}\,x)}{10} \leq \text{int}\,x + \text{fra}\,x < \text{int}\,x + \frac{\text{int}(10\,\text{fra}\,x) + 1}{10}.$$

We let q_1 denote the rational number $\text{int}\,x + \frac{\text{int}(10\,\text{fra}\,x)}{10}$. Then the above inequalities are equivalent to $q_1 \leq x \leq q_1 + 1/10$, proving that $1 \in A$. Now suppose that $n \in A$. Then let q_n be a rational such that $q_n \leq x \leq q_n + 1/10^n$. Consider the real number $x_n = x - q_n$. Then

$$\text{int}\left(10^{n+1}\,\text{fra}\,x_n\right) \leq 10^{n+1}\,\text{fra}\,x_n < \text{int}\left(10^{n+1}\,\text{fra}\,x_n\right) + 1,$$

and

$$\text{int}\,x_n + \frac{\text{int}\left(10^{n+1}\,\text{fra}\,x_n\right)}{10^{n+1}} \leq \text{int}\,x_n + \text{fra}\,x_n < \text{int}\,x_n + \frac{\text{int}\left(10^{n+1}\,\text{fra}\,x_n\right) + 1}{10^{n+1}}.$$

The number $q_n' = \text{int}\,x_n + \frac{\text{int}\left(10^{n+1}\,\text{fra}\,x_n\right)}{10^{n+1}}$ is rational and the above inequalities translate to $q_n' \leq x_n < q_n' + 1/10^{n+1}$. The rational number $q_{n+1} = q_n + q_n'$ satisfies $q_{n+1} \leq x < q_{n+1} + 1/10^{n+1}$. Hence $n + 1 \in A$. The principle of mathematical induction now concludes our proof. $\qquad\square$

Algebraic and transcendental numbers   Recall that for a given integer n, a **polynomial** of degree n is an expression of the form

$$a_n x^n + a_{n-1} x^{n-1} + \cdots + a_1 x + a_0$$

where $a_0, a_1, \ldots, a_n$ are real numbers, called **coefficients**, with $a_n \neq 0$ and x is a quantity that can take on all numerical values. A **zero** or a **root** of a polynomial is a numerical value x for which

$$a_n x^n + a_{n-1} x^{n-1} + \cdots + a_1 x + a_0 = 0.$$

Example 1.46

Let $P(x) = a_n x^n + a_{n-1} x^{n-1} + \cdots + a_1 x + a_0$ be a polynomial with integer coefficients, i.e. $a_n, a_{n-1}, \ldots, a_0 \in \mathbb{Z}$, and where $a_0 \neq 0$ and $a_n \neq 0$. Show that if $P(x)$ has to a rational zero, say p/q, then p must divide a_0 and q must divide a_n.

Solution

Since p/q is a zero, we have

$$a_n \left(\frac{p}{q}\right)^n + a_{n-1} \left(\frac{p}{q}\right)^{n-1} + \cdots + a_1 \left(\frac{p}{q}\right) + a_0 = 0. \qquad (1.2)$$

Multiplying (1.2) by q^n, dividing by p, and then transposing the last term, we obtain

$$a_n p^{n-1} + a_{n-1} p^{n-2} q + \cdots + a_1 q^{n-1} = -\frac{a_0 q^n}{p}. \qquad (1.3)$$

Since the left side of (1.3) is an integer, so is $-\frac{a_0 q^n}{p}$. Since p does not divide q^n, p must divide a_0.

In a similar manner, by multiplying (1.2) by q^n, dividing by q, and then transposing the first term, we obtain

$$-\frac{a_n p^n}{q} = a_{n-1} p^{n-1} + \cdots + a_1 p q^{n-2} + a_0 q^{n-1}. \qquad (1.4)$$

Since the left side of (1.4) is an integer, so is $-\frac{a_n p^{n-1}}{q}$. Since q does not divide p^n, p must divide a_0. $\qquad \square$

A real number x is said to be an **algebraic number** if it is a zero of a polynomial with integer coefficients. A nonalgebraic real number is called a **transcendental number**.

For example, a rational number $r = \frac{p}{q}$ satisfies the polynomial equation $qx - p = 0$. Therefore any rational number is an algebraic number. The converse does not hold.

Example 1.47

Show that $\sqrt{2} + \sqrt{3}$ is an irrational algebraic number.

Solution

Let $x = \sqrt{2} + \sqrt{3}$. Then $x^2 = 5 + 2\sqrt{6}$ and thus $x^2 - 5 = 2\sqrt{6}$. Squaring both sides, we get $x^4 - 10x^2 + 1 = 0$. Thus $\sqrt{2} + \sqrt{3}$ is indeed an algebraic number. However, according to Example 1.46, the only possible rational solutions of $x^4 - 10x^2 + 1 = 0$ are ± 1. Thus $\sqrt{2} + \sqrt{3}$ cannot be rational. $\qquad \square$

1.3 Variables and Functions

One of the most prevalent ideas in Mathematics is that of a function. In this section we review some features of functions which we will need in the subsequent chapters.

First we recall that a **variable** is a quantity that can take on various numerical values. Variables are usually designated by letters such as x, y, z, t,..... Sometimes two or more variables are related to one another by a well-defined rule.

Definition 1.48

A **function** f is a correspondence that assigns to each value of a variable x in a given set, say X, exactly one value of a variable y in another set, say Y.

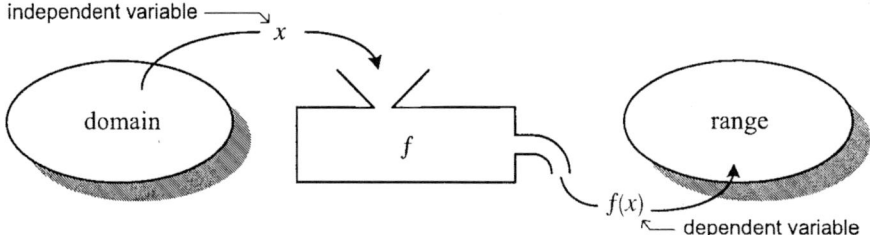

Figure 1.3   Function

Figure 1.3 is helpful to visualize the general idea behind the definition of a function. The notation for a function is usually as follows

$$f : X \to Y : x \longmapsto y = f(x).$$

However, it is a common practice to define a function by specifying a formula for finding $f(x)$ without mentioning its domain. Thus "the function $f(x) = \sqrt{x-2}$" should be understood as "the function $f : [2, +\infty) \longrightarrow \mathbb{R}; \; x \longmapsto \sqrt{x-2}$". Similarly, "the function $y = \ln x$" is understood as "the function $f : (0, +\infty) \longrightarrow \mathbb{R}; \; x \longmapsto \ln x$".

The set A of all x for which $f(x)$ is well defined, that is, $f(x)$ has a definite value, is called the **domain** of f and is denoted by dom f. For example, if $f(x) = x^2 - 2$, then dom $f = (-\infty, \infty)$ because $x^2 - 2$ is defined for all values of x. The function $f(x) = (x+2)/(x-2)$ is defined for all values of x with the exception of $x = 2$, hence dom $f = (-\infty, 2) \cup (2, \infty)$. For the function $f(x) = \sqrt{x-2}$, dom $f = [2, \infty)$ and so on.

For any subset $B \subset Y$, the **inverse image** of B is defined to be the set

$$f^{-1}(B) = \{x \in X : f(x) \in B\}.$$

In particular, if $f : X \to Y$, then we always have $f^{-1}(Y) = \text{dom} f$. Note that $f^{-1}(B)$ may be empty even when $B \neq \varnothing$: if $f(x) = \sqrt{x-2}$, then $f^{-1}([-2, -1]) = \varnothing$ as f does not take any negative value.

If $x \in \text{dom} f$, the element $f(x)$ of Y, read "f of x", is called the **value** of f at x. The subset $\{f(x) : x \in \text{dom} f\} \subset Y$ is called the **set of values** or the **range** of f denoted by ran f. Also, for any subset $A \subset X$, the **image** of A is defined to be the set

$$f(A) = \{f(x) : x \in A\}.$$

The function f is said to be **onto** or **surjective** if $f(X) = \text{ran} f = Y$.

Given a function f, each value of the variable $y = f(x) \in \text{ran} f$ depends on each given value of the variable $x \in \text{dom} f$. For this reason $y = f(x)$ is said to be the **dependent variable** for the function f. On the other hand, the choice of the variable x in dom f is free; x is called the **independent variable** or the **argument** of the function f.

The function f is said to be **one-to-one** or **injective** if for each value of the dependent variable $y \in \text{ran} f$ there corresponds exactly one value $x \in \text{dom} f$ of the independent variable such that $y = f(x)$. In other words, f is one-to-one if for every $x_1 \neq x_2$ in dom f, $f(x_1) \neq f(x_2)$ in ran f. For example the function $f(x) = x/(x+3)$ is defined for all values of the variable x, with the exception of $x = -3$, and f takes on all the values of the real numbers; hence dom $f = \mathbb{R} \setminus \{-3\}$ and ran $f = \mathbb{R}$. It is clear that if $x_1/(x_1+3) = x_2/(x_2+3)$, then $x_1 x_2 + 3x_1 \neq x_2 x_1 + 3x_2$ and hence $x_1 = x_2$. Therefore f is onto and one-to-one. The function $g(x) = \sin x$ is defined for all values of x; therefore dom $g = \mathbb{R}$; since $-1 \leq \sin x \leq 1$ for all $x \in \text{dom} g$, ran $g = [-1, 1]$. Then g is not onto $\mathbb{R}$. However, it is onto $[-1, 1]$. It is clear that $\sin 0 = \sin \pi = 0$. Thus g is not one-to-one on $\mathbb{R}$.

Again, we are mostly interested in functions whose domain and range are both subsets of $\mathbb{R}$. Such functions are called **real-valued functions of a real variable** or simply **real functions**. The interesting case of real-valued functions of an integer variable will be discussed in Chapter 2.

We recall a few basic elementary functions:

- **Power function:** $f(x) = x^\alpha$ where $\alpha \in \mathbb{R} \setminus \{0\}$. If $\alpha \geq 0$, then dom $f = \mathbb{R}$. If $\alpha < 0$, then dom $f = \mathbb{R} \setminus \{0\}$.

- **Exponential function:** $f(x) = a^x$ where $a \in \mathbb{R}_+ \setminus \{1\}$; dom $f = \mathbb{R}$.

- **Logarithmic function:** $f(x) = \log_a x$ where $a \in \mathbb{R}_+ \setminus \{1\}$; dom $f = \mathbb{R}_+$.

- **Trigonometric functions:**

$$
\begin{array}{ll}
f(x) = \sin x & \text{dom } f = \mathbb{R} \\
f(x) = \cos x & \text{dom } f = \mathbb{R} \\
f(x) = \tan x \left(= \frac{\sin x}{\cos x}\right) & \text{dom } f = \mathbb{R} \setminus \left\{(2k+1)\frac{\pi}{2} : k \in \mathbb{Z}\right\} \\
f(x) = \cot x \left(= \frac{\cos x}{\sin x}\right) & \text{dom } f = \mathbb{R} \setminus \{k\pi : k \in \mathbb{Z}\} \\
f(x) = \sec x \left(= \frac{1}{\cos x}\right) & \text{dom } f = \mathbb{R} \setminus \left\{(2k+1)\frac{\pi}{2} : k \in \mathbb{Z}\right\} \\
f(x) = \csc x \left(= \frac{1}{\cos x}\right) & \text{dom } f = \mathbb{R} \setminus \{k\pi : k \in \mathbb{Z}\}
\end{array}
$$

Other examples of functions can be obtained by means of a finite number of operations of addition, subtraction, multiplication, division and by taking the "function of a function". Given two real-valued functions f and g,

- the **sum** $f + g$ is defined on dom $f \cap$ dom g by the rule

$$(f + g)(x) = f(x) + g(x);$$

- the **product** $f \cdot g$ is defined on dom $f \cap$ dom g by the rule

$$(f \cdot g)(x) = f(x) \cdot g(x);$$

- the **quotient** $\left(\frac{f}{g}\right)$ is defined on $\{x \in \text{dom } f \cap \text{dom } g : g(x) \neq 0\}$ by the rule

$$\left(\frac{f}{g}\right)(x) = \frac{f(x)}{g(x)};$$

- the **composition** $g \circ f$ is defined on $\{x \in \text{dom } f : f(x) \in \text{dom } g\}$ by

$$(g \circ f)(x) = g(f(x));$$

- the **maximum** $\max\{f, g\}$ (or $f \vee g$) is defined on dom $f \cap$ dom g by

$$\max\{f, g\}(x) = \max\{f(x), g(x)\};$$

- the **minimum** $\min\{f, g\}$ (or $f \wedge g$) is defined on dom $f \cap$ dom g by

$$\min\{f, g\}(x) = \min\{f(x), g(x)\}.$$

For example, a **polynomial function**

$$f(x) = a_n x^n + a_{n-1} x^{n-1} + \cdots + a_1 x + a_0$$

is a finite sum of multiples of power functions. A **rational function**

$$f(x) = \frac{a_n x^n + a_{n-1} x^{n-1} + \cdots + a_1 x + a_0}{b_m x^m + b_{m-1} x^{m-1} + \cdots + b_1 x + b_0}$$

is the quotient of two polynomial functions. More generally, a function $y = f(x)$ which satisfies an equation of the form

$$P_n(x) y^n + P_{n-1}(x) y^{n-1} + \cdots + P_1(x) y + P_0(x) = 0$$

where $P_n(x), P_{n-1}(x), \cdots, P_1(x), P_0(x)$ are polynomials in x, is called an **algebraic function**.

Example 1.49

Show that $f(x) = \frac{1+\sqrt{x}}{3+x}$, $x \neq -3$, is an algebraic function.

Solution

If $y = \frac{1+\sqrt{x}}{3+x}$, then $(3+x) y - 1 = \sqrt{x}$. Squaring, we have $(3+x)^2 y^2 - 2(3+x) y + 1 = x$ and $(3+x)^2 y^2 - 2(3+x) y + 1 - x = 0$. Thus f is an algebraic function.   □

A function which is not algebraic is called **transcendental**. Examples of transcendental functions are $y = \sin x$, $y = 2^x$, and so on.

If f is a real function defined on a subset S of $\mathbb{R}$, then we say that f is

- **nondecreasing** on S if $f(x) \leq f(y)$ whenever $x \leq y$ in S;

- **increasing**[2] on S if $f(x) < f(y)$ whenever $x < y$ in S;

- **nonincreasing** on S if $f(x) \geq f(y)$ whenever $x \leq y$ in S;

- **decreasing** on S if $f(x) > f(y)$ whenever $x < y$ in S.

A function which is either nondecreasing or nonincreasing on S is termed a **monotone** or **monotonic function**.

Finally, a real function $y = f(x)$ may be graphically represented on a plane by locating in a rectangular coordinate system the points defined by the pairs $(x, f(x))$ (Figure 1.4). The **graph** of a function $y = f(x)$ is defined to be the collection, in a rectangular coordinate system, of all the points of the form $(x, f(x))$ for all $x \in \text{dom } f$.

---

[2] The terminology *strictly increasing* and *strictly decreasing* is also used in the literature instead of *increasing* and *decreasing* respectively.

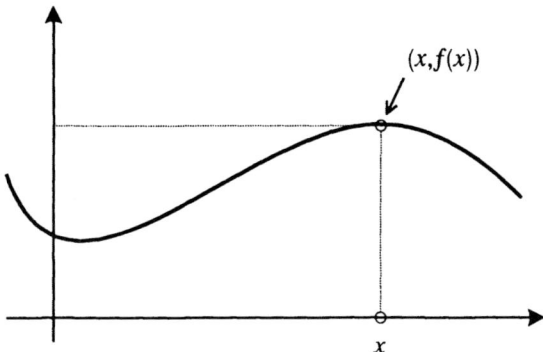

Figure 1.4   Graph of a function

EXERCISES

1.1 Prove (4) and (5) of Example 1.2, page 2.

1.2 Verify that the real number 0 is unique.

1.3 Verify that the opposite of a real number t is uniquely determined by $-t$ and $-(-t) = t$.

1.4 Show that $\mathbb{N}$ does not satisfy either **A3** or **A4** of Axiom 1.1. Show that $\mathbb{Z}$ satisfies all of the axioms of Axiom 1.1 except **A8**. Verify that $\mathbb{Q}$ satisfies all the axioms of a field.

1.5 Show that the set $\left\{a + b\sqrt{2} : a, b \in \mathbb{R}\right\}$ is a field under ordinary addition and multiplication.

1.6 Show that if $a^2 + b^2 = 0$ in $\mathbb{R}$, then $a = b = 0$.

1.7 Indicate whether the given set is bounded (above/below) and determine its supremum, infimum, maximum, and minimum provided they exist.

(a) $[0,1)$ (d) $\bigcap_{n=1}^{\infty} \left[\frac{-1}{n}, 1 + \frac{1}{n}\right]$ (g) $\left\{\sqrt{x} : x \in \mathbb{R}_+\right\}$

(b) $\left\{(-2)^n : n \in \mathbb{N}\right\}$ (e) $\bigcup_{n=1}^{\infty} [\pi n, \pi n + 1]$ (h) $\left\{\frac{n^2}{n^2+1} : n \in \mathbb{N}\right\}$

(c) $\left\{\frac{\cos \pi n}{n} : n \in \mathbb{N}\right\}$ (f) $\left\{1 - \frac{1}{n} : n \in \mathbb{N}\right\}$ (i) $\left\{\ln \frac{1}{n} : n \in \mathbb{N}\right\}$

1.8 Prove the following statements.

(1) $1 + 2^2 + \cdots + n^2 = n(n + 1)(2n + 1)/6$ for all $n \in \mathbb{N}$.

(2) $1 + 2^3 + \cdots + n^3 = (n(n + 1)/2)^2$ for all $n \in \mathbb{N}$.

(3) $1 + 2^{-1} + 2^{-2} + \cdots + 2^{-n} = 2 - 2^{-n}$ for all $n \in \mathbb{N}$.

(4) $1 + a + a^2 + \cdots + a^n = \frac{1 - a^{n+1}}{1 - a}$ for $a \neq 1$ and for all $n \in \mathbb{N}$.

(5) $|a_1 + a_2 + \cdots + a_n| \leq |a_1| + |a_2| + \cdots + |a_n|$ for all $n \in \mathbb{N}$, where $a_i \in \mathbb{R}$, $i = 1, 2, \cdots, n$.

(6) 2 divides $n(n+1)$ for all $n \in \mathbb{N}$.

(7) 7 divides $3^{2n+1} + 2^{n+2}$ for all $n \in \mathbb{N}$.

(8) $\ln(x^n) = n \ln x$ for $x > 0$ and for all $n \in \mathbb{N}$.

1.9   Show that $x - y$ divides $x^n - y^n$ for all $n \in \mathbb{N}$.

1.10 Show that $2 \cos nx \sin(x/2) = \sin[(n + 1/2)x] - \sin[(n - 1/2)x]$ for all $n \in \mathbb{N}$. Deduce that for $x \neq 2k\pi, k = 0, \pm 1, \pm 2, \cdots$,

$$|\cos x + \cos 2x + \cdots + \cos nx| \leq \frac{1}{\left|\sin \frac{1}{2}x\right|} \quad \text{for all } n.$$

1.11 Use induction to prove the **binomial formula**: given two real numbers a and b, then for all $n \in \mathbb{N}$

$$(a + b)^n = \binom{n}{0}a^n + \binom{n}{1}a^{n-1}b + \binom{n}{2}a^{n-2}b^2 + \cdots$$
$$+ \binom{n}{n-1}ab^{n-1} + \binom{n}{n}b^n$$

where $\binom{n}{k} = \frac{n!}{k!(n-k)!}$ for $k = 1, 2, \cdots, n$. (Hint: $\binom{n}{k} + \binom{n}{k-1} = \binom{n+1}{k}$.)

1.12 Show that the equation $q^2 = 3$ is not satisfied by any rational.

1.13 Show that $\sqrt{2} + \sqrt{3} + \sqrt{5}$ is an irrational algebraic number.

1.14 Show that the square root of any odd integer is odd.

1.15 Show that $\operatorname{int} x + \frac{\operatorname{int}(n \operatorname{fra} x)}{n} \leq x < \operatorname{int} x + \frac{\operatorname{int}(n \operatorname{fra} x) + 1}{n}$ for all $n \in \mathbb{N}$.

1.16 Show that $|a/b| = |a| / |b|$ if $a, b \in \mathbb{R}$, $b \neq 0$ and $|abc| = |a| |b| |c|$ for all $a, b, c \in \mathbb{R}$.

1.17 Show that in $\mathbb{R}$, y is between x and z if and only if $|x - y| + |y - z| = |x - z|$.

1.18 Show that for every $\varepsilon > 0$

(1) $|a - b| < \varepsilon$ if and only if $b - \varepsilon < a < b + \varepsilon$;

(2) if $|a - b| < \varepsilon$, then $|a| < |b| + \varepsilon$ and $|b| < |a| + \varepsilon$.

1.19   Show that if $a, b \in \mathbb{R}$, then $\sqrt{|ab|} \leq \frac{|a| + |b|}{2}$.

1.20   Show that if $a > 0$, then the number $b = \sup \left\{ x \in \mathbb{R} : x \geq 0, x^2 \leq a \right\}$ exists and that $b^2 = a$.

1.21   Show that given $a > 0$, and two integers $m, n \in \mathbb{Z}$, then the number $b = \sup \left\{ x \in \mathbb{R} : x^n \leq a^m \right\}$ exists.

1.22   Let $x_1, x_2, \ldots, x_n$ be real numbers. Show that $x_1^2 + x_2^2 + \cdots + x_n^2$ is a square.

1.23   Prove the results of Example 1.6.

1.24   Show that an inductive subset of $\mathbb{R}$ cannot be bounded above.

1.25   Show that if a set A has an upper bound, then it has infinitely many upper bounds.

1.26   Write out the proof of Theorem 1.24.

1.27   Let $a, b \in \mathbb{R}$. Show that if every number greater than b is greater than a, then $b \geq a$.

1.28   Let $q \in \mathbb{Q}$ and $\varepsilon \in \mathbb{R}_+$. Show that there exists an irrational number x such that $|x - q| < \varepsilon$.

1.29   Show that if A and B are subsets of $\mathbb{R}$ which are both bounded below, then $A + B$ is also bounded below and $\inf (A + B) = \inf A + \inf B$.

1.30   Let A and B be nonempty bounded subsets of $\mathbb{R}$. Show that

   (1)   if $A \subset B$, then $\inf B \leq \inf A \leq \sup A \leq \sup B$;

   (2) $\sup (A \cup B) = \max \left\{ \sup A, \sup B \right\}$;

   (3) $\inf A = -\sup (-A)$ where $-A = \left\{ x \in \mathbb{R} : -x \in A \right\}$.

1.31   For all a and b in $\mathbb{R}$ verify that

   (1) $a + b = \max \left\{ a, b \right\} + \min \left\{ a, b \right\}$;

   (2) $|a - b| = \max \left\{ a, b \right\} - \min \left\{ a, b \right\}$;

   (3) $\max \left\{ a, b \right\} = \left[(a + b) + |a - b| \right] / 2$;

   (4) $\min \left\{ a, b \right\} = \left[(a + b) - |a - b| \right] / 2$;

   (5) $|a| \leq \max \left\{ |a + b|, |a - b| \right\}$.

1.32   Write out the proof of Remark 1.36.

1.33   Prove Theorem 1.43.

1.34 Fill in the table with appropriate and meaningful formulas.

$f(x)$	$g(x)$	$(f+g)(x)$	$(fg)(x)$	$f \circ g(x)$	$g \circ f(x)$
$\sqrt{3-x}$	x^2				
	$\ln x$	$\ln\left(\frac{x+1}{x}\right)$			
$\sqrt{1-x}$	$\sqrt{x^2-4}$				
$\frac{x}{x-2}$					$\tan^{-1}\frac{x}{x-2}$
$\cos 3x$				$\cos\left(x^2+1\right)$	
	$\sin 2x$		$\cos x$		
	x	$\mathrm{fra}\, x$			
$\ln x$	x^2+2				

1.35 Give the domain of all the functions of the previous exercises.

1.36 Consider the functions $f^+ = \max\{f,0\}$ and $f^- = -\min\{f,0\}$. Show that $f = f^+ - f^-$ and $|f| = f^+ + f^-$.

1.37 Show that $\max\{f,g\} = (f-g)^+ + g$ and $\min\{f,g\} = f - (g-f)^-$.

1.38 Given $f(x) = \ln x^2$ and $g(x) = \ln \sqrt{x}$, write a formula for $\max\{f,g\}$, $\min\{f,g\}$, f^+, f^-, g^+, and g^-, specifying their domain.

1.39 Let $f : \mathbb{R}_+ \to \mathbb{R}$ be a nonnegative function satisfying $f(xy) = f(x) + f(y)$.

   (1) Show that $f(1) = 0$.

   (2) Let $a \in \mathbb{R}_+$. Write $f\left(a^2\right), f\left(a^3\right)$, and more generally $f(a^n)$ for all $n \in \mathbb{N}$ in terms of $f(a)$.

1.40 Let $g : \mathbb{R} \to \mathbb{R}_+$ be a function satisfying $g(x+y) = g(x) \cdot g(y)$.

   (1) Let $a \in \mathbb{R}$. Evaluate $g(2a), g(3a)$, and more generally $g(na)$ for all $n \in \mathbb{N}$.

   (2) Compare $g(-x)$ and $g(x)$. Write $g(ma)$ in terms of $g(a)$ and m, where $m \in \mathbb{Z}$.

   (3) Write $g(ra)$ in terms of $g(a)$ and r, where $r \in \mathbb{Q}$.

1.41 Let $f : X \to Y$ be a function. Show that

   (1) if $A \subset X$, then $A \subset f^{-1}(f(A))$;

   (2) if $B \subset Y$, then $f\left(f^{-1}(B)\right) \subset B$;

   (3) if $B \subset Y$, then $f^{-1}(Y \setminus B) = X \setminus f^{-1}(B)$.

1.42 Let $f : X \to Y$ be a function. Let $\{A_i : i \in I\}$ and $\{B_i : i \in I\}$ respectively be a family of subsets of X and Y. Show that

   (1) $f\left(\bigcup_{i \in I} A_i\right) = \bigcup_{i \in I} f(A_i)$;

(2) $f\left(\bigcap_{i\in I} A_i\right) \subset \bigcap_{i\in I} f\left(A_i\right);$

(3) $f^{-1}\left(\bigcup_{i\in I} B_i\right) = \bigcup_{i\in I} f^{-1}\left(B_i\right);$

(4) $f^{-1}\left(\bigcap_{i\in I} B_i\right) = \bigcap_{i\in I} f^{-1}\left(B_i\right).$

2
Sequences

Functions whose domains are subsets of the integers present particularly important and interesting features. The argument of such functions does not take on values in a *continuous* way but rather in a *sequential* manner.

2.1 Definition of a Sequence

Definition 2.1

An (**infinite**) **sequence** is a function whose domain is a set of the form $\{n \in \mathbb{N}_0 : n \geq m\}$ for some fixed $m \in \mathbb{N}_0$.

Hence, a function $a : \{m, m+1, \ldots\} \longrightarrow X : n \longmapsto a(n)$ is a sequence of elements of the set X. To each $n \in \{m, m+1, \ldots\}$, the sequence a associates an element of the set X denoted by $a(n)$.

Notation

It is standard practice to use subscript notation for the values of a given sequence: $a(n)$ is written a_n and is called the n-th **term** of the sequence. It is then usual to write a sequence as $(a_n)_{n=m}^{\infty}$, or in the case where $m = 1$ (resp. $m = 0$) $(a_n)_{n \in \mathbb{N}}$ (resp. $(a_n)_{n \in \mathbb{N}_0}$). When the domain of the sequence is understood from the context we simply write (a_n).

In this chapter, we will mostly deal with sequences of real numbers, i.e. sequences with values in $\mathbb{R}$. Thus the range of the sequence $(a_n)_{n=m}^{\infty}$ also called the **set of values** or **range** of $(a_n)_{n=m}^{\infty}$ is the subset of $\mathbb{R}$: $\{a_n : n = m, m + 1, \ldots\}$.

Note

One should be careful not to confuse the range of a sequence $(a_n)_{n=m}^{\infty}$ with the sequence itself. For example $((-1)^n)_{n\in\mathbb{N}}$ is a sequence while its range is the set $\{-1, 1\}$.

An important notion in mathematics is contained in the following definition.

Definition 2.2

A nonempty subset A of a set X is said to be **countable** if it is the range of some sequence.

According to this definition, every finite set is countable. A set which is not countable is said to be **uncountable**.

Example 2.3

Show that the set of all rational numbers in $[0, 1]$ is countable.

Solution

The rational numbers in $[0, 1]$ can be obtained by writing all fractions with denominator 1, then 2, then 3, ... considering equivalent fractions no more than once. Thus we can write

$$\mathbb{Q} \cap [0, 1] = \left\{0, 1, \frac{1}{2}, \frac{1}{3}, \frac{2}{3}, \frac{1}{4}, \frac{3}{4}, \frac{1}{5}, \frac{2}{5}, \frac{3}{5}, \frac{4}{5}, \cdots\right\}.$$

The set $\mathbb{Q} \cap [0, 1]$ is then clearly seen to be the range of the sequence

n :	1	2	3	4	5	6	7	8	$\cdots$
a_n :	0	1	$\frac{1}{2}$	$\frac{1}{3}$	$\frac{2}{3}$	$\frac{1}{4}$	$\frac{3}{4}$	$\frac{1}{5}$	$\cdots$.

$\square$

Example 2.4

Show that the set of all real numbers in $[0, 1]$ is uncountable.

Solution[1]

We write each real number in $[0,1]$ in its decimal expansion $0.d_1 d_2 d_3 \ldots$ where $d_1, d_2, \ldots$ are any of the digits $0, 1, 2, \ldots, 9$. For definiteness, we agree that a terminating decimal expansion will be represented by an infinite string of 9's. For example 0.5365 is represented by $0.5364999999\ldots$. Let (x_n) be a sequence in $[0,1]$. We write (x_n) as

$$
\begin{array}{ll}
n \;\mapsto\; & x_n \\
1 & 0.d_{11} d_{12} d_{13} \ldots \\
2 & 0.d_{21} d_{22} d_{23} \ldots \\
3 & 0.d_{31} d_{32} d_{33} \ldots \\
\vdots & \vdots
\end{array}
$$

We consider the number $a = 0.a_1 a_2 a_3 \ldots$ where $a_1 \neq d_{11}, a_2 \neq d_{22}, a_3 \neq d_{33}, \ldots$ and where there exists no m such that $a_n = 9$ for all $n > m$. Then clearly $a \in [0,1]$ but $a \neq x_n$ for all n. This shows that the range of any sequence in $[0,1]$ cannot cover the whole of the interval $[0,1]$ and proves that $[0,1]$ is uncountable. $\qquad\square$

Example 2.5

Show that every infinite set contains an infinite countable subset.

Solution

It suffices to show that an infinite set A contains (the range of) an infinite sequence. Since A is not empty, there exists an element $a_1 \in A$. Since A is infinite the set $A_1 = A \setminus \{x_1\}$ is nonempty. There exists $a_2 \in A_1$. Again since A is infinite $A_2 = A \setminus \{x_1, x_2\}$ is nonempty. There exists $a_3 \in A_2$. Proceeding in this way, we obtain the infinite subset $\{x_1, x_2, x_3, \ldots\}$ of A, which by construction is the range of the sequence (x_n). $\qquad\square$

Recall that a subset S of $\mathbb{R}$ is said to be bounded if there exists $M > 0$ such that $|s| \leq M$ for all s in S (Remark 1.16).

Definition 2.6

A sequence (a_n) is said to be **bounded** if there exists $M > 0$ such that $|a_n| \leq M$ for all n.

[1] The technique used here is known as the *Cantor Diagonal Method*.

It is easy to realize that a sequence (a_n) is bounded if and only if its range is a bounded subset of $\mathbb{R}$. Let us look at few examples.

- Since for each $n \in \mathbb{N}$, $\left|\frac{n+1}{n}\right| = 1 + \frac{1}{n} \leq 1 + 1 = 2$, the sequence $\left(\frac{n+1}{n}\right)_{n \in \mathbb{N}}$ is bounded.

- Since for any $M > 0$, there is n large enough such that $M < 2^n$ (it suffices for example to take $n > M$), the sequence $((2)^n)_{n=0}^{\infty}$ is not bounded.

- For any $M > 0$, by the Archimedean property of $\mathbb{R}$, there is n such that $M < n$. Thus for such n, we also have $M < n + \frac{1}{n}$. Hence the sequence $\left(\left(n + \frac{1}{n}\right)\right)_{n \in \mathbb{N}}$ is not bounded.

- Since for each $n \in \mathbb{N}$, $-1 \leq \cos \frac{n\pi}{4} \leq 1$, we have $\left|2 + \cos \frac{n\pi}{4}\right| \leq 2 + \left|\cos \frac{n\pi}{4}\right| \leq 3$. Thus $\left(2 + \cos \frac{n\pi}{4}\right)_{n=0}^{\infty}$ is a bounded sequence.

Definition 2.7

A sequence (a_n) is called **nondecreasing** (resp. **increasing**) if $a_n \leq a_{n+1}$ (resp. $a_n < a_{n+1}$) for all n and (a_n) is called **nonincreasing** (resp. **decreasing** if $a_n \geq a_{n+1}$ (resp. $a_n > a_{n+1}$ for all n. A sequence that is nondecreasing or nonincreasing is called a **monotone sequence** or a **monotonic sequence**.

For example, the sequences defined respectively by $a_n = 1 + \frac{1}{n}$, $b_n = \frac{1}{\sqrt{n}}$, and $c_n = \ln \frac{1}{n}$ are nonincreasing sequences, while the sequences defined respectively by $d_n = -2^{-n}$, $e_n = n^{2/3}$, and $f_n = 2$ are nondecreasing sequences. Thus each of these sequences is a monotonic sequence. There are sequences which are neither nonincreasing nor nondecreasing, i.e. which are not monotonic: $u_n = (-1)^n n$, $v_n = \cos \frac{n\pi}{3}$, $w_n = \frac{(-2)^n}{n^2}$, and so on.

Example 2.8

A sequence (u_n) is defined by the recursion formula $u_1 = 1$, $2u_{n+1} = 3 + u_n$ for all $n \in \mathbb{N}$. Show that (u_n) is increasing.

Solution

For each n, subtracting the following equalities

$$2u_{n+1} = 3 + u_n$$
$$2u_n = 3 + u_{n-1}$$

we have $2(u_{n+1} - u_n) = u_n - u_{n-1}$. It follows that for each n,

$$u_{n+1} - u_n = 2^{n-2}(u_2 - u_1) = 2^{n-2}(2 - 1) > 0.$$

This shows that (u_n) is increasing. □

Example 2.9

A sequence (u_n) is defined by the recursion formula $u_1 = 3$, $u_{n+1} = \sqrt{2 + u_n}$ for all $n \in \mathbb{N}$.

(1) Show that (u_n) is bounded.

(2) Show that (u_n) is nonincreasing.

Solution

(1) We first show by induction that $2 \leq u_n \leq 3$ for all n which implies that the sequence (u_n) is bounded.

Let $A = \{n \in \mathbb{N} : 2 \leq u_n \leq 3\}$. It is clear that $1 \in A$. Suppose that $k \in A$. Then $4 \leq 2 + u_k \leq 5$. Taking square roots, we have

$$2 \leq \sqrt{2 + u_k} < \sqrt{5}$$

and thus $2 \leq u_{k+1} \leq 3$, i.e. $k + 1 \in A$. By the principle of mathematical induction we have $A = \mathbb{N}$, and hence our claim is verified.

(2) For each n, since $u_n \geq 2$, we have

$$u_{n+1} - u_n = \sqrt{2 + u_n} - u_n$$
$$= \frac{2 + u_n - u_n^2}{\sqrt{2 + u_n} + u_n} \leq 0.$$

This shows that the sequence is nonincreasing. □

2.2 Convergence and Limits

In this section, we consider sequences whose n-th term *approaches* a single value as n gets larger and larger.

Definition 2.10

A sequence $(a_n)_{n=m}^{\infty}$ of real numbers is said to **converge** to $a \in \mathbb{R}$, if

for every $\varepsilon > 0$, there exists $N \in \mathbb{N}$ such that
$n > N$ implies $|a_n - a| < \varepsilon$.

If such a number a does not exist, the sequence is said to **diverge**.

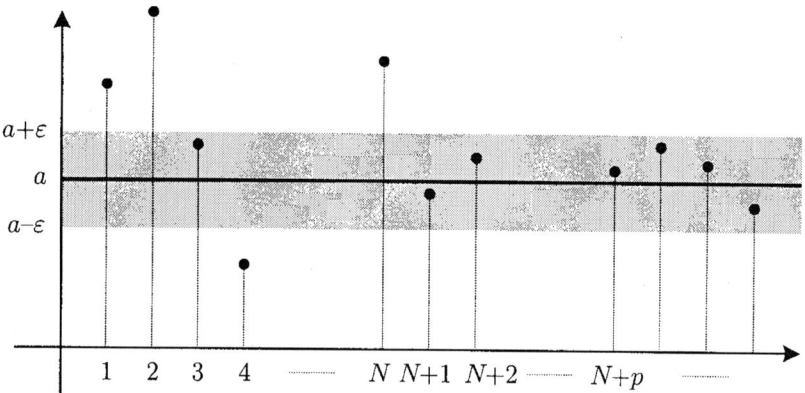

Figure 2.1   Limit of a sequence

In other words, after choosing $\varepsilon > 0$, no matter how small, we can find large enough $N \in \mathbb{N}$ so that all the terms of (a_n) coming after the N-th term are within distance ε of a. See Figure 2.1. Notice that the condition $|a_n - a| < \varepsilon$ is equivalent to $a - \varepsilon < a_n < a + \varepsilon$, or $a_n \in (a - \varepsilon, a + \varepsilon)$ and means that a_n is at a distance less than ε of a. An interval of the form $(a - \varepsilon, a + \varepsilon)$ is often called an ε-**neighborhood** of a and will be denoted by $N(a, \varepsilon)$. The definition can be restated as follows.

Theorem 2.11

A sequence $(a_n)_{n=m}^{\infty}$ of real numbers converges to $a \in \mathbb{R}$ if and only if for every ε-neighborhood $N(a, \varepsilon)$ of a, there exists $N \in \mathbb{N}$ such that $a_n \in N(a, \varepsilon)$ for all $n > N$.

If a sequence $(a_n)_{n=m}^{\infty}$ converges to some real number, then we say that $(a_n)_{n=m}^{\infty}$ is **convergent**; otherwise we say that the sequence is **divergent**.

Notation

The notation $a_n \to a$ is used to indicate that the sequence (a_n) converges to a. The number a is then called the **limit of the sequence** (a_n) and we write $\lim_{n\to\infty} a_n = a$ (read "limit of a_n as $n \to \infty$") or simply $\lim a_n = a$.

Example 2.12

Prove that the sequence $\left(a_n = 1 + (-1)^n \frac{1}{n}\right)$ converges to 1.

Solution

We first notice that for each $n \in \mathbb{N}$,

$$|a_n - 1| = \left| \left(1 + (-1)^n \frac{1}{n} \right) - 1 \right| = \frac{1}{n}.$$

Let $\varepsilon > 0$. By the Archimedean property of $\mathbb{R}$, there is N in $\mathbb{N}$ such that $N\varepsilon > 1$, i.e. $\frac{1}{N} < \varepsilon$. Thus $|a_n - 1| < \varepsilon$ is satisfied whenever $n > N$. Hence $\lim \left(1 + (-1)^n \frac{1}{n} \right) = 1$. $\square$

Example 2.13

Prove that the sequence $\left(a_n = \frac{2n-3}{5n+1} \right)$ converges to $\frac{2}{5}$.

Solution

Let $\varepsilon > 0$. We want to show that there is N in $\mathbb{N}$ large enough such that

$$\left| \frac{2n-3}{5n+1} - \frac{2}{5} \right| < \varepsilon \tag{2.1}$$

for all $n > N$. Since

$$\left| \frac{2n-3}{5n+1} - \frac{2}{5} \right| = \left| \frac{-17}{25n+5} \right| = \frac{17}{25n+5} < \frac{17}{25n}$$

the inequality (2.1) is satisfied if $\frac{17}{25n} < \varepsilon$, or equivalently if $n > \frac{17}{25\varepsilon}$. By the Archimedean property of $\mathbb{R}$, we can chose an integer $N \geq \frac{17}{25\varepsilon}$. The inequality (2.1) then holds provided $n > N$. This proves that $\lim \frac{2n-3}{5n+1} = \frac{2}{5}$. $\square$

Example 2.14

Prove that $\lim \left[\sqrt{n^2 + n} - n \right] = 1/2$.

Solution

Let $\varepsilon > 0$. We want to show that there is N in $\mathbb{N}$ large enough such that for $n > N$,

$$\frac{1}{2} - \varepsilon < \sqrt{n^2 + n} - n < \frac{1}{2} + \varepsilon. \tag{2.2}$$

We first notice that

$$\sqrt{n^2 + n} - n = \left(\sqrt{n^2 + n} - n \right) \frac{\sqrt{n^2 + n} + n}{\sqrt{n^2 + n} + n} = \frac{1}{\sqrt{1 + \frac{1}{n}} + 1}.$$

It follows that the inequalities in (2.2) are equivalent to

$$\frac{1}{2} - \varepsilon < \frac{1}{\sqrt{1 + \frac{1}{n}} + 1} < \frac{1}{2} + \varepsilon. \tag{2.3}$$

Since for every n, $1/\left(\sqrt{1 + \frac{1}{n}} + 1\right) < 1/2$, the second inequality in (2.3) is obvious. It is also clear that the first inequality of (2.3) holds if $\varepsilon \geq 1/2$. For $0 < \varepsilon < 1/2$, solving this inequality for n, we have $n > \frac{1}{\left(\frac{2}{1-2\varepsilon} - 1\right)^2 - 1}$. Thus (2.2) holds for all

$$n > \mathrm{int}\left(\frac{1}{\left(\frac{2}{1-2\varepsilon} - 1\right)^2 - 1}\right) + 1.$$

Hence $\lim\left[\sqrt{n^2 + n} - n\right] = 1/2$.   □

Definition 2.15

Let $(a_n)_{n=m}^{\infty}$ be a sequence of real numbers

(1) $(a_n)_{n=m}^{\infty}$ is said to **diverge to** ∞ if for every $M > 0$ there exists $N \in \mathbb{N}$ such that $n > N$ implies $a_n > M$.

(2) $(a_n)_{n=m}^{\infty}$ is said to **diverge to** $-\infty$ if for every $M > 0$ there exists $N \in \mathbb{N}$ such that $n > N$ implies $a_n < -M$.

In other words, the sequence $(a_n)_{n=m}^{\infty}$ diverges to ∞ (resp. $-\infty$) if for every preassigned $M > 0$, no matter how large, it is possible to find N such that, beginning with $n = N + 1$, all the subsequent terms of the sequence satisfy the inequality $a_n > M$ (resp. $a_n < -M$).

Notation

If $(a_n)_{n=m}^{\infty}$ diverges to $\pm\infty$, we write $a_n \to \pm\infty$, or $\lim a_n = \pm\infty$.

Example 2.16

Show that $\lim \sqrt{n + 5} = +\infty$.

Solution

For each $M > 0$, consider $N = \mathrm{int}\left(M^2 - 5\right)$. Then $n > N$ implies $n > \left(M^2 - 5\right)$; hence $\sqrt{n + 5} > M$ and so $\lim \sqrt{n + 5} = +\infty$.   □

Example 2.17

Show that $\lim (-2^n) = -\infty$.

Solution

For each $M > 0$, consider $N = \text{int}(\log_2 M)$. Then $n > N$ implies $n > \log_2 M$, hence $2^n > M$ and thus $-2^n < -M$. $\quad\square$

One of the first important results about convergent sequences is the uniqueness of their limits. The key idea is that the terms of a convergent sequence cannot be arbitrarily close to two distinct numbers.

Theorem 2.18

Let (a_n) be a sequence of real numbers. Suppose that $\lim a_n = a$ and $\lim a_n = b$. Then $a = b$.

Proof

Let (a_n) be a sequence with $\lim a_n = a$ and $\lim a_n = b$. Suppose to the contrary that $a \neq b$. Consider $\varepsilon = |a - b|$. Since $a_n \to a$, there is a $N_1 \in \mathbb{N}$ such that

$$|a_n - a| < \varepsilon/2 \quad \text{for} \quad n > N_1.$$

Similarly, since $a_n \to b$, there is a $N_2 \in \mathbb{N}$ such that

$$|a_n - b| < \varepsilon/2 \quad \text{for} \quad n > N_2.$$

Thus for a fixed $n > \max\{N_1, N_2\}$,

$$\varepsilon = |a - b| \leq |a - a_n| + |a_n - b| < \frac{\varepsilon}{2} + \frac{\varepsilon}{2} = \varepsilon,$$

a contradiction. Hence $a = b$. $\quad\square$

One should particularly notice the use of the triangle inequality in the above proof. Similar arguments will be used in many proofs involving limits.

Example 2.19

Show that convergent sequences are bounded.

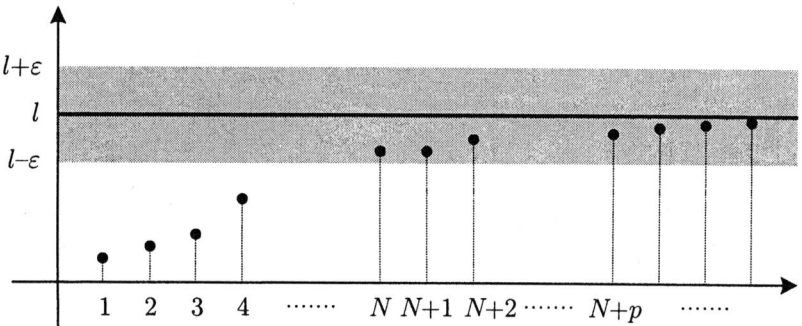

Figure 2.2 Bounded increasing sequence

Solution

Let (a_n) be a convergent sequence and let $\lim a_n = a$. For each n, we have

$$|a_n| = |a_n - a + a| \le |a_n - a| + |a|.$$

Consider $\varepsilon = 1$. There is N in $\mathbb{N}$ such that $|a_n - a| < 1$ for $n > N$. Therefore $|a_n| < |a| + 1$, for all $n > N$. It follows that if

$$M = \max\{|a_1|, |a_2|, \ldots, |a_N|, |a| + 1\},$$

then $|a_n| \le M$ for all $n \in \mathbb{N}$. $\square$

Note

Bounded sequences *may* or *may not* converge.

The sequence $((-1)^n)_{n \in \mathbb{N}}$ is an example of a sequence which is bounded but not convergent. So boundedness is not appropriate for testing the convergence of sequences. However, the following theorem is rather interesting. See Figure 2.2.

Theorem 2.20

Every bounded, monotonic sequence of real numbers converges.

Proof

Let $(a_n)_{n \in \mathbb{N}}$ be a bounded, monotonic sequence. Let us suppose that $(a_n)_{n \in \mathbb{N}}$ is nondecreasing. Since $(a_n)_{n \in \mathbb{N}}$ is a bounded sequence, its set of values $A =$

$\{a_n : n \in \mathbb{N}\}$ is bounded. By the completeness axiom of $\mathbb{R}$, $l = \sup A$ exists. We will show that $\lim a_n = l$. Let $\varepsilon > 0$. Since l is a supremum, there exists an $M \in \mathbb{N}$ such that $l - \varepsilon < a_M \leq l$. Since (a_n) is nondecreasing, $n > M$ implies $l - \varepsilon < a_n \leq l$. Therefore $\lim a_n = l$.

The nonincreasing case uses similar arguments with obvious changes and is left as an exercise. $\qquad\square$

Example 2.21

Consider the two sequences (u_n) and (v_n) defined for each n by:

$$u_n = 1 + \frac{1}{1!} + \frac{1}{2!} + \cdots + \frac{1}{n!}$$
$$v_n = u_n + \frac{1}{n\,(n!)}.$$

(1) Show that (u_n) is increasing, (v_n) is decreasing and $u_n < v_m$ for all $n, m \in \mathbb{N}$.

(2) Show that both sequences are bounded.

(3) Show that the two converge to the same limit which we denote by e.

(4) Show that e is not rational.

Solution

(1) For each n, we have

$$u_{n+1} - u_n = \frac{1}{(n+1)!} > 0.$$

Thus $u_n < u_{n+1}$ for all n, i.e. (u_n) is increasing. Also, for each n, we have

$$
\begin{aligned}
v_{n+1} - v_n &= u_{n+1} - u_n + \frac{1}{(n+1)\,((n+1)!)} - \frac{1}{n\,(n!)} \\
&= \frac{1}{(n+1)!} + \frac{1}{(n+1)\,((n+1)!)} - \frac{1}{n\,(n!)} \\
&= \frac{1}{(n+1)!}\left[1 + \frac{1}{n+1} - \frac{n+1}{n}\right] \\
&= \frac{-1}{n\,(n+1)\,((n+1)!)} < 0.
\end{aligned}
$$

Therefore $v_{n+1} < v_n$ for all n and (v_n) is decreasing. Finally, since

$$v_n - u_n = \frac{1}{n\,(n!)} > 0$$

for each n, we have $u_n < v_n$ for all n, and if $n < m$ in $\mathbb{N}$, then $u_n < u_m < v_m < v_n$. Hence $u_n < v_m$ for all $n, m \in \mathbb{N}$.

(2) We infer from the previous part of the solution that each term of the sequence (v_n) is an upper bound for the increasing sequence (u_n). In particular, we have

$$u_n < v_1 = 3 \quad \text{for each } n.$$

Thus the sequence $\lim_{n \to \infty} u_n = u$ exists.

Similarly, each term of the sequence (u_n) is a lower bound for the decreasing sequence (v_n). In particular, we have

$$u_1 = 2 < v_n \quad \text{for each } n.$$

Thus both sequences are bounded.

(3) Since $\lim_{n \to \infty} (v_n - u_n) = \lim_{n \to \infty} \frac{1}{n(n!)} = 0$, we must have

$$u = \lim_{n \to \infty} u_n = \lim_{n \to \infty} v_n = v = e.$$

(4) Suppose that e is rational. We write $e = \frac{p}{q}$ where $p, q \in \mathbb{N}$. Then since $u_q < e < v_q$, we have

$$1 + \frac{1}{1!} + \frac{1}{2!} + \cdots + \frac{1}{q!} < \frac{p}{q} < 1 + \frac{1}{1!} + \frac{1}{2!} + \cdots + \frac{1}{q!} + \frac{1}{q\,(q!)}.$$

Multiplying these inequalities by $q\,(q!)$, we obtain

$$q\,(q!) + \cdots + q < p\,(q!) < q\,(q!) + \cdots + q + 1.$$

We then notice that the number $N = q\,(q!) + \cdots + q \in \mathbb{N}$, and that the natural number $p\,(q!)$ is such that

$$N < p\,(q!) < N + 1.$$

Such a double inequality is impossible. Hence e cannot be rational. $\qquad\square$

Example 2.22 (Nested Intervals)

Let

$$I_n = \{x \in \mathbb{R} : a_n \leq x \leq b_n\}, \quad n = 1, 2, \ldots,$$

be a sequence of closed bounded intervals such that

$$I_1 \supset I_2 \supset I_3 \supset \cdots \supset I_n \supset I_{n+1} \supset \cdots.$$

Suppose that $\lim (b_n - a_n) = 0$. Show that the I_n have exactly one common element.

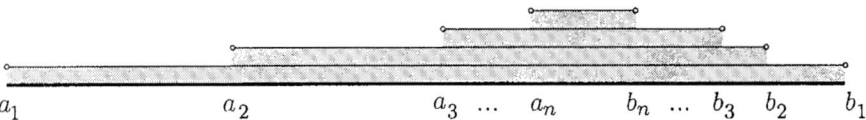

a_1 $\qquad\qquad$ a_2 $\qquad\qquad$ a_3 ... a_n $\qquad$ b_n ... b_3 b_2 $\qquad$ b_1

Figure 2.3   Nested intervals

Solution

The sequence (a_n) is nondecreasing and bounded above by b_1 (Figure 2.3). In view of Theorem 2.20, $a = \lim a_n$ exists as a real number. Since for each n and each p

$$a_n \le a_{n+p} \le b_{n+p} \le b_n,$$

we have $a \le b_p$ for every p. Therefore $a \in I_p$ for all p. Hence $\bigcap_{n=1}^{\infty} I_n \ni a$. Suppose that $b \in \bigcap_{n=1}^{\infty} I_n$. Then $b \in [a_n, b_n]$ for each n. Thus $0 \le b - a_n \le b_n - a_n$. It follows from $\lim (b_n - a_n) = 0$ that $b = \lim a_n = a$. This completes our proof. $\qquad\square$

The following example makes the result of the Density Theorem 1.44 more precise.

Example 2.23

Show that any real number can be expressed as the limit of a sequence of rational numbers.

The readers should convince themselves that the construction in Example 1.45 (page 22) provides a quick solution to this example. Here is another construction.

Solution

Fix $a \in \mathbb{R}$. Then by the density theorem, there is a rational number r_1 with

$$a - 1/1 < r_1 < a.$$

Now applying the density theorem to $\max \{r_1, a - 1/2\} < a$, there exists rational number r_2 with

$$\max \{r_1, a - 1/2\} < r_2 < a.$$

In particular, we have

$$r_1 < r_2 < a \text{ and } |a - r_2| < 1/2.$$

Similarly, we apply the density theorem to $\max\{r_2, a - 1/3\} < a$, and we can find another rational number r_3 such that

$$r_2 < r_3 < a \text{ and } |a - r_3| < 1/3.$$

Continuing in this fashion, we inductively construct an increasing sequence (r_n) of rational numbers such that

$$|a - r_n| < 1/n \text{ for each } n.$$

Since the sequence (r_n) is bounded above by a, Theorem 2.20 informs us that $r = \lim r_n$ exists. We wish to show that $a = r$. To see this, fix $\varepsilon > 0$. We can choose $N_1 \in \mathbb{N}$ large enough so that

$$|r - r_n| < \frac{\varepsilon}{2} \text{ whenever } n > N_1.$$

On the other hand, by the Archimedean property, we can choose N_2 large enough so that $1 < n\varepsilon/2$ for $n > N_2$. It follows that for $n > \max\{N_1, N_2\}$, we have

$$|r - a| \le |r - r_n| + |r_n - a|$$
$$\le \frac{\varepsilon}{2} + \frac{1}{n} < \varepsilon.$$

Since $\varepsilon > 0$ is arbitrary, we must have $|r - a| = 0$ or equivalently $a = r$. $\square$

Later we will find a critically important test for convergence of sequences. For now, we establish a few basic rules for dealing with the limit of convergent sequences.

Proposition 2.24[2]

Suppose that (a_n) and (b_n) are convergent sequences in $\mathbb{R}$, and let $k \in \mathbb{R}$. Then the sequences (ka_n), $(a_n + b_n)$, and $(a_n b_n)$ are all convergent and

(1) $\lim (ka_n) = k \lim (a_n)$;

(2) $\lim (a_n + b_n) = \lim a_n + \lim b_n$;

(3) $\lim (a_n b_n) = \lim a_n \cdot \lim b_n$;

(4) $\lim \left(\frac{a_n}{b_n}\right) = \frac{\lim a_n}{\lim b_n}$, provided $b_n \ne 0$ for all n and $\lim b_n \ne 0$.

---

[2] One should notice that these results do not apply to infinite limits.

Proof

For simplicity, we let $\lim a_n = a$ and $\lim b_n = b$.

(1) Let $\varepsilon > 0$. If $k \neq 0$, there is N in $\mathbb{N}$ such that $|a_n - a| < \varepsilon / |k|$ for all $n > N$. Then

$$|ka_n - ka| = |k| |a_n - a| < \varepsilon,$$

for all $n > N$. Hence $\lim (ka_n) = ka = k \lim (a_n)$ as desired. The case $k = 0$ is trivial.

(2) Again let $\varepsilon > 0$. We want to show that for large enough n

$$|(a_n + b_n) - (a + b)| < \varepsilon.$$

Since $a_n \to a$, there is N_1 in $\mathbb{N}$ such that

$$|a_n - a| < \frac{\varepsilon}{2} \text{ for all } n > N_1.$$

Similarly since $b_n \to b$, there is N_2 in $\mathbb{N}$ such that

$$|b_n - b| < \frac{\varepsilon}{2} \text{ for all } n > N_2.$$

It follows that for all $n > \max \{N_1, N_2\}$,

$$\begin{aligned}
|(a_n + b_n) - (a + b)| &= |(a_n + a) - (b_n + b)| \\
&\leq |a_n - a| + |b_n - b| \\
&< \frac{\varepsilon}{2} + \frac{\varepsilon}{2} = \varepsilon,
\end{aligned}$$

as expected.

(3) Let $\varepsilon > 0$. We need to prove that $|a_n b_n - ab| < \varepsilon$ for large enough n. By Example 2.19, since the sequence (a_n) converges, it is bounded, i.e. there is a constant $M_0 > 0$ such that $|a_n| < M_0$ for all n. Let $M = \max \{M_0, b\}$. Since $a_n \to a$, there is N_1 in $\mathbb{N}$ such that

$$|a_n - a| < \frac{\varepsilon}{2M} \text{ for all } n > N_1.$$

Similarly since $b_n \to b$, there is N_2 in $\mathbb{N}$ such that

$$|b_n - b| < \frac{\varepsilon}{2M} \text{ for all } n > N_2.$$

From the triangle inequality, we have for $n > \max \{N_1, N_2\}$

$$\begin{aligned}
|a_n b_n - ab| &= |a_n b_n - a_n b + a_n b - ab| \\
&\leq |a_n| |b_n - b| + |a_n - a| |b| \\
&\leq \frac{\varepsilon M_0}{2M} + \frac{\varepsilon |b|}{2M} < \varepsilon,
\end{aligned}$$

as desired.

(4) It is sufficient to show that $\lim \frac{1}{b_n} = \frac{1}{b}$, for once this is done the desired result may be obtained by applying (3) to the $a_n \cdot \frac{1}{b_n}$. Let $\varepsilon > 0$. Since $b_n \neq 0$ and $\lim b_n = b \neq 0$, there is $m > 0$ such that $|b_n| > m$ for all n. Also there exists N in $\mathbb{N}$ such that

$$n > N \text{ implies } |b - b_n| < \varepsilon m \, |b| \, .$$

Then for $n > N$, we have

$$\left| \frac{1}{b_n} - \frac{1}{b} \right| = \frac{|b - b_n|}{|b_n b|} \leq \frac{|b - b_n|}{m \, |b|} < \varepsilon.$$

The proof is complete.   □

The above result also implies that the difference of two convergent sequences is convergent. Both of the sequences $(\sqrt{n})$ and $(\sqrt{n+1})$ are unbounded, and therefore both are divergent; as the next example shows, their difference is, nevertheless, a convergent sequence.

Example 2.25

Show that the sequence $(\sqrt{n+1} - \sqrt{n})$ is convergent.

Solution

First, we write

$$0 \leq \sqrt{n+1} - \sqrt{n}$$

$$= (\sqrt{n+1} - \sqrt{n}) \frac{(\sqrt{n+1} + \sqrt{n})}{(\sqrt{n+1} + \sqrt{n})}$$

$$= \frac{1}{\sqrt{n+1} + \sqrt{n}} < \frac{1}{2\sqrt{n}}.$$

To finish the proof, for a given $\varepsilon > 0$, we can apply the Archimedean property to 1 and $4\varepsilon^2$ to find $N \in \mathbb{N}$ so that $1 < 2\sqrt{N}\varepsilon$. Hence for $n > N$, we have $\left| \sqrt{n+1} - \sqrt{n} - 0 \right| < \varepsilon$. The sequence $(\sqrt{n+1} - \sqrt{n})$ converges to 0.   □

2.3 Subsequences

Suppose that $(a_n)_{n \in \mathbb{N}}$ is a sequence and consider an ordered subset of $\mathbb{N}$, say $S = \{n_1, n_2, n_3, \ldots\}$ where $n_1 < n_2 < n_3 < \cdots$. Then we define a new sequence by considering only the terms $a_{n_1}, a_{n_2}, a_{n_3}, \ldots$. Since the terms of the new sequence $(a_{n_k})_{k \in \mathbb{N}}$ are selected from the original sequence $(a_n)_{n \in \mathbb{N}}$, the set of values of the sequence $(a_{n_k})_{k \in \mathbb{N}}$ is contained in the set of values of the original sequence $(a_n)_{n \in \mathbb{N}}$. The new sequence $(a_{n_k})_{k \in \mathbb{N}}$ is called a *subsequence* of the sequence $(a_n)_{n \in \mathbb{N}}$. This concept is more concisely defined in

Definition 2.26

Let $a = (a_n)_{n \in \mathbb{N}}$ be a sequence and $f : \mathbb{N} \longrightarrow \mathbb{N}$ be an increasing function (i.e. $f(n) < f(n+1)$ for all $n \in \mathbb{N}$). Then $a \circ f$ is called a **subsequence** of the sequence a. We denote $a(f(k))$ by a_{n_k}.

For example, consider the sequence $(a_n)_{n \in \mathbb{N}}$ defined by $a_n = (-1)^n \frac{1}{n}$, and let $f : \mathbb{N} \longrightarrow \mathbb{N}$ be the function defined by $f(k) = 2k$. It is clear that f is an increasing function. Hence $a \circ f$ defines a subsequence of (a_n) namely the sequence (a_{n_k}) defined by $a_{n_k} = (-1)^{2k} \frac{1}{2k} = \frac{1}{2k}$ for each $k \in \mathbb{N}$.

Remark 2.27

Let $(a_n)_{n \in \mathbb{N}}$ be a sequence. If $(a_{n_k})_{k \in \mathbb{N}}$ is a subsequence of $(a_n)_{n \in \mathbb{N}}$, then $n_k \geq k$ for all $k \in \mathbb{N}$.

Proof

The proof is by induction. Let $A = \{k \in \mathbb{N} : n_k \geq k\}$. Since $n_1 \in \mathbb{N}$, $n_1 \geq 1$. Hence $1 \in A$. Suppose that $k \in A$. Then $n_{k+1} > n_k \geq k$. Hence $n_{k+1} \geq k+1$, and thus $k+1 \in \mathbb{N}$. By the principal of mathematical induction $A = \mathbb{N}$. □

Theorem 2.28

Let $(a_n)_{n \in \mathbb{N}}$ be a convergent sequence. Then every subsequence (a_{n_k}) of (a_n) is a convergent sequence and

$$\lim a_{n_k} = \lim a_n.$$

Proof

Suppose that $\lim a_n = a$. Let $\varepsilon > 0$. There exists N in $\mathbb{N}$ such that $n > N$ implies $|a_n - a| < \varepsilon$. If $k > N$, then by Remark 2.27 $n_k > N$ which implies that $|a_{n_k} - a| < \varepsilon$. Hence $\lim a_{n_k} = \lim a_n$. □

In other words, if a sequence converges, then every subsequence converges to the same limit.

Example 2.29

Determine whether the sequence $a_n = \sin\left(\frac{n\pi}{2}\right) + \cos n\pi$ converges.

Solution

For each $n \in \mathbb{N}$, we have

$$a_{4n} = \sin\left(\frac{4n\pi}{2}\right) + \cos 4n\pi = 0 + 1 = 1,$$
$$a_{4n+1} = \sin\left(\frac{(4n+1)\pi}{2}\right) + \cos(4n+1)\pi = 1 - 1 = 0.$$

Thus the subsequence (a_{4n}) converges to 1 while the subsequence (a_{4n+1}) converges to 0. If (a_n) were convergent, then by Theorem 2.28, it would converge to 1 and to 0. Such a situation is impossible according to the uniqueness Theorem 2.18; therefore we conclude that the sequence (a_n) does not converge.   □

In the above example, we infer that infinitely many values of the sequence (a_n) cluster around the number 0 or the number 1. We say that 0 and 1 are cluster points for the sequence (a_n). Formally, a number a is called a **cluster point** of a sequence (a_n) if it can be expressed as the limit of some subsequence of (a_n).

So, for example, combining the result of Theorem 2.18 with that of Theorem 2.28, we see that a sequence is convergent if and only if it has a unique cluster point.

Another fundamental property of sequences of real numbers is

Theorem 2.30 (Bolzano–Weierstrass)

Every bounded sequence of real numbers has a convergent subsequence.

Another way of stating the Bolzano–Weierstrass Theorem is: a bounded sequence of real numbers has at least one cluster point.

Proof

Suppose that (a_n) is a bounded sequence. We will construct a subsequence and show that it converges. Since for each $m \in \mathbb{N}$ the set $\{a_n : n \geq m\}$ is bounded, we can define a sequence (b_m) by $b_m = \sup\{a_n : n \geq m\}$. Then (b_m)

is bounded. On the other hand, since for each $m \in \mathbb{N}$ we have

$$\{a_n : n \geq m + 1\} \subset \{a_n : n \geq m\},$$

then $b_{m+1} \leq b_m$, i.e. (b_m) is a nonincreasing sequence. Hence $\lim b_n = b$ exists.

We now construct a subsequence (a_{n_k}) of (a_n) converging to b. First, take $\varepsilon = 1/2$. Since $b_m \to b$, we can choose m_1 large enough so that $|b_{m_1} - b| < 1/2$. For such m_1, there is $n_1 \geq m_1$ such that $b_{m_1} - 1/2 < a_{n_1} < b_{m_1}$. Thus by the triangle inequality,

$$|a_{n_1} - b| \leq |a_{n_1} - b_{m_1}| + |b_{m_1} - b|$$
$$< \frac{1}{2} + \frac{1}{2} = 1.$$

Now take $\varepsilon = 1/4$. We can find $m_2 > n_1$ such that $|b_{m_2} - b| < 1/4$. Again for such m_2 there is $n_2 \geq m_2$ such that $b_{m_2} - 1/4 < a_{n_1} < b_{m_2}$, and thus

$$|a_{n_2} - b| \leq |a_{n_2} - b_{m_2}| + |b_{m_2} - b|$$
$$< \frac{1}{4} + \frac{1}{4} = \frac{1}{2}.$$

Suppose that we have chosen $n_1 < n_2 < n_3 < \cdots < n_k$ such that

$$|a_{n_i} - b| < \frac{1}{2i} \text{ for } 1 \leq i \leq k.$$

Take $\varepsilon = 1/(2(k+1))$. Choose $m_{k+1} > n_k$ such that $\left| b_{m_{k+1}} - b \right| < \frac{1}{2(k+1)}$. For such m_{k+1} there is $n_{k+1} \geq m_{k+1}$ such that $b_{m_{k+1}} - \frac{1}{2(k+1)} < a_{n_{k+1}} < b_{m_{k+1}}$. Therefore

$$\left| a_{n_{k+1}} - b \right| \leq \left| a_{n_{k+1}} - b_{m_{k+1}} \right| + \left| b_{m_{k+1}} - b \right|$$
$$< \frac{1}{2(k+1)} + \frac{1}{2(k+1)} = \frac{1}{k+1}.$$

By the principle of mathematical induction for all $k \in \mathbb{N}$ we have $n_{k+1} > n_k$ and $|a_{n_k} - b| < \frac{1}{k}$. Hence (a_{n_k}) is a subsequence of the sequence (a_n) which converges to b. $\qquad\square$

2.4 Upper and Lower Limits

Let (a_n) be a sequence of real numbers. For each fixed n consider the set

$$\{a_{n+p} : p \in \mathbb{N}\} = \{a_{n+1}, a_{n+2}, a_{n+3}, \ldots\}.$$

Let (u_n) and (v_n) denote the sequences defined respectively by

$$u_n = \inf \{a_{n+p} : p \in \mathbb{N}\} \text{ and } v_n = \sup \{a_{n+p} : p \in \mathbb{N}\} .$$

Since $\{a_{(n+1)+p} : p \in \mathbb{N}\} \subset \{a_{n+p} : p \in \mathbb{N}\}$ we have

$$u_1 \leq u_2 \leq u_3 \leq \cdots \text{ and } v_1 \geq v_2 \geq v_3 \geq \cdots .$$

If the sequence (a_n) is bounded, then so are the sequences (u_n) and (v_n). It follows from Theorem 2.20 that both $\lim u_n = u$ and $\lim v_n = v$ exist. If the sequence (a_n) is not bounded below (resp. above), we simply agree to write $\lim u_n = -\infty$ (resp. $\lim v_n = +\infty$).

Definition 2.31

Let (a_n) be a sequence of real numbers. We define the **lower limit** and the **upper limit** of (a_n) respectively by

(1) $\liminf a_n = \lim_{n \to \infty} (\inf \{a_{n+p} : p \in \mathbb{N}\})$;

(2) $\limsup a_n = \lim_{n \to \infty} (\sup \{a_{n+p} : p \in \mathbb{N}\})$.

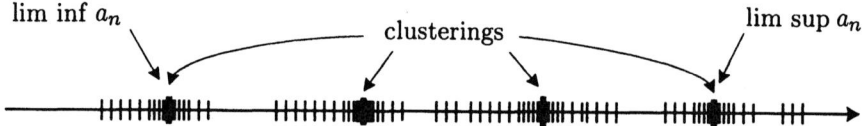

Figure 2.4   Upper and lower limits

If (a_n) is a sequence of real numbers, then the following facts are obvious:

$$\inf \{a_p : p \in \mathbb{N}\} \leq \liminf a_n \leq \limsup a_n \leq \sup \{a_p : p \in \mathbb{N}\} .$$

Theorem 2.32

Let (a_n) be a sequence of real numbers. Then $\lim a_n$ is defined if and only if $\liminf a_n = \limsup a_n$. Moreover, in such a case $\lim a_n = \liminf a_n = \limsup a_n$.

Proof

Suppose first that $\lim a_n = +\infty$. Let $M > 0$. Then there exists n in $\mathbb{N}$ such that $a_{n+p} > M$ for all $p \in \mathbb{N}$. Then $u_n = \inf \{a_{n+p} : p \in \mathbb{N}\} \geq M$. This shows that $\lim u_n = +\infty$, i.e. $\liminf a_n = +\infty$. Hence $\liminf a_n = \limsup a_n = +\infty$. The case $\lim a_n = -\infty$ is left as an exercise.

Now suppose that $\lim a_n = a$. Let $\varepsilon > 0$. There is $n \in \mathbb{N}$ such that $|a_{n+p} - a| < \varepsilon$ for all $p \in \mathbb{N}$. Thus $a_{n+p} < a + \varepsilon$ for all p in $\mathbb{N}$ and so

$$v_n = \sup\{a_{n+p} : p \in \mathbb{N}\} \leq a + \varepsilon.$$

Since (v_n) is a nonincreasing sequence, we have for $m > n$, $v_m \leq a + \varepsilon$. So $\limsup a_n = \lim v_n \leq a + \varepsilon$. Since $\varepsilon > 0$ was chosen arbitrarily, we get $\limsup a_n \leq a = \lim a_n$. A similar argument shows that $\lim a_n \leq \liminf a_n$. Finally since $\liminf a_n \leq \limsup a_n$, we conclude that

$$\lim a_n = \liminf a_n = \limsup a_n.$$

Conversely, suppose that $\liminf a_n = \limsup a_n = a$, and let $\varepsilon > 0$. There exists n_0 such that

$$|\sup\{a_{n_0+p} : p \in \mathbb{N}\} - a| < \varepsilon.$$

Thus $\sup\{a_{n_0+p} : p \in \mathbb{N}\} < a + \varepsilon$ and so

$$a_{n_0+p} < a + \varepsilon \text{ for all } p \in \mathbb{N}. \tag{2.4}$$

Similarly, there exists n_1 such that $|\inf\{a_{n_1+p} : p \in \mathbb{N}\} - a| < \varepsilon$, so

$$a - \varepsilon < a_{n_1+p} \text{ for all } p \in \mathbb{N}. \tag{2.5}$$

Combining (2.4) and (2.5), we have

$$a - \varepsilon < a_n < a + \varepsilon \text{ for all } n > \max\{n_0, n_1\}.$$

This proves that $\lim a_n = a$. $\qquad\square$

The limit $\limsup a_n$ ($\liminf a_n$) can be thought of as the supremum (infinimum) of all cluster points of the sequence (a_n). See Figure 2.4.

Example 2.33

Let (a_n) be a bounded sequence of real numbers. Show that

(1) $\limsup a_n$ is the largest cluster point of (a_n);

(2) $\liminf a_n$ is the smallest cluster point of (a_n).

Solution

We only prove (1). The proof of (2) uses similar arguments.

Since (a_n) is bounded, $b = \limsup (a_n)$ exists as a real number. We first choose n_1 such that

$$b - 1 < a_{n_1} < b + 1.$$

Let $\varepsilon = 1/2$. There is $N_1 \in \mathbb{N}$ large enough so that $a_{N_1+p} < b + 1/2$ for all p. Thus we can choose $n_2 \geq \max \{n_1, N_1 + 1\}$ such that

$$b - \frac{1}{2} < a_{n_2} < b + \frac{1}{2}.$$

Now let $\varepsilon = 1/2^2$. There is $N_2 > N_1$ in $\mathbb{N}$ large enough so that $a_{N_2+p} < b + 1/2^2$ for all p. Then we can choose $n_3 \geq \max \{n_2, N_2 + 1\}$ such that

$$b - \frac{1}{2^2} < a_{n_3} < b + \frac{1}{2^2}.$$

Continuing in this fashion, we construct a subsequence (a_{n_k}) of (a_n) which converges to b. Thus we have shown that b is indeed a cluster point.

Now suppose that c is another cluster point of (a_n), say $c = \lim \left(a_{f(n)} \right)$ where $f : \mathbb{N} \to \mathbb{N}$ is an increasing function. We define

$$c_m = \sup \left\{ a_{f(n)} : n \geq m \right\}.$$

Since $m \leq f(m)$, we have $c_m \leq \sup \{a_n : n \geq m\}$. It follows that

$$c \leq \lim c_m \leq \limsup (a_n) = b.$$

Hence every cluster point of (a_n) is at most equal to b. We have finished the proof.   □

Example 2.34

Let (a_n) be a bounded sequence of real numbers. Show that

(1) $a = \limsup a_n$ if and only if for every $\varepsilon > 0$ there is only a finite number of n's such that $a_n > a + \varepsilon$, and there exist infinitely many n's such that $a_n > a - \varepsilon$;

(2) $a = \liminf a_n$ if and only if for every $\varepsilon > 0$ there is only a finite number of n's such that $a_n < a - \varepsilon$, and there exist infinitely many n's such that $a_n < b + \varepsilon$.

Solution

Again, we only prove (1).

Suppose that $a = \limsup a_n$. Let $\varepsilon > 0$. There exists n_0 such that

$$|\sup \{a_{n_0+p} : p \in \mathbb{N}\} - a| < \varepsilon.$$

Thus $\sup \{a_{n_0+p} : p \in \mathbb{N}\} > a - \varepsilon$ and so

$$a_{n_0+p} > a - \varepsilon \text{ for all } p \in \mathbb{N}.$$

That is, there exist infinitely many n's such that $a_n > a - \varepsilon$.

Now suppose that for some $\varepsilon_0 > 0$, there exist infinitely many n's such that $a_n > a + \varepsilon_0$. Let $(a_{n'})$ be those a_n with the property that $a_n > a + \varepsilon_0$. Then by the Bolzano–Weierstrass Theorem, $(a_{n'})$ would have a subsequence converging to some number $c \geq a + \varepsilon_0$. In other words c is a cluster point for (a_n) and $c > a$. This would contradict the fact established in the previous example. Thus there can only be a finite number of n's such that $a_n > a + \varepsilon$. $\qquad\square$

2.5 Cauchy Criterion

We now introduce a concept which is of great practical and theoretical importance in classical analysis. To see whether a sequence is convergent or not, it is not always necessary to find the actual limit. The *Cauchy criterion* offers a critical tool for a test for convergence. First we prove the following result.

Theorem 2.35

Let (a_n) be a convergent sequence. Then for every $\varepsilon > 0$ there exists a number $N \in \mathbb{N}$ such that $|a_n - a_m| < \varepsilon$ for all $n, m \in \mathbb{N}$.

Proof

Suppose that $\lim a_n = a$. Let $\varepsilon > 0$. There is an N in $\mathbb{N}$ such that

$$n > N \text{ implies } |a_n - a| < \varepsilon/2.$$

Thus for $n, m > N$ we have

$$|a_n - a_m| \leq |a_n - a| + |a - a_m|$$
$$< \frac{\varepsilon}{2} + \frac{\varepsilon}{2} = \varepsilon,$$

as desired. $\qquad\square$

Loosely speaking, the terms of a convergent sequence can be made arbitrarily close to each other. This concept is defined more precisely as follows.

Definition 2.36

A sequence (a_n) of real numbers is said to be **Cauchy,** or to satisfy the **Cauchy criterion**, if for every $\varepsilon > 0$ there exists a number $N \in \mathbb{N}$ such that $|a_n - a_m| < \varepsilon$ for all $n, m \geq N$.

Sometimes it is more convenient to write the Cauchy criterion in the following equivalent form:

A sequence (a_n) is Cauchy if and only if for every $\varepsilon > 0$ there exists a number $N \in \mathbb{N}$ such that $|a_{n+p} - a_n| < \varepsilon$ for all $n > N$, and for all $p \in \mathbb{N}$.

Example 2.37

Let (x_n) be the sequence in $\mathbb{R}$ defined recursively by

$$x_1 = 1, \ x_2 = 2, \ldots, x_{n+1} = \frac{1}{2}(x_n + x_{n-1})$$

for $n > 2$. Show that (x_n) satisfies the Cauchy criterion.

Solution

We first notice that for each n, the equality $x_{n+1} = \frac{1}{2}(x_n + x_{n-1})$ implies

$$x_{n+1} - x_n = -\frac{1}{2}(x_n - x_{n-1}).$$

It follows that for each n,

$$
\begin{aligned}
|x_{n+1} - x_n| &= \frac{1}{2}|x_n - x_{n-1}| \\
&= \frac{1}{2^2}|x_{n-1} - x_{n-2}| = \cdots \\
&= \frac{1}{2^{n-1}}|x_2 - x_1| = \frac{1}{2^{n-1}}.
\end{aligned}
$$

Thus for $p, n \in \mathbb{N}$, we have

$$
\begin{aligned}
|x_{n+p} - x_n| &\leq |x_{n+p} - x_{n+p-1}| + \cdots + |x_{n+1} - x_n| \\
&= \frac{1}{2^{n+p-1}} + \cdots + \frac{1}{2^{n-1}} \\
&= \frac{1}{2^{n-1}} \left(\frac{1}{2^p} + \frac{1}{2^{p-1}} + \cdots + \frac{1}{2} + 1 \right) \\
&< \frac{1}{2^{n-2}}.
\end{aligned}
$$

It follows that for any given $\varepsilon > 0$, we can choose N large enough so that $1/2^{n-2} < \varepsilon$ for all $n > N$. Therefore

$$
|x_{n+p} - x_n| < \varepsilon \text{ whenever } n > N \text{ and } p \in \mathbb{N},
$$

i.e. the sequence (x_n) satisfies the Cauchy criterion.   □

According to the above definition, Theorem 2.35 can be restated as: "Every convergent sequence is Cauchy". It turns out that the converse is also true; technically, we say $\mathbb{R}$ is **complete**. First we have the following lemma.

Lemma 2.38

Every Cauchy sequence is bounded.

Proof

Let (a_n) be a Cauchy sequence. Consider $\varepsilon = 1$. There is N in $\mathbb{N}$ such that

$$
|a_{n+p} - a_n| < 1 \text{ for all } n \geq N \text{ and for all } p \in \mathbb{N}.
$$

In particular, $|a_{N+p}| < |a_{N+1}| + 1$, for all $p \in \mathbb{N}$. It follows that if

$$
M = \max \{ |a_1|, |a_2|, \ldots, |a_N|, |a_{N+1}| + 1 \},
$$

then $|a_n| \leq M$ for all $n \in \mathbb{N}$.   □

Theorem 2.39

A sequence of real numbers is convergent if and only if it is Cauchy.

Proof

We already saw the necessity (Theorem 2.35). For the sufficiency, suppose that (a_n) is a Cauchy sequence and let $\varepsilon > 0$. There exists N in $\mathbb{N}$ such that

$$|a_{n+p} - a_n| < \varepsilon \text{ for all } n \geq N \text{ and for all } p \in \mathbb{N}.$$

In particular $a_N - \varepsilon < a_{N+p} < a_N + \varepsilon$, for all $p \in \mathbb{N}$. Thus

$$
\begin{aligned}
a_N - \varepsilon &\leq \inf \{a_{N+p} : p \in \mathbb{N}\} \\
&\leq \liminf a_n \leq \limsup a_n \\
&\leq \sup \{a_{N+p} : p \in \mathbb{N}\} \leq a_N + \varepsilon.
\end{aligned}
$$

It follows that $0 \leq \limsup a_n - \liminf a_n < 2\varepsilon$. Since $\varepsilon > 0$ was arbitrary, we conclude that $\limsup a_n = \liminf a_n$, so by Theorem 2.32 (a_n) is convergent. $\qquad\square$

Example 2.40

Show that the sequence (a_n), defined by:

(1) $a_n = 1 + \frac{1}{2} + \cdots + \frac{1}{n}$ for $n = 1, 2, \cdots$, does not converge;

(2) $a_n = 1 + \frac{1}{1!} + \cdots + \frac{1}{n!}$ for $n = 1, 2, \cdots$, converges.

Solution

(1) For each n, we notice that

$$a_{2n} - a_n = \frac{1}{n+1} + \frac{1}{n+2} + \cdots + \frac{1}{n+n} > \frac{1}{2n} + \frac{1}{2n} + \cdots + \frac{1}{2n} = \frac{1}{2}.$$

Thus the sequence (a_n) cannot be Cauchy and therefore it does not converge.

(2) Utilizing the result of Example 2.21, we have for each n, and for each p,

$$
\begin{aligned}
a_{n+p} - a_n &= \frac{1}{(n+1)!} + \frac{1}{(n+2)!} + \cdots + \frac{1}{(n+p)!} \\
&= \frac{1}{(n+1)!} \left[1 + \frac{(n+1)!}{(n+2)!} + \cdots + \frac{(n+1)!}{(n+p)!} \right] \\
&\leq \frac{1}{(n+1)!} \left[1 + \frac{1}{2!} + \cdots + \frac{1}{p!} \right] \\
&< \frac{1}{(n+1)!} e.
\end{aligned}
$$

It follows that for a given $\varepsilon > 0$, we can choose $N \in \mathbb{N}$ large enough so that $e/(n+1)! < \varepsilon$ whenever $n > N$. Therefore,

$$|a_{n+p} - a_n| < \varepsilon \text{ whenever } n > N \text{ and for every } p \in \mathbb{N}.$$

This shows that the sequence (a_n) is Cauchy, and as such, it converges. □

EXERCISES

2.1 Show that every subset of a countable set is countable.

2.2 Let $(A_n)_{n \in \mathbb{N}}$ be a family of countable sets. Show that the union $S = \bigcup_{n=1}^{\infty} A_n$ is countable.

2.3 Let $A = \{(x_n) : x_n \in \{0, 1\} \text{ for all } n\}$. Show that A is uncountable.

2.4 Show that the set of all finite subsets of a countable set is countable.

2.5 Show that a set A is countable if and only if there exists a one-to-one correspondence from A into $\mathbb{N}$.

2.6 Write out (in terms of $\varepsilon > 0$ and $\delta > 0$) the negation of "$\lim a_n = a$".

2.7 Determine whether each given sequence is monotonic, bounded, or convergent.

(a) $\frac{1}{n}$ (c) $\frac{\cos n}{n}$ (e) $\frac{n!}{e^n}$ (g) $(-2)^{-n}$

(b) $(-2)^n$ (d) $n \sin \frac{1}{n}$ (f) $\frac{n^2}{n^2+1}$ (h) $n^{1/n}$

2.8 Let (a_n) be the sequence defined by $a_1 = 1/2$ and $a_{k+1} = \frac{k+1}{k} a_k$ for $k \geq 1$.

(1) Find a_1, a_2, and a_3. Show that $a_n > 0$ for all n.

(2) Show that (a_n) is increasing and that it diverges.

2.9 Let (a_n) be the sequence defined by $a_1 = a$ and $a_{k+1} = r a_k$ for $k \geq 1$, where $a \in \mathbb{R}$ and $r > 0$.

(1) Show that $a_n = r^{n-1} a$ for all $n \in \mathbb{N}$.

(2) Deduce that $a_n \to 0$ if $0 < r < 1$, $a_n \to a$ for $r = 1$, and $a_n \to \infty$ for $r > 1$.

(3) Determine $\lim a_n$.

2.10 Let (a_n) be the sequence defined by $a_1 = 1$ and $a_{k+1} = \sqrt{a_k + 1}$ for $k \geq 1$.

(1)   Show that (a_n) is increasing (i.e. $a_{n+1} > a_n$ for all n).

(2)   Show that $a_n < 2$ for all n.

(3)   Determine $\lim a_n$.

2.11   Let (a_n) be the sequence defined by $a_1 = 1$ and $a_{k+1} = \frac{1}{2}\left(a_k + \frac{1}{a_k}\right)$ for $k \geq 1$.

(1)   Show that (a_n) is convergent.

(2)   Determine $\lim a_n$.

2.12   Show that $\lim_{n \to \infty} x^n = 0$ if $|x| < 1$.

2.13   Show that $\lim_{n \to \infty} n^{1/n} = 1$.

2.14   Show that $\lim_{n \to \infty} \left(1 + n + n^2\right)^{1/n} = 1$.

2.15   Show that the sequence $\left(\left(1 + \frac{1}{n}\right)^n\right)_{n \in \mathbb{N}}$ is increasing and bounded. (The limit is the number e.)

2.16   Prove that $n! > (n/e)^n$ for all $n \in \mathbb{N}$. (Hint: use the result in the previous problem.)

2.17   Prove that $\lim_{n \to \infty} \frac{1}{n}\left(1 + \frac{1}{2} + \frac{1}{3} + \cdots + \frac{1}{n}\right) = 0$.

2.18   **Césaro summability**. A sequence (a_n) is said to be Césaro summable or C1-summable if $\lim_{n \to \infty} \frac{a_1 + a_2 + \cdots + a_n}{n}$ exists. Show that if (a_n) is convergent, then it is C1-summable and $\lim_{n \to \infty} a_n = \lim_{n \to \infty} \frac{a_1 + a_2 + \cdots + a_n}{n}$. Show that the converse is not true.

2.19   **Abel summability**. A sequence (a_n) is said to be Abel summable if for any sequence (r_k) in the interval $(0, 1)$ converging to 1, one has for each k, $\sigma_k = \lim_{n \to \infty} \left(r_k a_1 + r_k^2 a_2 + \cdots + r_k^n a_n\right)$ exists as a real number, and the sequence (σ_k) converges. Show that if (a_n) is convergent, then it is Abel summable and $\lim_{n \to \infty} a_n = \lim_{k \to \infty} \sigma_k$. Show that the converse is not true.

2.20   Let (a_n) be the **Fibonacci** sequence defined by $a_1 = a_2 = 1$ and $a_{k+1} = a_k + a_{k-1}$ for $k > 1$.

(1)   Observe that $a_n > 0$ for all n, and show that (a_n) is increasing (i.e. $a_{n+1} > a_n$ for all n).

(2)   Consider the sequence (**golden ratio**) defined by $r_k = a_{k+1}/a_k$ for each $k \in \mathbb{N}$. Show that (r_n) is increasing and $1 < r_n < 2$, for all n.

(3)   Determine $\lim_{n \to \infty} r_n$.

2.21 Determine

    (1) $\lim_{n\to\infty} a^{\frac{1}{n}}$ for $a > 0$;

    (2) $\lim_{n\to\infty} n\left(a^{\frac{1}{n}} - 1\right)$ for $a > 0$.

2.22 Let $0 < a \leq b$. Show that the sequence (x_n) defined by $x_n = (a^n + b^n)^{1/n}$ converges to b.

2.23 Verify that $\lim_{n\to\infty} \frac{1+\sqrt{2}+\sqrt[3]{3}+\cdots+\sqrt[n]{n}}{n} = 1$.

2.24 Determine (a) $\lim_{n\to\infty} \frac{1+2+\cdots+n}{n^2}$; (b) $\lim_{n\to\infty} \frac{1+2^2+\cdots+n^2}{n^3}$.

2.25 Complete the proof of Theorem 2.20, by dealing with the case where (a_n) is nonincreasing.

2.26 Complete the proof of Theorem 2.32, by dealing with the case $\lim a_n = -\infty$.

2.27 Let (a_n) and (b_n) be sequences of real numbers. Show that

    (1) if $a_n \to a$, then $|a_n| \to |a|$;

    (2) if $a_n \to a$, and $|a_n - b_n| \to 0$, then $b_n \to a$;

    (3) if $a_n \to a$, and $b_n \to b$, then $|a_n - b_n| \to |a - b|$;

    (4) if $a_n \to a$, then $|a_n - b| \to |a - b|$.

2.28 Write out the proof of the fact that a subsequence of a Cauchy sequence is a Cauchy sequence.

2.29 Let (a_n) and (b_n) be Cauchy sequences of real numbers. Show that the sequence $(c_n = |a_n - b_n|)$ converges.

2.30 **Sandwich theorem.** Let (a_n), (b_n), and (c_n) be sequences of real numbers. Suppose that for each $n \in \mathbb{N}$, $a_n \leq c_n \leq b_n$ and that both (a_n) and (b_n) converge to the same limit a. Show that (c_n) also converges to the same limit a.

2.31 Give an example of a convergent sequence (a_n) in a set A where $\lim a_n \notin A$.

2.32 Suppose that (a_n) is a sequence of nonnegative real numbers converging to a. Show that $a \geq 0$.

2.33 Let (a_n) and (b_n) be sequences of real numbers. Suppose that for each $n \in \mathbb{N}$, $a_n \geq b_n$. Show that $\lim a_n \geq \lim b_n$.

2.34 Show that $\lim_{n\to\infty} \left(1 + \frac{1}{n}\right)^n \leq \lim_{n\to\infty} \left(1 + \frac{1}{2!} + \frac{1}{3!} + \cdots + \frac{1}{n!}\right)$.

2.35 **Euler's constant.**

(1) Show that if $x > -1$, $x \neq 0$, then

$$\frac{x}{1+x} < \ln(1+x) < x.$$

(2) Let $\gamma_n = \left(1 + \frac{1}{2} + \cdots + \frac{1}{n}\right) - \ln n$. Show that (γ_n) is a decreasing sequence.

(3) Show that (γ_n) converges. ($\lim \gamma_n = \gamma$ is known as Euler's constant.)

(4) Prove that $\lim \frac{1 + 1/2 + \cdots + 1/n}{\ln n} = 1$.

2.36 Determine and compare $\liminf a_n$, $\limsup a_n$, $\lim a_n$ each given sequence (a_n).

(a) $a_n = (-1)^n \left(1 + \frac{1}{n}\right)$ (c) $a_n = \frac{2 + \cos n}{n}$ (e) $a_n = (-n)^n \sin \frac{\pi}{n}$
(b) $a_n = 2 + \cos^n(n\pi)$ (d) $a_n = \frac{n^n}{n!}$ (f) $a_n = \ln \frac{1}{n}$

2.37 For each n, let $f_n : [-1, 1] \to \mathbb{R}$ be defined by $f_n(t) = t^n$. Determine $\limsup f_n$ and $\liminf f_n$.

2.38 Let (a_n) and (b_n) be sequences of real numbers. Show that

(1) $\limsup(-a_n) = -\liminf(a_n)$;

(2) $\limsup c a_n = c \limsup a_n$ for any $c > 0$;

(3) $\limsup(a_n + b_n) \leq \limsup a_n + \limsup b_n$.

2.39 Let (a_n) be a sequence. Show that

$$\limsup \left(|n a_n|\right)^{1/n} = \limsup \left(|a_n|\right)^{1/n}.$$

2.40 Let (a_n) be a sequence of nonnegative real numbers. For each $n \in \mathbb{N}$, define $\sigma_n = \frac{a_1 + a_2 + \cdots + a_n}{n}$. Show that

$$\liminf a_n \leq \liminf \sigma_n \leq \limsup \sigma_n \leq \limsup a_n.$$

(Compare with Exercise 2.18.)

3
Series

This chapter focuses on sequences whose terms are defined as sums of terms of another sequence. Such sequences are referred to as series. We will study some standard tests for the convergence of series.

3.1 Infinite Series

Given a sequence (a_n) of real numbers, one can define a new sequence (s_n) by

$$s_n = a_1 + a_2 + \cdots + a_n$$

for each $n \in \mathbb{N}$. If the new sequence (s_n) converges, then the sequence (a_n) is said to be **summable** and $s = \lim s_n$ is called the **infinite sum** of the sequence (a_n). In practice, sequences of the form (s_n) are of particular importance.

Definition 3.1

Let $(a_k)_{k=1}^{\infty}$ be a sequence of real numbers. An **(infinite) series** is an expression of the form

$$a_1 + a_2 + \cdots + a_n + \cdots = \sum_{k=1}^{\infty} a_k.$$

The number a_n is called the **n-th term** of the series.

For each n the number $s_n = a_1 + a_2 + \cdots + a_n = \sum_{k=1}^{n} a_k$ is called the **n-th partial sum** of the infinite series $\sum_{k=1}^{\infty} a_k$.

Series	10th term	4th partial sum
$\sum_{n=1}^{\infty} \frac{1}{n}$	$\frac{1}{10}$	$1 + \frac{1}{2} + \frac{1}{3} + \frac{1}{4} = \frac{25}{12}$
$\sum_{n=2}^{\infty} \ln n$	$\ln 10$	$\ln 2 + \ln 3 + \ln 4 + \ln 5 = \ln 120$
$\sum_{n=1}^{\infty} (-1)^n$	$(-1)^{10} = 1$	$(-1) + 1 + (-1) + 1 = 0$
$\sum_{n=0}^{\infty} \sin n\frac{\pi}{3}$	$\sin 10\frac{\pi}{3} = -\frac{\sqrt{3}}{2}$	$0 + \frac{\sqrt{3}}{2} + \frac{\sqrt{3}}{2} + 0 = \sqrt{3}$

We note that an expression of the form $\sum_{k=m}^{\infty} b_k$ for any $m \in \mathbb{N}$ represents a series; it can be understood as the series $\sum_{k=1}^{\infty} a_k$ where $a_k = 0$ if $k \leq m-1$, and $a_k = b_k$ if $k \geq m$. Similarly, an expression of the form $\sum_{k=m}^{\infty} b_k$ where $m \in \mathbb{Z}$ can be considered as the series $\sum_{k=1}^{\infty} b_{k-(1-m)}$. We shall often write $\sum a_n$ instead of $\sum_{k=m}^{\infty} a_k$ when m is either understood or irrelevant.

Definition 3.2

A series $\sum_{k=1}^{\infty} a_k$ is said **to be convergent**, or **to converge**, if the sequence (s_n) of its partial sums converges. A series that does not converge is said **to be divergent**, or **to diverge**.

In other words, the series $\sum_{k=1}^{\infty} a_k$ is convergent if the sequence (a_n) is summable. If the series $\sum_{k=1}^{\infty} a_k$ converges to a number s, we write $\sum_{k=1}^{\infty} a_k = s$. The number s is then called the **sum** of the series. If $\lim_{n\to\infty} \sum_{k=1}^{n} a_k = \pm\infty$, we write $\sum_{k=1}^{\infty} a_k = \pm\infty$.

Example 2.40 (page 60) shows that the series $\sum_{n=1}^{\infty} \frac{1}{n}$ (known as a **harmonic series**) is divergent.

Example 3.3

Show that the series $\sum_{n=1}^{\infty} \frac{1}{n(n+1)}$ converges.

Solution

First we notice that $\frac{1}{n(n+1)} = \frac{1}{n} - \frac{1}{n+1}$. It follows that for each n, the n-th partial sum is

$$s_n = \left(1 - \frac{1}{2}\right) + \left(\frac{1}{2} - \frac{1}{3}\right) + \cdots + \left(\frac{1}{n} - \frac{1}{n+1}\right) = 1 - \frac{1}{n+1}.$$

Thus $\lim s_n = 1$, i.e. the series converges and $\sum_{n=1}^{\infty} \frac{1}{n(n+1)} = 1$. $\square$

Example 3.4

Show that the series $\sum_{k=0}^{\infty} r^k$ (known as a **geometric series**) converges if $|r| < 1$ and diverges if $|r| \geq 1$.

Solution

If $r = 1$, the partial sum $s_n = 1 + 1 + \cdots + 1 = n$. Since $\lim s_n = +\infty$, the series diverges.

If $r = -1$, the partial sum $s_n = 1$ if n is odd and $s_n = 0$ if n is even. Thus $\lim s_n$ does not exist and therefore the series diverges.

If $|r| \neq 1$, then

$$
\begin{aligned}
s_n &= 1 + r + r^2 + \cdots + r^n, \\
r s_n &= r + r^2 + r^3 + \cdots + r^{n+1}.
\end{aligned}
\tag{3.1}
$$

Subtracting the second line of (3.1) from the first line, we have $(1 - r) s_n = 1 - r^{n+1}$. Hence

$$
s_n = \frac{1 - r^{n+1}}{1 - r}.
$$

So if $|r| < 1$, $\lim s_n = \frac{1}{1-r}$ and if $|r| \geq 1$, $\lim s_n$ does not exist. $\square$

If $s_n = \sum_{k=1}^{n} a_k$ is the partial sum of the series $\sum_{k=1}^{\infty} a_k$, then saying that the series $\sum_{k=1}^{\infty} a_k$ converges means that there exists s such that $\lim_n (s - s_n) = \lim_{n \to \infty} \sum_{k=n}^{\infty} a_k = 0$. More precisely we have the following result.

Proposition 3.5

A series $\sum_{k=m}^{\infty} a_k$ converges if and only if $\lim_n \sum_{k=n}^{\infty} a_k = 0$.

In the light of Proposition 3.5, we notice the following fact:

Remark 3.6

Changing a finite number of terms of a series has no effect on its convergence or its divergence.

For example, the series $\sum_{k=100}^{\infty} a_n$ and $\sum_{n=1}^{\infty} a_n$ either both converge or both diverge.

A necessary condition for convergence of a series is given in the following proposition.

Proposition 3.7

If $\sum a_n$ converges, then $\lim a_n = 0$.

Proof

Suppose that the series $\sum a_n$ is convergent. Let (s_n) be the sequence of its partial sums. Then $\lim s_n = s$ exists as a real number. Hence

$$\lim_{n \to \infty} a_n = \lim_{n \to \infty} (s_n - s_{n-1}) = \lim_{n \to \infty} s_n - \lim_{n \to \infty} s_{n-1} = s - s = 0.$$

$\square$

The result of the above proposition is mainly used as a test for divergence: if $\lim a_n \neq 0$, then the series $\sum a_n$ diverges. For example, the series $\sum \frac{n}{2n+1}$ diverges, since $\lim \frac{n}{2n+1} = \frac{1}{2} \neq 0$.

Warning

If $\lim a_n = 0$, the series $\sum a_n$ *may* or *may not* converge.

The series $\sum_{n=1}^{\infty} \frac{1}{n}$ of Example 2.40 diverges even though $\lim \frac{1}{n} = 0$.

Example 3.8

Consider the series $\sum_{n=1}^{\infty} \left(\sqrt{n+1} - \sqrt{n}\right)$. Show that

(1) $\lim_{n \to \infty} \left(\sqrt{n+1} - \sqrt{n}\right) = 0$;

(2) $\sum_{n=1}^{\infty} \left(\sqrt{n+1} - \sqrt{n}\right)$ diverges.

Solution

For the proof of (1), see Example 2.25, page 50.

(2) The n-th partial sum of $\sum_{n=1}^{\infty} \left(\sqrt{n+1} - \sqrt{n}\right)$ can be written as

$$s_n = \left(\sqrt{2} - \sqrt{1}\right) + \left(\sqrt{3} - \sqrt{2}\right) + \cdots$$
$$+ \left(\sqrt{n+1} - \sqrt{n}\right)$$
$$= \sqrt{n+1} - 1.$$

Since $\left(\sqrt{n+1} - 1\right)$ diverges to ∞, $\sum_{n=1}^{\infty} \left(\sqrt{n+1} - \sqrt{n}\right)$ is a divergent series.

$\square$

The next result is both a necessary and sufficient condition for convergence of a series.

Theorem 3.9

A series $\sum a_n$ converges if and only if for every $\varepsilon > 0$, there exists N in $\mathbb{N}$ such that for every $n > N$ and for all $p \in \mathbb{N}$

$$\left| \sum_{k=n+1}^{n+p} a_k \right| < \varepsilon.$$

Proof

Let $s_n = \sum_{k=1}^{n} a_k$ be the n-th partial sum of the series $\sum a_n$. If n and p are natural numbers, then $|s_{n+p} - s_n| = \left| \sum_{k=n+1}^{n+p} a_k \right|$. Thus the stated condition is nothing but the Cauchy criterion for the sequence (s_n).   □

The condition of the above theorem is also referred to as the Cauchy criterion for the series $\sum a_n$.

Proposition 3.10

If the series $\sum a_n$ and $\sum b_n$ converge, then so do $\sum (a_n + b_n)$ and $\sum c a_n$ for every $c \in \mathbb{R}$, and we have

$$\sum (a_n + b_n) = \sum a_n + \sum b_n, \quad \sum c a_n = c \sum a_n.$$

Proof

Apply Proposition 2.24 to the sequences of partial sums (Exercise 3.2).   □

For example, since both of the series $\sum \frac{1}{n(n+1)}$ and $\sum \frac{1}{3^n}$ converge, so does the series $\sum \left[\frac{5}{n(n+1)} - \frac{\pi}{3^n} \right]$.

Corollary 3.11

If the series $\sum a_n$ converges and the series $\sum b_n$ diverges, then the series $\sum (a_n + b_n)$ diverges.

Proof

Suppose to the contrary that the series $\sum (a_n + b_n)$ converges. Then by Proposition 3.10, the series $\sum [(a_n + b_n) - a_n] = \sum b_n$ converges. Contradiction! Thus the series $\sum (a_n + b_n)$ diverges. $\square$

For example, the series $\sum \left[\frac{1}{2^n} - \frac{1}{n} \right]$ diverges because $\sum \frac{1}{2^n}$ is convergent while $\sum \frac{1}{n}$ diverges.

Proposition 3.12

Suppose that $a_n \geq 0$ for all $n \in \mathbb{N}$. Then either $\sum a_n$ converges or $\sum a_n = +\infty$.

Proof

If $a_n \geq 0$, then (s_n) is a nondecreasing sequence. If (s_n) is bounded, since it is monotonic, it is convergent by Theorem 2.20. Otherwise it is clear that $\sum a_n = +\infty$. $\square$

3.2 Conditional Convergence

Definition 3.13

A sequence (b_n) is said to be a **rearrangement** of another sequence (a_n) if there exists a one-to-one mapping σ of $\mathbb{N}$ onto $\mathbb{N}$ (called a **permutation**) such that $b_n = a_{\sigma(n)}$ for all $n \in \mathbb{N}$.

As an example, consider the permutation $\sigma : \mathbb{N} \to \mathbb{N}$ defined by

$$\sigma(n) = \begin{cases} 2k - 1 & \text{if} \quad n = 2k, \\ 2k & \text{if} \quad n = 2k - 1. \end{cases}$$

Thus,   if $(a_n) = \left(1, \ \frac{1}{2}, \ \frac{1}{3}, \ \frac{1}{4}, \ \cdots, \ \frac{1}{2n-1}, \ \frac{1}{2n}, \cdots \right)$,

then $\left(a_{\sigma(n)}\right) = \left(\frac{1}{2}, \ 1, \ \frac{1}{4}, \ \frac{1}{3}, \ \cdots, \ \frac{1}{2n}, \ \frac{1}{2n-1}, \cdots \right)$.

A rearrangement of a given sequence (a_n) affects only the relative order of the a_n but not their occurrence nor the possible reoccurrence of their numerical values. Thus if two sequences are a rearrangement of each other, then necessarily they have the same set of values. That is to say

$$\{a_n : n \in \mathbb{N}\} = \left\{a_{\sigma(n)} : n \in \mathbb{N}\right\}.$$

The question that we are naturally led to ask is: If $(a_{\sigma(n)})$ is a rearrangement of (a_n), what can one say about the nature of the series $\sum a_{\sigma(n)}$ as compared with that of $\sum a_n$? We begin by introducing the following definition.

Definition 3.14

A series $\sum a_n$ is said to be **unconditionally convergent** if for any rearrangement $(a_{\sigma(n)})$ of the sequence (a_n) the series $\sum a_{\sigma(n)}$ converges. Otherwise the series is said to be **conditionally convergent**.

Example 3.15

Show that if (a_n) is a sequence of nonnegative real numbers, then $\sum a_n = \sum a_{\sigma(n)}$ for any permutation $\sigma : \mathbb{N} \to \mathbb{N}$.

Solution

Let (a_n) be a sequence of nonnegative real numbers and let $\sigma : \mathbb{N} \to \mathbb{N}$ be any permutation. For simplicity, we set $a = \sum a_n$ and $b = \sum a_{\sigma(n)}$. (Notice that a and b could be ∞.) Let $n \in \mathbb{N}$. If

$$q_n = \max \{\sigma(1), \sigma(2), \ldots, \sigma(n)\},$$

then

$$\sum_{i=1}^{n} a_{\sigma(i)} \leq \sum_{k=1}^{q_n} a_k \leq a.$$

Since n is arbitrary, it ensues that $b \leq a$. A similar argument shows that $a \leq b$. This completes the proof. $\qquad\square$

It is clear from the definition that an unconditionally convergent series is convergent. Our next question is: Given an unconditionally convergent series $\sum a_n$, and $(a_{\sigma(n)})$ a rearrangement of the sequence (a_n), when do we have

$$\sum a_n = \sum a_{\sigma(n)}?$$

Example 3.15 implies in particular that a convergent series with nonnegative terms is unconditionally convergent and all the rearrangements of such series converge to the same limit.

Definition 3.16

A series $\sum a_n$ is said to be **rearrangement invariant** if it is unconditionally convergent and if for any rearrangement $\left(a_{\sigma(n)}\right)$ of the sequence (a_n),

$$\sum a_n = \sum a_{\sigma(n)}.$$

For example, we infer from Example 3.15 that the series $\sum 1/n^2$ is rearrangement invariant. Another type of convergence, which is often of great importance in treating series, is contained in the following

Definition 3.17

Let $\sum a_n$ be a series. Then $\sum a_n$ is said to be

(1) **absolutely convergent** if the series $\sum |a_n|$ converges;

(2) **nonabsolutely convergent** if $\sum a_n$ is convergent and $\sum |a_n|$ is divergent.

Since the series $\sum \left| \frac{(-1)^n}{3^n} \right| = \sum \frac{1}{3^n}$ is a convergent geometric series, the series $\sum \frac{(-1)^n}{3^n}$ is absolutely convergent. We will see later that the series $\sum \frac{(-1)^n}{n}$ is convergent; however, the series $\sum \left| \frac{(-1)^n}{n} \right| = \sum \frac{1}{n}$ diverges. Thus the series $\sum \frac{(-1)^n}{n}$ is nonabsolutely convergent.

Example 3.18

Show that every absolutely convergent series is convergent.

Solution

Suppose that $\sum a_n$ is absolutely convergent. For all natural numbers n and p, from the triangle inequality we have

$$\left| \sum_{k=n+1}^{n+p} a_k \right| \leq \sum_{k=n+1}^{n+p} |a_k|.$$

Since $\sum |a_n|$ is convergent, it satisfies the Cauchy criterion. It follows from the above inequality that $\sum a_n$ also satisfies the Cauchy criterion and thus it is convergent. $\qquad\square$

It turns out that nonabsolute convergence implies conditional convergence or equivalently unconditional convergence implies absolute convergence. This remarkable observation is due to Riemann.

Theorem 3.19 (Riemann)

If a series $\sum a_n$ is nonabsolutely convergent and if c is an arbitrary real number, then there exists a rearrangement $(a_{\sigma(n)})$ of the sequence (a_n) such that $\sum a_{\sigma(n)} = c$.

Proof

For each n, let $a_n^+ = a_n \vee 0$ and $a_n^- = -a_n \vee 0$. Then $a_n^+ \geq 0$ and $a_n^- \geq 0$ for all n. We claim that both series $\sum a_n^+$ and $\sum a_n^-$ diverge.

Indeed, if both are convergent, then $\sum (a_n^+ + a_n^-) = \sum |a_n|$ would converge. Contradiction! On the other hand, if one of them is convergent and the other is divergent, by Corollary 3.11, $\sum a_n = \sum (a_n^+ - a_n^-)$ diverges. Another contradiction which proves our claim.

Without loss of generality, we may assume that $c \geq 0$. Now let n_1 be the smallest integer for which

$$c < \sum_{k=1}^{n_1} a_k^+.$$

Then there exists a smallest integer n_2 such that

$$\sum_{k=1}^{n_1} a_k^+ - \sum_{k=1}^{n_2} a_k^- < c < \sum_{k=1}^{n_1} a_k^+.$$

Again let n_3 be the smallest integer such that

$$\sum_{k=1}^{n_1} a_k^+ - \sum_{k=1}^{n_2} a_k^- < c < \sum_{k=1}^{n_1} a_k^+ - \sum_{k=1}^{n_2} a_k^- + \sum_{k=n_1+1}^{n_3} a_k^+ x.$$

Then there exists a smallest integer n_4 such that

$$\sum_{k=1}^{n_1} a_k^+ - \sum_{k=1}^{n_2} a_k^- + \sum_{k=n_1+1}^{n_3} a_k^+ x - \sum_{k=n_1+1}^{n_4} a_k^+ x < c$$

$$< \sum_{k=1}^{n_1} a_k^+ - \sum_{k=1}^{n_2} a_k^- + \sum_{k=n_1+1}^{n_3} a_k^+ x.$$

Continuing in this manner, we see that the number c is sandwiched between two partial sums of the same series. Since the terms of this series are obviously a rearrangement of the a_n, and since $\lim a_n = 0$, we conclude that the series is convergent and has c as its sum. $\qquad \square$

The converse of the above result is contained in the following so-called **rearrangement rule.**

Theorem 3.20

If a series $\sum a_n$ is absolutely convergent, then for any rearrangement $\left(a_{\sigma(n)}\right)$ of (a_n) the series $\sum a_{\sigma(n)}$ is absolutely convergent and

$$\sum a_n = \sum a_{\sigma(n)}.$$

Proof

Let $\sum a_n = a$ and let $\left(a_{\sigma(n)}\right)$ be any rearrangement of (a_n). Since the series $\sum |a_n|$ is convergent, the sequence of its partial sums is bounded. Let M be an upper bound for the sequence $\left(\sum_{k=1}^n |a_k|\right)_{n\in\mathbb{N}}$. Then it is clear that M is also an upper bound for the sequence $\left(\sum_{k=1}^n \left|a_{\sigma(k)}\right|\right)_{n\in\mathbb{N}}$. It follows that this latter sequence, being nondecreasing and bounded above, is convergent.

Now let $b = \sum a_{\sigma(n)}$. We still have to show that $b = a$. Let $\varepsilon > 0$. Then there exists $N > 0$ such that $m > n \geq N$ implies

$$\left|a - \sum_{k=1}^n a_k\right| < \frac{\varepsilon}{3}, \quad \left|b - \sum_{k=1}^n a_{\sigma(k)}\right| < \frac{\varepsilon}{3} \text{ and } \left|\sum_{k=n+1}^m a_k\right| < \frac{\varepsilon}{3}.$$

For each $k \in \mathbb{N}$ there is a unique $n_k \in \mathbb{N}$ such that $k = \sigma(n_k)$. Let $q = \max\{n_1, n_2, \ldots, n_k\}$. Then

$$\{1, 2, \ldots, k\} \subset \{\sigma(1), \sigma(2), \ldots, \sigma(q)\}.$$

Choose $r \in \mathbb{N}$ large enough so that

$$\left\{a_{\sigma(1)}, a_{\sigma(2)}, \ldots, a_{\sigma(q)}\right\} \subset \{a_1, a_2, \ldots, a_r\}.$$

Then if $r \geq n \geq N$, we have

$$\left|\sum_{k=1}^n a_k - \sum_{k=1}^n a_{\sigma(k)}\right| \leq \left|\sum_{k=n+1}^r a_k\right| < \frac{\varepsilon}{3}.$$

It follows that

$$|a - b| \leq \left|a - \sum_{k=1}^n a_k\right| + \left|\sum_{k=1}^n a_k - \sum_{k=1}^n a_{\sigma(k)}\right| + \left|\sum_{k=1}^n a_{\sigma(k)} - b\right|$$
$$< \frac{\varepsilon}{3} + \frac{\varepsilon}{3} + \frac{\varepsilon}{3} = \varepsilon.$$

Since $\varepsilon > 0$ is arbitrary, we have $a = b$ as desired. The proof is complete. $\square$

In summary, we have established in this section the following theorem.

Theorem 3.21

Let a_n be a sequence of real numbers. Then the following conditions are equivalent.

(1) the series $\sum a_n$ is rearrangement invariant;

(2) the series $\sum a_n$ is absolutely convergent;

(3) the series $\sum a_n$ is unconditionally convergent.

The rearrangement invariance, the absolute convergence, and the unconditional convergence of series in $\mathbb{R}$ are equivalent properties.

Double Sequences

Let $a = (a_{n,m})$ be a **double sequence**. That is to say a is a real-valued mapping with domain $\mathbb{N} \times \mathbb{N}$. It may be helpful to think of a double sequence as an array of numbers.

$$
\begin{array}{cccc}
a_{1,1} & a_{1,2} & \cdots & a_{1,m} & \cdots \\
a_{2,1} & a_{2,2} & \cdots & a_{2,m} & \cdots \\
\vdots & \vdots & & \vdots \\
a_{n,2} & a_{n,2} & & a_{n,m} & \cdots \\
\vdots & \vdots & & \vdots
\end{array}
$$

We say that the double sequence $(a_{n,m})$ converges to the number l if

$$\text{for every } \varepsilon > 0, \text{ there exists } N > 0 \text{ such that}$$

$$n, m > N \quad \text{implies} \quad |a_{n,m} - l| < \varepsilon.$$

Much of the theory on limits of sequences carries over, with obvious small changes, to double sequences.

To each double sequence $(a_{n,m})$, one can associate two **iterated series**, namely

$$\sum_{n=1}^{\infty} \sum_{m=1}^{\infty} a_{n,m} = \lim_{k \to \infty} \sum_{n=1}^{k} \sum_{m=1}^{\infty} a_{n,m}$$

and

$$\sum_{m=1}^{\infty} \sum_{n=1}^{\infty} a_{n,m} = \lim_{k \to \infty} \sum_{m=1}^{k} \sum_{n=1}^{\infty} a_{n,m}.$$

On the other hand, we define the partial sums of $(a_{n,m})$ to be the double sequence $s_{n,m}$ defined by

$$s_{n,m} = \sum_{i=1}^{n} \sum_{j=1}^{m} a_{i,j},$$

and we say that the **double series** $\sum a_{n,m}$ converges if the double sequence $(s_{n,m})$ converges. We also say that the double series is absolutely convergent if the series $\sum |a_{n,m}|$ converges. The natural question that arises is: When do all of these series converge to the same limit?

We first notice that (prove this!) the double series $\sum a_{n,m}$ is absolutely convergent if and only if the set

$$\left\{ \sum_{i=1}^{n} \sum_{j=1}^{m} |a_{i,j}| : n, m \in \mathbb{N} \right\}$$

is bounded.

The next two examples establish two elementary facts.

Example 3.22

Show that if the iterated series $\sum_{m=1}^{\infty} \sum_{n=1}^{\infty} |a_{n,m}|$ converges, then the double series $\sum a_{n,m}$ is absolutely convergent.

Solution

Since the iterated series $\sum_{m=1}^{\infty} \sum_{n=1}^{\infty} |a_{n,m}|$ converges, for each m the series $\sum_{n=1}^{\infty} |a_{n,m}|$ converges to, say a_m, and the series $\sum a_m$ converges to a positive number a. Clearly, for every $n, m \in \mathbb{N}$ we have

$$\sum_{i=1}^{n} \sum_{j=1}^{m} |a_{i,j}| \le a,$$

thereby establishing that the double series $\sum a_{n,m}$ is absolutely convergent. □

Example 3.23

Show that if $a_{n,m} \ge 0$, then the order of summation does not affect the sum of the series, i.e.

$$\sum_{n=1}^{\infty} \sum_{m=1}^{\infty} a_{n,m} = \sum_{m=1}^{\infty} \sum_{n=1}^{\infty} a_{n,m}.$$

Solution

Let $a = \sum_{n=1}^{\infty} \sum_{m=1}^{\infty} a_{n,m}$ and $b = \sum_{m=1}^{\infty} \sum_{n=1}^{\infty} a_{n,m}$. (Both a and b can be $+\infty$.) Let p and q be arbitrary, fixed in $\mathbb{N}$. Then

$$\sum_{n=1}^{p} \sum_{m=1}^{q} a_{n,m} = \sum_{m=1}^{q} \sum_{n=1}^{p} a_{n,m} \leq \sum_{n=1}^{q} \sum_{m=1}^{\infty} a_{n,m} \leq \sum_{m=1}^{\infty} \sum_{n=1}^{\infty} a_{n,m} = b.$$

Whence $a \leq b$. In a similar fashion, we have $b \leq a$. The proof is complete. $\square$

It is an exercise to show that if a double series is absolutely convergent, then it is convergent.

Theorem 3.24

Suppose that the double series $\sum a_{n,m}$ is absolutely convergent. Then

$$\sum_{n=1}^{\infty} \sum_{m=1}^{\infty} a_{n,m} = \sum_{m=1}^{\infty} \sum_{n=1}^{\infty} a_{n,m} = \sum a_{n,m}.$$

Proof

Since $\sum a_{n,m}$ is absolutely convergent, both $a = \sum a_{n,m}$ and

$$A = \sup \left\{ \sum_{i=1}^{n} \sum_{j=1}^{m} |a_{i,j}| : n, m \in \mathbb{N} \right\}$$

exist as real numbers. It follows that:

(1) if m is fixed, we have for every n

$$\sum_{i=1}^{n} |a_{i,m}| \leq \sum_{i=1}^{n} \sum_{j=1}^{m} |a_{i,j}| \leq A;$$

(2) for each $\varepsilon > 0$, there exists $N > 0$ such that $n, m > N$ implies

$$|s_{n,m} - a| = \left| \sum_{i=1}^{n} \sum_{j=1}^{m} a_{i,j} - a \right| < \varepsilon.$$

The statement in (1) implies that for each m, the series $\sum_{i=1}^{\infty} a_{i,m}$ is absolutely convergent to an element, say a_m in $\mathbb{R}$.

In (2), taking the limit as $n \to \infty$, we have

$$\left| \left(\sum_{i=1}^{\infty} a_{i,1} + \sum_{i=1}^{\infty} a_{i,2} + \cdots + \sum_{i=1}^{\infty} a_{i,m} \right) - a \right| < \varepsilon,$$

hence

$$|(a_1 + a_2 + \cdots + a_m) - a| = \left| \sum_{k=1}^{m} a_k - a \right| < \varepsilon.$$

Since m is arbitrary, we infer that the iterated series $\sum_{m=1}^{\infty} \sum_{n=1}^{\infty} a_{n,m}$ converges and $a = \sum_{m=1}^{\infty} \sum_{n=1}^{\infty} a_{n,m}$. A similar argument applies to the other iterated series. The proof is complete.   □

3.3 Comparison Tests

Term Comparison Test

Theorem 3.25

If $0 \le a_n \le b_n$ for each $n \in \mathbb{N}$, and if $\sum b_n$ converges, then $\sum a_n$ converges.

Proof

Let $s_n = \sum_{k=1}^{n} a_k$ and $t_n = \sum_{k=1}^{n} b_k$ be the n-th partial sums of $\sum a_n$ and $\sum b_n$ respectively. Suppose that $\sum b_n$ converges and has sum t. Then for each n, $s_n \le t_n \le t$. Thus (s_n) is a bounded monotonic sequence; it converges. Thus the series $\sum a_n$ converges.   □

This next result is an equivalent statement of the above theorem. The proof is left as an exercise (Exercise 3.3).

Corollary 3.26

If $0 \le a_n \le b_n$ for each $n \in \mathbb{N}$, and if $\sum a_n$ diverges, then $\sum b_n$ diverges.

For example, since $\frac{\text{fra}(2^{-n})}{1+2^n} \le \frac{1}{2^n}$ for every n, and since the series $\sum \frac{1}{2^n}$ is a convergent geometric series $\left(r = \frac{1}{2} \right)$, the series $\sum \frac{\text{fra}(2^{-n})}{1+2^n}$ converges. On the other hand, since $\frac{1}{\sqrt{n}} > \frac{1}{n}$ and since the harmonic series $\sum \frac{1}{n}$ diverges, the series $\sum \frac{1}{\sqrt{n}}$ diverges.

Limit Comparison Test

Theorem 3.27

If $0 < a_n, b_n$ for each $n \in \mathbb{N}$, and if $\lim \frac{a_n}{b_n} = c > 0$ $(c \neq \pm\infty)$, then the series $\sum a_n$ and $\sum b_n$ are either both convergent or both divergent.

Proof

Suppose that $\lim \frac{a_n}{b_n} = c > 0$. There exists N in $\mathbb{N}$ such that

$$n > N \text{ implies } \left| \frac{a_n}{b_n} - c \right| < \frac{c}{2},$$

or equivalently

$$n > N \text{ implies } \frac{c}{2} < \frac{a_n}{b_n} < \frac{3c}{2}.$$

Thus for $n > N$ we have $\frac{c}{2} b_n < a_n < \frac{3c}{2} b_n$. If $\sum a_n$ converges, then $\sum \frac{c}{2} b_n$ also converges by Theorem 3.25. Hence $\sum b_n = \sum \frac{2}{c} \frac{c}{2} b_n$ converges by Proposition 3.10. Similarly, if $\sum a_n$ diverges, then $\sum \frac{3c}{2} b_n$ diverges by Corollary 3.26, and hence so does $\sum b_n = \sum \frac{2}{3c} \frac{3c}{2} b_n$ by Proposition 3.10. So $\sum a_n$ converges if and only if $\sum b_n$ converges. Consequently we also have that $\sum a_n$ diverges if and only if $\sum b_n$ diverges. $\qquad\square$

For example, since $\lim \left(\frac{2n^2}{3n^3+1} \right) / \left(\frac{1}{n} \right) = 2/3 > 0$ and since the series $\sum \frac{1}{n}$ diverges, the series $\sum \frac{2n^2}{3n^3+1}$ diverges. Also since $\lim \left(\frac{3n+5}{n2^n} \right) / \left(\frac{1}{2^n} \right) = 3 > 0$, and since the geometric series $\sum \frac{1}{2^n}$ converges, the series $\sum \frac{3n+5}{n2^n}$ converges.

Integral Comparison Test

Theorem 3.28

Let $\sum a_n$ be a series of real numbers such that

(1) $a_n \geq 0$ for all $n \in \mathbb{N}$;

(2) $a_1 \geq a_2 \geq a_3 \geq \cdots$;

(3) there exists a nonnegative nonincreasing continuous function such that $f(n) = a_n$ for all $n \in \mathbb{N}$.

Then $\sum a_n$ converges if and only if $\lim_{a \to \infty} \int_1^a f(x)\, dx$ exists as a real number.

Proof

First, we notice that since f is nonincreasing and continuous, and $f(n) = a_n \geq 0$ for all $n \in \mathbb{N}$, then $f(x) \geq 0$ for all $x \in [1,\infty)$. Thus for $k \geq 2$, the integral $\int_{k-1}^{k} f(x)\, dx$ represents the area of the region delimited by $y = f(x)$, $y = 0$, $x = k - 1$, and $x = k$ on the coordinate plane. It follows that

$$f(k)[k - (k-1)] \leq \int_{k-1}^{k} f(x)\, dx \leq f(k-1)[k - (k-1)].$$

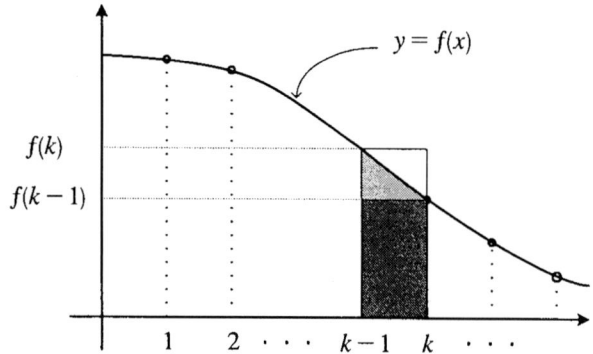

Therefore,

$$\sum_{k=2}^{n+1} a_k = \sum_{k=2}^{n+1} f(k) = \sum_{k=2}^{n+1} f(k)[k - (k-1)]$$

$$\leq \int_{1}^{n+1} f(x)\, dx$$

$$\leq \sum_{k=2}^{n+1} f(k-1)[k - (k-1)] = \sum_{k=1}^{n} a_k.$$

Whence, if s_n denotes the n-th partial sum of the series $\sum a_n$,

$$s_{n+1} - a_1 \leq \int_{1}^{n+1} f(x)\, dx \leq s_n. \tag{3.2}$$

Assume that $\lim_{a \to \infty} \int_{1}^{a} f(x)\, dx = I$ exists as a real number. Then (3.2) implies that (s_n) is bounded, namely for all $n \in \mathbb{N}$

$$s_{n+1} \leq I + a_1.$$

On the other hand, since the a_n are nonnegative, the sequence (s_n) is nondecreasing. By virtue of Theorem 2.20, we have that $\lim s_n = s$ exists and the series converges.

Assume next that $\lim_{a \to \infty} \int_1^a f(x)\, dx$ does not exist as a real number. Thus $\lim_{a \to \infty} \int_1^a f(x)\, dx = \infty$. By virtue of inequality (3.2), $\lim s_n = \infty$ and the series diverges. The proof is complete. $\qquad\square$

Example 3.29

Use the integral test to show that the series $\sum_{n=3}^{\infty} \left(\frac{1}{n-2} - \frac{1}{n} \right)$ converges.

Solution

Consider $f(x) = \frac{1}{x-2} - \frac{1}{x}$. Then f is nonnegative, continuous, and $f'(x) = \frac{-1}{(x-2)^2} + \frac{1}{x^2} > 0$ for $x > 3$. Thus the integral test applies. But for each $N > 3$, we have

$$\int_3^N f(x)\, dx = \int_3^N \left(\frac{1}{x-2} - \frac{1}{x} \right) dx$$

$$= \left[\ln \frac{x-2}{x} \right]_3^N = \ln \frac{N-2}{N}.$$

Hence

$$\lim_{N \to \infty} \int_3^N f(x)\, dx = \lim \ln \frac{N-2}{N} = 0.$$

This shows that the series converges. $\qquad\square$

Example 3.30

Use the integral test to show that the series $\sum_{n=27}^{\infty} \frac{1}{n \ln n \ln(\ln n)}$ diverges.

Solution

It is easy to see that the function $f(x) = \frac{1}{x \ln x \ln(\ln x)}$ is nonnegative, continuous, and decreasing on $[27, \infty)$. Also

$$\lim_N \int_{27}^N \frac{1}{x \ln x \ln(\ln x)}\, dx = \lim_N \int_{27}^N \frac{1}{\ln(\ln x)}\, d(\ln(\ln x))$$

$$= \lim_N \left[\ln |\ln(\ln x)| \right]_{27}^N = \infty.$$

Thus the series $\sum_{n=27}^{\infty} \frac{1}{n \ln n \ln(\ln n)}$ diverges by the integral test. $\qquad\square$

3.4 Root and Ratio Tests

Root Test

Theorem 3.31 (Cauchy)

Let $\sum a_n$ be a series and let $\alpha = \limsup |a_n|^{1/n}$. The series $\sum a_n$

(1) absolutely converges if $\alpha < 1$;

(2) diverges if $\alpha > 1$ or $+\infty$.

Note

If $\limsup |a_n|^{1/n} = 1$, we say that the root test fails. In such a case, the series $\sum a_n$ *may* or *may not* converge.

Proof

Suppose that $\alpha = \limsup |a_n|^{1/n} < 1$ and select $\varepsilon > 0$ such that $\alpha + \varepsilon < 1$. Then by Definition 2.31, there is $N \in \mathbb{N}$ such that

$$\alpha - \varepsilon < \sup \left\{ |a_n|^{1/n} : n > N \right\} < \alpha + \varepsilon.$$

It follows that $|a_n| < (\alpha + \varepsilon)^n$ for $n > N$. Since $0 < \alpha + \varepsilon < 1$, $\sum_{n=N}^{\infty} (\alpha + \varepsilon)^n$ is a convergent geometric series and hence by the comparison test (Theorem 3.25), the series $\sum_{n=N}^{\infty} a_n$ converges. Consequently, by Remark 3.6, the series $\sum a_n$ converges.

Now suppose that $\alpha = \limsup |a_n|^{1/n} > 1$. Then a subsequence of $|a_n|^{1/n}$ has limit $\alpha > 1$. It follows that $\lim a_n$ cannot be zero and the series diverges. $\square$

For example, for the harmonic series $\sum 1/n$, which we know is divergent, we have $\lim \sqrt[n]{1/n} = 1$. We also have $\lim \sqrt[n]{1/n^2} = \lim \sqrt[n]{1/n} \sqrt[n]{1/n} = 1$ but we will see later that the series $\sum 1/n^2$ converges.

Example 3.32

Test the series $\sum_{n=1}^{\infty} \left(\frac{n}{2n+1} \right)^n$ for convergence.

Solution

Since

$$\lim \sqrt[n]{\left(\frac{n}{2n+1}\right)^n} = \lim \frac{n}{2n+1} = \frac{1}{2},$$

the series $\sum_{n=1}^{\infty} \left(\frac{n}{2n+1}\right)^n$ converges. □

Ratio Test

Theorem 3.33 (d'Alembert)

Let $\sum a_n$ be a series of nonzero terms. Then the series $\sum a_n$

(1) converges absolutely if $\limsup \left|\frac{a_{n+1}}{a_n}\right| < 1$;

(2) diverges if $\liminf \left|\frac{a_{n+1}}{a_n}\right| > 1$ or $+\infty$.

Note

If $\liminf \left|\frac{a_{n+1}}{a_n}\right| \le 1 \le \limsup \left|\frac{a_{n+1}}{a_n}\right|$, we say that the ratio test fails. In such case, the series $\sum a_n$ *may* or *may not* converge.

The proof of Theorem 3.33 follows from the following

Lemma 3.34

Let (a_n) be a sequence of nonzero real numbers. Then we have

$$\liminf \left|\frac{a_{n+1}}{a_n}\right| \le \liminf |a_n|^{1/n} \le \limsup |a_n|^{1/n} \le \limsup \left|\frac{a_{n+1}}{a_n}\right|.$$

We notice that according to this lemma if $\lim \left|\frac{a_{n+1}}{a_n}\right|$ exists, then so does $\lim |a_n|^{1/n}$. Also this lemma implies that if the root test fails, then so does the ratio test. We now prove the lemma.

Proof (of Lemma 3.34)

We only need to prove the first and the third inequalities. We will prove the third inequality; the first one can be proved in a similar fashion.

Let $\alpha = \limsup |a_n|^{1/n}$ and $r = \limsup |a_{n+1}/a_n|$. If $r = +\infty$, there is nothing to prove. Suppose that $r < +\infty$. Let $\varepsilon > 0$. Then there is N in $\mathbb{N}$ such that

$$|a_{N+p}/a_{N+p-1}| < r + \varepsilon \text{ for } p \in \mathbb{N}.$$

Hence for each $p \in \mathbb{N}$, we have

$$|a_{N+p}| = \left|\frac{a_{N+p}}{a_{N+p-1}}\right| \cdot \left|\frac{a_{N+p-1}}{a_{N+p-2}}\right| \cdot \ldots \cdot \left|\frac{a_{N+2}}{a_{N+1}}\right| \cdot |a_{N+1}| < (r+\varepsilon)^p |a_{N+1}|,$$

and thus for every $p \in \mathbb{N}$

$$|a_{N+p}|^{1/(N+p)} < (r+\varepsilon)^{p/(N+p)} |a_{N+1}|^{1/(N+p)}.$$

Since

$$\lim_p (r+\varepsilon)^{p/(N+p)} |a_{N+1}|^{1/(N+p)} = (r+\varepsilon),$$

we have

$$\alpha = \limsup |a_n|^{1/n} \leq (r+\varepsilon).$$

Since $\varepsilon > 0$ is arbitrary, we conclude that $\alpha \leq r$ as desired. $\square$

Proof (of Theorem 3.33)

Let $\alpha = \limsup |a_n|^{1/n}$. By the lemma,

$$\liminf \left|\frac{a_{n+1}}{a_n}\right| \leq \alpha \leq \limsup \left|\frac{a_{n+1}}{a_n}\right|.$$

If $\limsup \left|\frac{a_{n+1}}{a_n}\right| < 1$, then $\alpha < 1$ and the series converges by the root test.

If $\liminf \left|\frac{a_{n+1}}{a_n}\right| > 1$, then $\alpha > 1$ and the series diverges again by the root test. $\square$

Example 3.35

Test the series $\sum \frac{\cos n\pi}{n!}$ for convergence.

Solution

Here $a_n = \frac{\cos n\pi}{n!}$. Thus

$$\frac{a_{n+1}}{a_n} = \frac{n! \cos (n+1)\pi}{(n+1)! \cos n\pi} = \frac{1}{n+1} \frac{\cos (n+1)\pi}{\cos n\pi}.$$

It follows that $\limsup \left|\frac{a_{n+1}}{a_n}\right| = \lim \frac{1}{n+1} = 0 < 1$ and hence the series converges. $\square$

Example 3.36

Test the series $\sum \frac{e^n}{n}$ for convergence.

Solution

Here $a_n = \frac{e^n}{n}$. Thus $a_{n+1} = \frac{e^{n+1}}{n+1}$ and $\frac{a_{n+1}}{a_n} = e\frac{n}{n+1}$. Hence

$$\lim \left| \frac{a_{n+1}}{a_n} \right| = \lim e\frac{n}{n+1} = e > 1.$$

The series diverges. $\square$

3.5 Further Tests

Raabe's Test

We saw that if $\lim \left| \frac{a_{n+1}}{a_n} \right| = 1$ or $\lim |a_n|^{1/n} = 1$, then neither the ratio test nor the root test can determine the nature of the corresponding series. In such cases, Raabe's test is often used.

Theorem 3.37 (Raabe)

Let $\sum a_n$ be a series of nonzero terms. Then the series $\sum a_n$

(1) converges absolutely if $\liminf n \left(1 - \left| \frac{a_{n+1}}{a_n} \right| \right) > 1$;

(2) is not absolutely convergent if $\limsup n \left(1 - \left| \frac{a_{n+1}}{a_n} \right| \right) < 1$.

Proof

Let $\sum a_n$ be a series of nonzero terms.

(1) Suppose that $\liminf n \left(1 - \left| \frac{a_{n+1}}{a_n} \right| \right) = \alpha > 1$. Let $0 < \varepsilon$ such that $\alpha - \varepsilon > 1$. Then there exists N such that

$$\alpha - \varepsilon < n \left(1 - \left| \frac{a_{n+1}}{a_n} \right| \right) \quad \text{for } n > N.$$

It follows that for $n > N$,

$$(n - 1)|a_n| - n|a_{n+1}| \geq (\alpha - \varepsilon - 1)|a_n| > 0.$$

Therefore the sequence $(n\,|a_{n+1}|)$ is decreasing for $n > N$. Also we have, for each $n > N$,

$$\sum_{k=N}^{n} [(k-1)\,|a_k| - k\,|a_{k+1}|] = (N-1)\,|a_N| - n\,|a_{n+1}| \geq (\alpha - 1) \sum_{k=N}^{n} |a_k|.$$

Therefore the sequence of partial sums of $\sum |a_n|$ is bounded and this shows that $\sum |a_n|$ is convergent.

(2) Suppose that $\limsup n \left(1 - \left|\frac{a_{n+1}}{a_n}\right|\right) = \alpha < 1$. Let $0 < \varepsilon$ such that $\alpha + \varepsilon \leq 1$. Then there exists N such that

$$n\left(1 - \left|\frac{a_{n+1}}{a_n}\right|\right) < \alpha + \varepsilon \text{ for } n > N.$$

Hence for $n > N$,

$$(n-1)\,|a_n| \leq (n - \alpha - \varepsilon)\,|a_n| \leq n\,|a_{n+1}|.$$

We then notice that the sequence $(n\,|a_{n+1}|)$ is increasing for $n \geq N$, and thus there exists $c > 0$ such that

$$|a_{n+1}| > c/n \text{ for } n > N.$$

Since the series $\sum 1/n$ diverges, the series $\sum |a_{n+1}|$ diverges as well. □

Alternating Series Test

The signs of the terms of the series such as $\sum (-1)^n$, $\sum \frac{\cos n\pi}{n}$, $\sum \frac{\sin((2n+1)\pi/2)}{n\ln(n+1)}$, and so on, alternate, i.e. two successive terms have opposite signs. The next theorem gives sufficient conditions for the convergence of such series.

Theorem 3.38 (Leibniz)

If (a_n) is a nonincreasing sequence, i.e. $a_1 \geq a_2 \geq \cdots \geq a_n \geq \cdots$ and $\lim a_n = 0$, then the series $\sum (-1)^n a_n$ converges.

Proof

It suffices to show that the series satisfies the Cauchy criterion. Fix $\varepsilon > 0$. Since $\lim a_n = 0$, there exists $N_1 \in \mathbb{N}$ such that

$$|a_n| < \varepsilon/3 \text{ for } n > N_1. \tag{3.3}$$

We also notice that for each n we have

$$0 \leq \sum_{k=1}^{n} (a_k - a_{k+1}) = (a_1 - a_2) + (a_3 - a_2) + \cdots + (a_n - a_{n+1}) < a_1,$$

thus the nondecreasing sequence $\left(\sum_{k=1}^{n} (a_k - a_{k+1})\right)$ converges, i.e. the series $\sum_{k=1}^{\infty} (a_k - a_{k+1})$ converges. Hence there exists N_2 in $\mathbb{N}$ such that

$$\left| \sum_{k=n}^{n+p} (a_k - a_{k+1}) \right| < \varepsilon/3 \text{ for } n > N_2 \text{ and for all } p \in \mathbb{N}. \tag{3.4}$$

Next for each k, let $b_k = (-1)^k$ and consider $s_n = \sum_{k=1}^{n} b_k$ so that for each k, $b_k = s_k - s_{k-1}$. Notice that $s_k = 0$ or -1 depending on whether k is even or odd. It follows that

$$\sum_{k=n}^{n+p} (-1)^k) a_k = \sum_{k=n}^{n+p} a_k (s_k - s_{k-1}) = \sum_{k=n}^{n+p} a_k s_k - \sum_{k=n}^{n+p} a_k s_{k-1}$$

$$= \sum_{k=n}^{n+p} a_k s_k - \sum_{k=n-1}^{n+p-1} a_{k+1} s_k$$

$$= \sum_{k=n}^{n+p-1} (a_k - a_{k+1}) s_k + a_{n+p} s_{n+p} - a_n s_{n-1}.$$

Hence for $n > \max\{N_1, N_2\}$ and for all $p \in \mathbb{N}$,

$$\left| \sum_{k=n}^{n+p} (-1)^k) a_k \right| \leq \sum_{k=n}^{n+p-1} |a_k - a_{k+1}| |s_k| + |a_{n+p}| |s_{n+p}| + |a_n| |s_{n-1}|$$

$$\leq \sum_{k=n}^{n+p-1} |a_k - a_{k+1}| + |a_{n+p}| + |a_n|$$

$$\leq \frac{\varepsilon}{3} + \frac{\varepsilon}{3} + \frac{\varepsilon}{3} = \varepsilon,$$

as desired. $\square$

For example, since $1 \geq \frac{1}{2} \geq \frac{1}{3} \geq \cdots$ and $\lim \frac{1}{n} = 0$, we see that the series $\sum (-1)^n / n$ converges. Note that since $\sum 1/n$ is a divergent series, the series $\sum (-1)^n / n$ is not absolutely convergent. Similarly, each of the series $\sum (-1)^n / \sqrt{n}$, $\sum (-1)^n / \ln(n+1)$, and $\sum (-1)^n / n^{0.99}$ are nonabsolutely convergent series. Since $\lim(\ln(n)) = \infty$, the alternating series $\sum (-1)^n \ln(n)$ diverges.

Our next example shows that the n-th partial sum of a convergent alternating series differs from its sum by no more than the absolute value of its $(n+1)$-th term.

Example 3.39

Show that if the alternating series $\sum_{k=1}^{\infty} (-1)^k a_k$ is convergent and its sum is s, then

$$\left| s - \sum_{k=1}^{n} (-1)^k a_k \right| \leq a_{n+1}.$$

Solution

Let s be the sum of the series $\sum_{k=1}^{\infty} (-1)^k a_k$ and for each n, let s_n denote its n-th partial sum. Then, if n is odd we have

$$s - s_n = (a_{n+1} - a_{n+2}) + (a_{n+3} - a_{n+4}) + \cdots$$
$$= a_{n+1} - (a_{n+2} - a_{n+3}) - (a_{n+4} - a_{n+5}) - \cdots \leq a_{n+1}.$$

Similarly if n is even, then

$$s_n - s = a_{n+1} - (a_{n+2} - a_{n+3}) - (a_{n+4} - a_{n+5}) - \cdots \leq a_{n+1}.$$

The desired estimate follows. $\square$

Cauchy's Condensation Test

Theorem 3.40

If (a_n) is a nonincreasing sequence, i.e. $a_1 \geq a_2 \geq \cdots \geq a_n \geq \cdots$ and $a_n > 0$ for each n, then the series $\sum a_n$ converges if and only if $\sum 2^n a_{2^n}$ converges.

Proof

Since $a_n > 0$, the partial sums s_n of $\sum a_n$ form an increasing sequence. Similarly the partial sums t_n of $\sum 2^n a_{2^n}$ also form an increasing sequence. Thus we are done if we show that the sequence (s_n) is bounded if and only if so is the sequence (t_n). First we notice that for k in $\mathbb{N}$

$$s_{2^k} - s_{2^{k-1}} = a_{2^{k-1}+1} + a_{2^{k-1}+2} + \cdots + a_{2^k}.$$

Since (a_n) is nonincreasing, we have for each k in $\mathbb{N}$

$$2^{k-1} a_{2^k} \leq s_{2^k} - s_{2^{k-1}} \leq 2^{k-1} a_{2^{k-1}}.$$

Adding, we obtain

$$\frac{1}{2} \sum_{k=1}^{n} 2^k a_{2^k} \leq s_{2^n} - s_1 \leq \sum_{k=1}^{n} 2^{k-1} a_{2^{k-1}},$$

hence $\frac{1}{2}t_n \le s_{2^n} \le t_n + a_1$ and so we see that (s_n) is bounded if and only if (t_n) is bounded.   □

The above test can be used to show the following classical results.

Example 3.41

Show that the p-series $\sum_{n=1}^{\infty} \frac{1}{n^p}$ converges if and only if $p > 1$.

Solution

If $p > 0$, $\left(\frac{1}{n^p}\right)$ is a nonincreasing sequence of positive terms and Cauchy's condensation test applies. Since $2^n \frac{1}{(2^n)^p} = 2^{(1-p)n}$, the series $\sum 2^n \frac{1}{(2^n)^p} = \sum \left(\frac{1}{2^{(p-1)}}\right)^n$ is a geometric series. It converges if and only if $\frac{1}{2^{(p-1)}} < 1$, i.e. if and only if $p > 1$.   □

Example 3.42

Determine for what values of $a > 1$ the series $\sum_{n=1}^{\infty} \frac{1}{a^{\ln n}}$ converges.

Solution

Since $a > 1$, the sequence $\left(\frac{1}{a^{\ln n}}\right)$ is a nonincreasing sequence of positive terms. Thus Cauchy's condensation test applies. The series

$$\sum 2^n \frac{1}{a^{\ln(2^n)}} = \sum \left(\frac{2}{a^{\ln 2}}\right)^n$$

is a geometric series with $r = \frac{2}{a^{\ln 2}}$. Thus the given series converges if and only if $\frac{2}{a^{\ln 2}} < 1$, that is, if and only if $a > e$.   □

EXERCISES

3.1 Write down the first four terms of the sequence of partial sums of each of the given series and determine which ones converge.

(a) $\sum_{n=0}^{\infty} \frac{n}{n^2+1}$ (c) $\sum_{n=2}^{\infty} \frac{(-1)^n}{\operatorname{int} \sqrt{n}}$

(b) $\sum_{n=0}^{\infty} \frac{2^n}{(2n)!}$ (d) $\sum_{n=0}^{\infty} \left(\sqrt{n+1} - \sqrt{n}\right)$

3.2 Complete the proof of Proposition 3.10.

3.3 Prove Corollary 3.26.

3.4 Show that $\sum_{n=0}^{\infty} \frac{1}{(x+n)(x+n+1)} = \frac{1}{x}$ if $x > 0$.

3.5 Show that $\sum_{n=0}^{\infty} \frac{4}{n(n+1)(n+2)} = 1$.

3.6 Test the following series for convergence.

(a) $\sum_{n=1}^{\infty} \frac{n+1}{n}$ (e) $\sum_{n=1}^{\infty} \frac{\cos n\pi}{n}$ (i) $\sum_{n=1}^{\infty} \sin \frac{n\pi}{2}$ (m) $\sum_{n=1}^{\infty} \frac{(-1)^n n}{n^2+2}$

(b) $\sum_{n=1}^{\infty} \frac{n}{n+\sqrt{n}}$ (f) $\sum_{n=0}^{\infty} \frac{n^4}{2^n}$ (j) $\sum_{n=0}^{\infty} \frac{\sin n-1}{2^n}$ (n) $\sum_{n=1}^{\infty} \frac{\ln n}{n^2}$

(c) $\sum_{n=2}^{\infty} \frac{(-1)^n}{\operatorname{int} \sqrt{n}}$ (g) $\sum_{n=1}^{\infty} \frac{\sqrt{n}}{(-n)^n}$ (k) $\sum_{n=1}^{\infty} \frac{\sqrt{(n+1)^3}}{n}$ (o) $\sum_{n=1}^{\infty} \frac{1}{n^{\ln n}}$

(d) $\sum_{n=0}^{\infty} \frac{n}{(n+1)^2}$ (h) $\sum_{n=1}^{\infty} \frac{n!}{n^n}$ (l) $\sum_{n=0}^{\infty} \frac{n+2}{n^3+5}$ (p) $\sum_{n=1}^{\infty} (\sqrt[n]{n} - 1)^n$

3.7 Test the following series for absolute/conditional convergence.

(a) $\sum_{n=1}^{\infty} \frac{(-1)^n}{n}$ (e) $\sum_{n=1}^{\infty} \frac{\operatorname{fra}(n\pi)}{n^2}$ (i) $\sum_{n=1}^{\infty} (\sqrt{n} \ln n)^{-1}$

(b) $\sum_{n=1}^{\infty} \frac{(-1)^n n}{n+\sqrt{n}}$ (f) $\sum_{n=0}^{\infty} \frac{(-1)^n n^4}{2^n}$ (j) $\sum_{n=0}^{\infty} \frac{\cos n-1}{2^n}$

(c) $\sum_{n=1}^{\infty} \frac{(-1)^n}{(2n-1)^2}$ (g) $\sum_{n=0}^{\infty} 2^{-n} \sin \frac{n}{3^n}$ (k) $\sum_{n=1}^{\infty} \left(\frac{1}{3^n} - \frac{2}{\sqrt{n}} \right)$

(d) $\sum_{n=1}^{\infty} \frac{(-n)^n}{(n+1)^2}$ (h) $\sum_{n=2}^{\infty} \frac{1}{n(\ln n)^2}$ (l) $\sum_{n=1}^{\infty} \sqrt{n} (-n)^{-n}$

3.8 Let (a_n) be a sequence of real numbers. Show that each of the following conditions is equivalent to the unconditional convergence of the series $\sum a_n$ in $\mathbb{R}$.

(1) For every sequence (ε_n) with range $\{-1, 1\}$, the series $\sum \varepsilon_n a_n$ converges in $\mathbb{R}$.

(2) For every subsequence (a_{n_k}) of (a_n) the series $\sum a_{n_k}$ converges in $\mathbb{R}$.

(3) For every $\varepsilon > 0$, there exists $N \in \mathbb{N}$ such that $S \subset \mathbb{N}$, S finite and $\min S > N$ implies $\left| \sum_{n \in S} a_n \right| < \varepsilon$.

3.9 Let $(a_{n,m})$ be the double sequence defined by

$$a_{n,m} = \begin{cases} 1 & \text{if } m - n = 1 \\ -1 & \text{if } m - n = -1 \\ 0 & \text{otherwise.} \end{cases}$$

(1) Show that both iterated series converge but the sums are different.

(2) Let $(s_{n,m})$ denote the sequence of partial sums of the $a_{n,m}$. Show that $\lim_{n\to\infty} s_{n,n}$ exists. What can you say about the double series $\sum a_{n,m}$?

3.10 Show that $\sum_{n=1}^{\infty}\sum_{m=1}^{\infty} 1/(n^2+m^2) = \infty$. However, if $p > 1$, show that the double series $\sum 1/(n^2+m^2)^p$ converges.

3.11 Show that if $p > 1$ and $q > 1$, then the double series $\sum \frac{1}{m^p n^q}$ converges.

3.12 Give examples of series $\sum a_n$ such that $\lim \frac{a_{n+1}}{a_n} = 1$ and

(1) $\sum a_n$ converges;

(2) $\sum a_n$ diverges.

3.13 Show that the series $\sum_{n=1}^{\infty} \frac{1}{n\sqrt[n]{n}}$ diverges.

3.14 Show that $\lim_{n\to\infty} n^p a^n = 0$, for every $p > 0$ provided $|a| < 1$.

3.15 Show that $\lim_{n\to\infty} \frac{2^n n!}{n^n} = 0$.

3.16 Consider the series $\sum_{n=1}^{\infty} \frac{x^n}{n+\sqrt{n}}$. Find out for which of the following values of the variable x the series converges.

(a) $x = 0$ (b) $x = 1$ (c) $x = 1/2$ (d) $x = -1$ (e) $x = -2$

3.17 Determine for which values of x each of the following series converges.

(a) $\sum_{n=1}^{\infty} \frac{x^n}{n}$ (b) $\sum_{n=1}^{\infty} x^k$ (c) $\sum_{n=1}^{\infty} \frac{(-x)^n}{n}$ (d) $\sum_{n=1}^{\infty} \frac{x^n}{n!}$

3.18 Determine for which values of α and β the series $\sum_{n=2}^{\infty} \frac{1}{n^\alpha (\ln n)^\beta}$ converges.

3.19 Determine for which values of β the series $\sum_{n=2}^{\infty} \frac{1}{n \ln n (\ln \ln n)^\beta}$ converges.

3.20 Let (a_n) be a sequence of real numbers. Suppose that $|a_{n+1} - a_n| < 2^{-n}$ for each $n \in \mathbb{N}$. Show that the sequence (a_n) is Cauchy.

3.21 Suppose that the series $\sum_{n=1}^{\infty} |a_{n+1} - a_n|$ converges for a given sequence of real numbers (a_n).

(1) Show that the sequence (a_n) is Cauchy.

(2) Is the converse true?

3.22 Let (a_n) be a sequence of real numbers. Suppose that there exist $M > 0$ and $0 < r < 1$ such that $|a_{n+1} - a_n| < Mr^n$ for each $n \in \mathbb{N}$. Show that the sequence (a_n) converges.

3.23 Let (a_n) be a sequence of real numbers. Show that if (a_n) is p-**summable**, that is to say, if $\sum |a_n|^p$ converges for some $p \geq 1$, then (a_n) is q-summing for every $q \geq p$.

3.24 Let (a_n) be a nonincreasing sequence of real numbers. Show that if $\sum a_n$ converges, then $\lim n a_n = 0$.

3.25 Show that if $\sum a_n$ diverges, then $\sum n a_n$ diverges.

3.26 Let (a_n) be a sequence of real numbers. For each $n \in \mathbb{N}$, let $b_n = \frac{|a_1| + |a_2| + \cdots + |a_n|}{n}$. Show that the series $\sum b_n$ always diverges.

3.27 Suppose that the sequence (a_n) is Césaro summable (see Exercise 2.18, page 62).

   (1) Show that if $\lim a_n = 0$, then $\lim \frac{1}{n} \sum_{k=1}^{n} a_k = 0$.

   (2) Show that the series $\sum a_n$ converges if and only if

$$\frac{1}{n} \sum_{k=1}^{n} k a_k = 0.$$

   (3) Show that if $\lim n a_n = 0$, then the series $\sum a_n$ converges.

3.28 Suppose that the sequence (a_n) is Abel summable (see Exercise 2.19, page 62). Show that if $\lim n a_n = 0$, then the series $\sum a_n$ converges.

3.29 Let (a_n) and (b_n) be sequences of real numbers such that both $\sum (a_n)^2$ and $\sum (b_n)^2$ converge. Show that $\sum a_n b_n$ converges absolutely.

3.30 Show that the convergence of $\sum a_n$, where the a_n are nonnegative, implies the convergence of $\sum \frac{\sqrt{a_n}}{n}$.

3.31 Show that if $\sum a_n$ converges and if (b_n) is monotonic and bounded, then $\sum a_n b_n$ converges.

3.32 Show that if $\sum a_n$ converges, where the a_n are nonnegative, and if (b_n) is bounded, then $\sum a_n b_n$ converges.

3.33 **Dirichlet's test.** Let (a_n) and (b_n) be sequences of real numbers. Suppose that the following conditions hold

   (1) $\lim a_n = 0$;

(2)   the series $\sum_{n=1}^{\infty} |a_{n+1} - a_n|$ converges;

(3)   the sequence $(\sum_{k=1}^{n} b_k)$ is bounded.

Show that the series $\sum a_n b_n$ converges.

3.34   Determine the nature of the series.

$$\text{(a) } \sum (1/n) \cos nx \qquad \text{(b) } \sum (1/n) \sin nx$$

3.35   **Abel's test.** Let (a_n) and (b_n) be sequences of real numbers. Suppose that the following conditions hold

(1)   the series $\sum_{n=1}^{\infty} |a_{n+1} - a_n|$ converges;

(2)   the sequence $(\sum_{k=1}^{n} b_k)$ is convergent.

Show that the series $\sum a_n b_n$ converges.

3.36   **Cauchy product.** Let (a_n) and (b_n) be sequences of nonnegative real numbers. Suppose that $\sum a_n = a$ and $\sum b_n = b$. Let

$$u_n = a_1 b_n + a_2 b_{n-1} + \cdots + a_n b_1 = \sum_{k=1}^{n} a_k b_{n-k}.$$

Show that

$$\sum u_n \le \left(\sum b_n \right) \left(\sum b_n \right) \le \sum u_{2n}$$

and hence that $\sum u_n = ab$. (The series $\sum u_n$ is called the Cauchy product of the series $\sum a_n$ with the series $\sum b_n$.)

3.37   **Merten's theorem.** Let (a_n) and (b_n) be sequences of real numbers. Suppose that $\sum a_n = a$ and $\sum b_n = b$ and that $\sum |a_n|$ converges. Show that the Cauchy product of the series $\sum a_n$ with the series $\sum b_n$ is convergent with sum equal to ab. (Hint: consider the series $\sum \beta_n$ where $\beta_n = b - \sum_{k=1}^{n} b_k$ and show that $\sum_{n=1}^{N} u_n = b \sum_{n=1}^{N} a_n + \sum_{k=0}^{N} a_k \beta_{N-k}$.)

3.38   Show that the sequence $(\sum_{k=1}^{n} \cos \frac{k\pi}{3})$ is bounded. Deduce that the series $\sum_{n=1}^{\infty} \frac{\cos \frac{n\pi}{3}}{\sqrt{n}}$ converges.

3.39   Test the series $\sum_{n=1}^{\infty} \frac{\sin \frac{n\pi}{2}}{n}$ for convergence.

3.40   Let (a_n) and (b_n) be sequences of real numbers. Suppose that $a_n > a_{n+1}$ for all n, $\lim a_n = 0$ and $(\sum_{k=1}^{n} b_k)$ is a bounded sequence. Show that the series $\sum a_n b_n$ converges.

3.41   **Infinite product**. We say that the infinite product

$$\prod_{n=1}^{\infty} (1 + a_n) = (1 + a_1)(1 + a_2)(1 + a_3) \cdots$$

converges if the series

$$S = \sum_{n=1}^{\infty} \ln (1 + a_n)$$

converges, in which case the value of the infinite product is e^S.

(1)   Prove that $\prod_{n=1}^{\infty} \left(1 + \frac{1}{n}\right)$ diverges.

(2)   Prove that $\prod_{n=1}^{\infty} \left(1 + \frac{1}{n^2}\right)$ converges.

(3)   Evaluate the infinite product $\prod_{n=2}^{\infty} \frac{n^2}{n^2-1}$.

3.42   Let (a_n) be the sequence defined by $a_n = \left(1 + \frac{1}{n}\right)^n$ for each $n \in \mathbb{N}$.

(1)   Show that for each $n \in \mathbb{N}$,

$$a_n = 1 + 1 + \frac{1}{2!}\left(1 - \frac{1}{n}\right) + \frac{1}{3!}\left(1 - \frac{1}{n}\right)\left(1 - \frac{2}{n}\right) + \cdots$$
$$+ \frac{1}{n!}\left(1 - \frac{1}{n}\right)\left(1 - \frac{2}{n}\right) \cdots \left(1 - \frac{n-1}{n}\right).$$

Deduce that the sequence (a_n) is increasing.

(2)   Show that for each $n \in \mathbb{N}$,

$$a_n < 1 + \sum_{k=1}^{n} \frac{1}{2^{n-1}}.$$

Deduce that the sequence (a_n) is bounded.

(3)   Show that $\lim a_n = e$ exists.

4
Limits and Continuity

The notion of limit is central in mathematical analysis. It is intimately bound with basic concepts such as the continuity of functions, the definitions of derivative and integral, and the convergence of sequences and series. It will play a fundamental role for the remainder of the course.

4.1 Limits of Functions

Given a function f, we are often interested in exploring the behavior of f near a given point a. Problems may arise if the function is defined on a neighborhood of the number a but not at a itself. In such a case, our interest is redirected to studying the variations of $f(x)$ when the variable x takes on values close to a. This idea is made more precise in the following definition.

Definition 4.1

Let f be a real-valued function of a real variable and let $A \subset \mathrm{dom}\,(f)$. Let a and l be two real numbers or the symbols $+\infty$ or $-\infty$. The statement

$$\lim_{x \to a^A} f(x) = l$$

(read "limit as x tends to a along A") means for every sequence (a_n) in A converging to a, the sequence $(f(a_n))$ converges to l, i.e. for every sequence

(a_n) in A

$$\lim a_n = a \text{ implies } \lim f(a_n) = l.$$

It should be noticed that in the above definition, a must be the limit of some sequence in A.

- If $a \in \mathbb{R}$, and $A = I \setminus \{a\}$ where $I = (a - \delta, a + \delta)$ is a δ-neighborhood of a, then $\lim_{x \to a^{I \setminus \{a\}}} f(x)$ is simply denoted by $\lim_{x \to a} f(x)$ and is called the **limit** of f at a.

- If $a \in \mathbb{R}$, and $A = (a - \delta, a)$ is an open interval, then $\lim_{x \to a^{(a-\delta,a)}} f(x)$ is denoted by $\lim_{x \to a^-} f(x)$ or by $f(a^-)$ and is called the **left-hand limit** of f at a.

- If $a \in \mathbb{R}$, and $A = (a, a + \delta)$ is an open interval, then $\lim_{x \to a^{(a,a+\delta)}} f(x) = l$ is denoted by $\lim_{x \to a^+} f(x)$ or $f(a^+)$ and is called the **right-hand limit** of f at a.

- If $A = (b, +\infty)$, then $\lim_{x \to +\infty^{(b,+\infty)}} f(x)$ is denoted by $\lim_{x \to +\infty} f(x)$.

- If $A = (-\infty, b)$, then $\lim_{x \to -\infty^{(-\infty,b)}} f(x)$ is denoted by $\lim_{x \to -\infty} f(x)$.

Example 4.2

Find $\lim_{x \to 3/2} \frac{4x^2 - 9}{2x - 3}$.

Solution

First we notice that if $x \neq 3/2$,

$$\frac{4x^2 - 9}{2x - 3} = \frac{(2x - 3)(2x + 3)}{2x - 3} = 2x + 3.$$

Thus for any sequence (a_n) converging to $3/2$, we have

$$\lim_{n \to \infty} \frac{4a_n^2 - 9}{2a_n - 3} = 2\left(\lim_{n \to \infty} a_n\right) + 3 = 2(3/2) + 3 = 6.$$

$\square$

Example 4.3

Show that

(1) $\lim_{x \to 2^-} \frac{1}{x - 2} = -\infty$;

(2) $\lim_{x \to 2^+} \frac{1}{x - 2} = +\infty$;

(3) $\lim_{x \to +\infty} \frac{1}{x-2} = 0$.

Solution

(1) Let (x_n) be a sequence converging to 2 such that $x_n < 2$. Let $M > 0$. Then there exists $N \in \mathbb{N}$ such that

$$2 - \frac{1}{M} < x_n < 2, \text{ for all } n > N.$$

Thus for $n > N$, $-\frac{1}{M} < x_n - 2$ or $\frac{1}{x_n - 2} < -M$. Hence $\lim_{x \to 2^-} \frac{1}{x-2} = -\infty$.

(2) Let (x_n) be a sequence converging to 2 such that $x_n > 2$. Let $M > 0$. Then there exists $N \in \mathbb{N}$ such that

$$2 < x_n < 2 + \frac{1}{M}, \text{ for all } n > N.$$

Thus for $n > N$, $\frac{1}{x_n - 2} > M$. Hence $\lim_{x \to 2^+} \frac{1}{x-2} = +\infty$.

(3) Let (x_n) be a sequence in $(2, +\infty)$ diverging to $+\infty$. Let $\varepsilon > 0$. Since we are interested in $\varepsilon > 0$ arbitrarily small, we may assume $0 < \varepsilon < 1/4$. There is $N \in \mathbb{N}$ such that $x_n > 2 + 1/\varepsilon$, for all $n > N$. Thus $x_n - 2 > 1/\varepsilon$, for $n > N$. That is, for $n > N$,

$$0 < \frac{1}{x_n - 2} < \varepsilon.$$

This shows that $\lim_{x \to +\infty} \frac{1}{x-2} = 0$. $\qquad \square$

Example 4.4

Let $f : \mathbb{R} \longrightarrow \mathbb{R}$ be defined by

$$f(t) = \begin{cases} 1 & \text{if } t \text{ is rational,} \\ 0 & \text{if } t \text{ is irrational.} \end{cases}$$

Show that for all a in $\mathbb{R}$, neither $f(a^+)$ nor $f(a^-)$ exists.

Solution

Let $a \in \mathbb{R}$. For every $\delta > 0$, the interval $(a, a + \delta)$ (resp. $(a - \delta, a)$) contains a sequence (a_n) of rationals as well as a sequence (b_n) of irrationals both converging to a. Since $\lim f(a_n) = 1$ while $\lim f(b_n) = 0$, $\lim_{x \to a^+} f(x)$ (resp. $\lim_{x \to a^-} f(x)$) does not exist. $\qquad \square$

Theorem 4.5

Let f and g be functions and let $A \subset \operatorname{dom}(f) \cap \operatorname{dom}(g)$. If $\lim_{x \to a^A} f(x)$ and $\lim_{x \to a^A} g(x)$ exist and are finite, then

(1) $\lim_{x \to a^A} (f + g)(x) = \lim_{x \to a^A} f(x) + \lim_{x \to a^A} g(x)$;

(2) $\lim_{x \to a^A} (f \cdot g)(x) = \lim_{x \to a^A} f(x) \cdot \lim_{x \to a^A} g(x)$;

(3) $\lim_{x \to a^A} \left(\frac{f}{g}\right)(x) = \frac{\lim_{x \to a^A} f(x)}{\lim_{x \to a^A} g(x)}$ if $\lim_{x \to a^A} g(x) \neq 0$ and $g(x) \neq 0$ for all $x \in A$.

Proof

Consider a sequence (a_n) in A converging to a. Then $\lim f(a_n) = \lim_{x \to a^A} f(x)$ and $\lim g(a_n) = \lim_{x \to a^A} f(x)$. By Proposition 2.24, we have

$$\lim (f + g)(a_n) = \lim f(a_n) + \lim g(a_n) = \lim_{x \to a^A} f(x) + \lim_{x \to a^A} g(x);$$
$$\lim (f \cdot g)(a_n) = \lim f(a_n) \cdot \lim g(a_n) = \lim_{x \to a^A} f(x) \cdot \lim_{x \to a^A} g(x)$$

and since $\lim g(a_n) \neq 0$ and $g(a_n) \neq 0$ for all n, we also have

$$\lim (f/g)(a_n) = \frac{\lim f(a_n)}{\lim g(a_n)} = \frac{\lim_{x \to a^A} f(x)}{\lim_{x \to a^A} g(x)}.$$

Since the sequence (a_n) is arbitrary chosen in A such that $a_n \to a$, we conclude that statements (1), (2) and (3) of the theorem hold. $\qquad \square$

Theorem 4.6

Let $A \subset \mathbb{R}$, and let $f : A \longrightarrow \mathbb{R}$. Let a be a real number such that it is the limit of some sequence in A, and let l be a real number. Then $\lim_{x \to a^A} f(x) = l$ if and only if

$$\begin{array}{r} \text{for every } \varepsilon > 0, \text{ there exists } \delta > 0 \text{ such that} \\ x \in A \text{ and } |x - a| < \delta \text{ implies } |f(x) - l| < \varepsilon. \end{array} \qquad (4.1)$$

Proof

Suppose that (4.1) holds and consider a sequence (a_n) in A converging to a. We show that $f(a_n) \to l$. Let $\varepsilon > 0$. By (4.1) there exists $\delta > 0$ such that

$$x \in A \text{ and } |x - a| < \delta \text{ implies } |f(x) - l| < \varepsilon$$

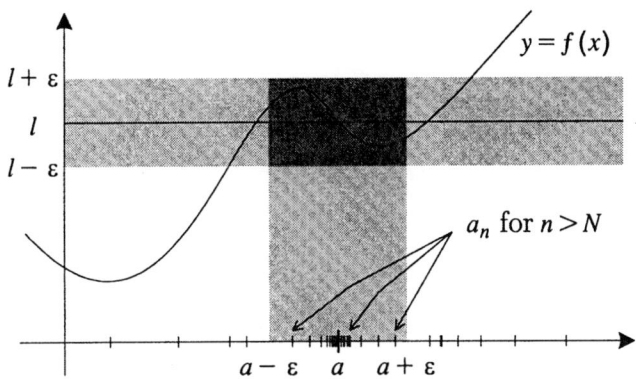

Figure 4.1 $\lim_{x \to a^A} f(x) = l$

(see Figure 4.1). Since $a_n \to a$, there exists N such that $n > N$ implies $|a_n - a| < \delta$. Since $a_n \in A$ for all n, we obtain

$$n > N \text{ implies } |f(a_n) - l| < \varepsilon,$$

as desired.

For the converse, suppose that (4.1) fails. Then for every $\delta > 0$ there exists an $\varepsilon > 0$ and $x \in A$ such that

$$|x - a| < \delta \text{ and } |f(x) - l| > \varepsilon.$$

So for each $n \in \mathbb{N}$, there exists $a_n \in A$ such that

$$|a_n - a| < \frac{1}{n} \text{ and } |f(a_n) - l| > \varepsilon.$$

Then (a_n) is a sequence in A, $a_n \to a$ but $f(a_n) \not\to l$. Thus $\lim_{x \to a^A} f(x) = l$ fails. This completes our proof. $\square$

Example 4.7

Use Theorem 4.6 to show that $\lim_{x \to 1} \frac{\sqrt{x}-1}{x-1} = \frac{1}{2}$.

Solution

Consider $\varepsilon > 0$. We notice that

$$\left| \frac{\sqrt{x}-1}{x-1} - \frac{1}{2} \right| = \left| \frac{1}{1+\sqrt{x}} - \frac{1}{2} \right| = \left| \frac{1-\sqrt{x}}{2(1+\sqrt{x})} \right|$$

$$= \left| \frac{(1-\sqrt{x})(1+\sqrt{x})}{2(1+\sqrt{x})^2} \right| = \frac{|1-x|}{2(1+\sqrt{x})^2} \leq \frac{|x-1|}{2}.$$

Choosing $\delta = 2\varepsilon$, we have

$$|x - 1| < \delta \text{ implies } \left| \frac{\sqrt{x} - 1}{x - 1} - \frac{1}{2} \right| \leq \frac{|x - 1|}{2} < \varepsilon.$$

The proof is complete.   □

Example 4.8

Use Theorem 4.6 to show that $\lim_{x \to 0} x \sin(1/x) = 0$.

Solution

Consider $\varepsilon > 0$. We notice that since $|\sin(1/x)| \leq 1$ for all $x \neq 0$,

$$|x \sin(1/x) - 0| = |x| \, |\sin(1/x)| \leq |x|.$$

Choosing $\delta = \varepsilon$, we have

$$|x - 0| < \delta \text{ implies } |x \sin(1/x) - 0| \leq |x - 0| < \varepsilon.$$

The proof is complete.   □

As corollaries of Theorem 4.6, for a function $f : A \to \mathbb{R}$ and for $a \in A$ and $l \in \mathbb{R}$, we have the following

$$\lim_{x \to a} f(x) = l \Leftrightarrow \begin{cases} \text{for every } \varepsilon > 0 \text{ there exists } \delta > 0 \text{ such that} \\ x \in A \text{ and } |x - a| < \delta \text{ implies } |f(x) - l| < \varepsilon. \end{cases} \quad (4.2)$$

$$\lim_{x \to a^+} f(x) = l \Leftrightarrow \begin{cases} \text{for every } \varepsilon > 0 \text{ there exists } \delta > 0 \text{ such that} \\ x \in A \text{ and } a < x < a + \delta \text{ implies } |f(x) - l| < \varepsilon. \end{cases} \quad (4.3)$$

$$\lim_{x \to a^-} f(x) = l \Leftrightarrow \begin{cases} \text{for every } \varepsilon > 0 \text{ there exists } \delta > 0 \text{ such that} \\ x \in A \text{ and } a - \delta < x < a \text{ implies } |f(x) - l| < \varepsilon. \end{cases} \quad (4.4)$$

Here the symbol "$\Leftrightarrow$" means "if and only if".

Since the condition $|x - a| < \delta$ is equivalent to $a - \delta < x < a + \delta$, it is clear that if (4.2) holds, then so do (4.3) and (4.4). Conversely, suppose that $\lim_{x \to a^+} f(x) = \lim_{x \to a^-} f(x) = l$. Let $\varepsilon > 0$. Then by (4.3) and (4.4) there exists $\delta_1 > 0$ and $\delta_2 > 0$ such that

$$a < x < a + \delta_1 \text{ implies } |f(x) - l| < \varepsilon$$

and

$$a - \delta_2 < x < a \text{ implies } |f(x) - l| < \varepsilon.$$

Thus if $\delta = \min\{\delta_1, \delta_2\}$, we have

$$|x - a| < \delta \text{ implies } |f(x) - l| < \varepsilon.$$

We have just proved the following fact.

Theorem 4.9

Let f be a function defined on an open interval containing a except possibly at a. Then $\lim_{x \to a} f(x)$ exists if and only if

(1) $f(a^+)$ exists;

(2) $f(a^-)$ exists; and

(3) $f(a^+) = f(a^-)$, in which case

$$f(a^+) = f(a^-) = \lim_{x \to a} f(x).$$

For example, if $f(x) = \text{fra}(x)$, then $\lim_{x \to 2} f(x)$ does not exist because $f(2^+) = 0$, and $f(2^-) = 2$. Similarly, if $g(x) = x/|x|$, then $\lim_{x \to 0} g(x)$ does not exist, because $g(0^-) = -1$, and $g(0^+) = 1$.

We now consider the case where a is $\pm\infty$. The proof of the next theorem is similar to that of Theorem 4.6.

Theorem 4.10

Let $A \subset \mathbb{R}$, and let $f : A \longrightarrow \mathbb{R}$. Let l be a real number. Then $\lim_{x \to +\infty} f(x) = l$ if and only if

$$\text{for every } \varepsilon > 0, \text{ there exists } N > 0 \text{ such that} \atop x \in A \text{ and } x > N \text{ implies } |f(x) - l| < \varepsilon. \tag{4.5}$$

Proof

Suppose that (4.5) holds and consider a sequence (a_n) in A converging to ∞. We show that $f(a_n) \to l$. Let $\varepsilon > 0$. By (4.5) there exists $M > 0$ such that

$$x \in A \text{ and } x > M \text{ implies } |f(x) - l| < \varepsilon.$$

Since $a_n \to \infty$, there exists N such that $n > N$ implies $a_n > \delta$. Since $a_n \in A$ for all n, we obtain

$$n > N \text{ implies } |f(a_n) - l| < \varepsilon,$$

as desired.

For the converse, suppose that (4.5) fails. Then for every $n \in \mathbb{N}$ there exists an $\varepsilon > 0$ and $x \in A$ such that

$$x > n \quad \text{and} \quad |f(x) - l| > \varepsilon.$$

So for each $n \in \mathbb{N}$ there exists $a_n \in A$ such that

$$a_n > n \quad \text{and} \quad |f(a_n) - l| > \varepsilon.$$

Then (a_n) is a sequence in A, $a_n \to \infty$ but $f(a_n) \nrightarrow l$. Thus $\lim_{x \to a^A} f(x) = l$ fails. This completes our proof.   □

Example 4.11

Use (4.5) to show that $\lim_{x \to +\infty}(e^{1-x} + 2) = 2$.

Solution

Let $\varepsilon > 0$. Consider $N = 1 + \ln \frac{1}{\min\{\varepsilon, 1\}}$. Then

$$x > N \text{ implies } 1 - x < -\ln \frac{1}{\min\{\varepsilon, 1\}}$$

and thus $e^{1-x} < \min\{\varepsilon, 1\}$. Hence for $0 < \varepsilon < 1$,

$$x > N \quad \text{implies} \quad \left|(e^{1-x} + 2) - 2\right| = e^{1-x} < \varepsilon.$$

This proves that $\lim_{x \to +\infty}(e^{1-x} + 2) = 2$.   □

In the same way, we have for any finite number l,

$$\lim_{x \to -\infty} f(x) = l \Leftrightarrow \left\{ \begin{array}{l} \text{for every } \varepsilon > 0 \text{ there exists } N > 0 \text{ such that} \\ x \in A \text{ and } x < -N \text{ implies } |f(x) - l| < \varepsilon. \end{array} \right.$$

Now for the case $l = \pm\infty$, we will state the results and leave the proofs as exercises.

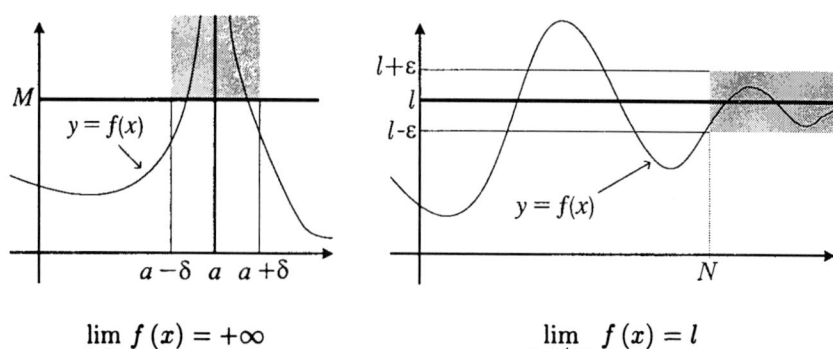

$$\lim_{x \to a} f(x) = +\infty \qquad\qquad \lim_{x \to +\infty} f(x) = l$$

Theorem 4.12

Let $A \subset \mathbb{R}$, and let $f : A \longrightarrow \mathbb{R}$. Then $\lim_{x \to a} f(x) = +\infty$ if and only if

for every $M > 0$ there exists $\delta > 0$ such that
$x \in A$ and $|x - a| < \delta$ implies $f(x) > M$.

Example 4.13

Show that $\lim_{x \to 1} \frac{1}{(1-x)^2} = +\infty$.

Solution

For any $M > 0$, we have $\frac{1}{(1-x)^2} > M$ provided that $(1 - x)^2 < \frac{1}{M}$; i.e. provided that

$$|1 - x| < \frac{1}{\sqrt{M}} = \delta.$$

Hence $\lim_{x \to 1} \frac{1}{(1-x)^2} = +\infty$. $\square$

The case $l = -\infty$ is similar and is obtained with obvious changes.

$$\lim_{x \to a} f(x) = -\infty \Leftrightarrow \left\{ \begin{array}{l} \text{for every } M > 0 \text{ there exists } \delta > 0 \text{ such that} \\ x \in A \text{ and } |x - a| < \delta \text{ implies } f(x) < -M. \end{array} \right.$$

Example 4.14

Show that $\lim_{x \to 1+} \ln(x - 1) = -\infty$.

Solution

Let $M > 0$. Consider $\delta = 1/e^M$. It follows that if $x > 1$ and $|x - 1| < \delta$, then

$$\ln(x - 1) < \ln \delta = \ln \left(\frac{1}{e^M} \right) = -M.$$

This proves that $\lim_{x \to 1+} \ln(x - 1) = -\infty$. $\square$

Finally if a and l are both infinite, we have the following result.

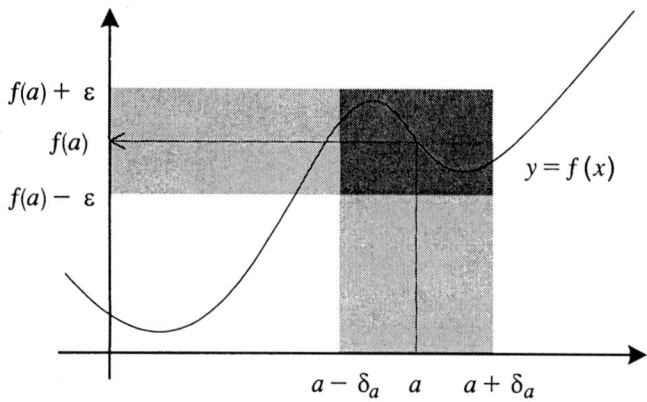

Figure 4.2   Continuity at the point a

Theorem 4.15

Let $A \subset \mathbb{R}$, and let $f : A \longrightarrow \mathbb{R}$. Then $\lim_{x \to -\infty} f(x) = +\infty$ if and only if

> for every $M > 0$ there exists $N > 0$ such that
> $x \in A$ and $x < -N$ implies $f(x) > M$.

4.2 Continuity of Functions

Continuity is another important notion in analysis. Loosely speaking, a function f is continuous if it preserves "closeness" in the sense that the variable $f(x)$ takes on values close to $f(a)$ whenever x is close enough to a (see Figure 4.2).

Definition 4.16

Let $A \subset \mathbb{R}$ and $f : A \to \mathbb{R}$. The function f is said to be **continuous** at $a \in A$ if

> for every $\varepsilon > 0$ there exists $\delta_a > 0$ such that
> $x \in A$ and $|x - a| < \delta_a$ implies $|f(x) - f(a)| < \varepsilon$.   (4.6)

If f is continuous at each point a of A, then we say f is continuous on A.

We have used the notation δ_a to emphasize the fact that the number δ_a depends not only on the given $\varepsilon > 0$, but also on the point a. If a function f is not continuous at a point a, then we say that "f is **discontinuous** at a" and that "a is a point of discontinuity of f".

Example 4.17

Show that the absolute value function $|\cdot| : \mathbb{R} \to \mathbb{R}$ is continuous.

Solution

Let $a \in \mathbb{R}$ and fix $\varepsilon > 0$. Since for every $x \in \mathbb{R}$, we have

$$x = (x - a) + a \le |x - a| + |a|$$

and

$$a = (a - x) + x \le |a - x| + |x|,$$

it follows that

$$||x| - |a|| \le |x - a|.$$

Thus if $\delta = \varepsilon$, we have

$$|x - a| < \delta \text{ implies } ||x| - |a|| < \varepsilon.$$

Hence $|\cdot| : \mathbb{R} \to \mathbb{R}$ is continuous. $\qquad\square$

It must be clear that for a function f to be continuous at $a \in A$, the definition requires that

(1) $f(a)$ exists as a real number and

(2) $\lim_{x \to a^A} f(x) = f(a)$.

Each of the following figures shows discontinuity at a because $f(a)$ is not defined.

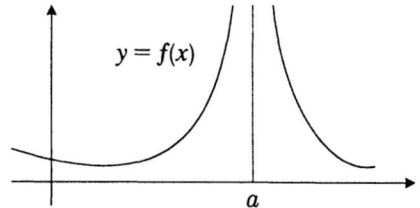 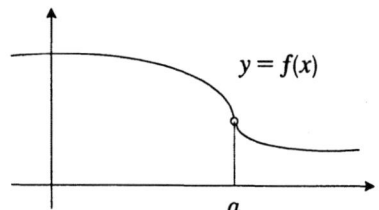

The next two figures show that even when f is defined at a point a, continuity is not realized if the limit condition is not satisfied.

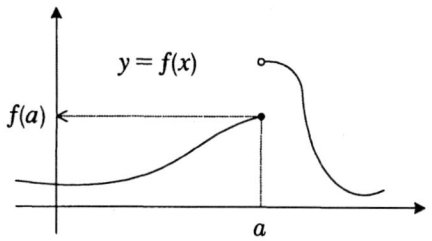 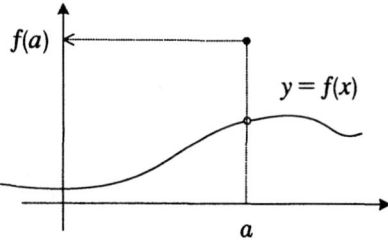

It should also be clear that according to the definition, a function f is continuous at a point $a \in A$ if and only if $f(a^+) = f(a^-) = f(a)$ as real numbers.

Finally, we notice that a function that is defined everywhere can be discontinuous at all points. Example 4.4 shows such a function: it is often referred to as **Dirichlet's discontinuous function**.

Example 4.18

Let $f : [a, b] \to \mathbb{R}$ be a nondecreasing function. Show that the set of all the points of discontinuity of f is countable.

Solution

Since f is nondecreasing, we first notice that for each $x \in (a, b)$, both

$$f\left(x^-\right) = \lim_{t \to x^-} f(x) \quad \text{and} \quad f\left(x^+\right) = \lim_{t \to x^+} f(x)$$

exist as real numbers. Let D be the set of all points of discontinuity of f on (a, b). Now if $x \in D$, then $f(x^-) < f(x^+)$. By the density theorem (page 22), we can choose a rational number $r(x)$ such that

$$f\left(x^-\right) < r(x) < f\left(x^+\right).$$

On the other hand, the monotonicity of f also implies that if $x < y$ in D, then

$$r(x) < f\left(x^+\right) < f\left(y^-\right) < r(y).$$

Thus, the correspondence $x \mapsto r(x)$ from the set D into $\mathbb{Q}$ is one-to-one . Since $\mathbb{Q}$ is countable, we conclude that D is also countable.   □

Since the expression $\lim_{x \to a^{\blacktriangle}} f(x)$ is defined in terms of convergence of sequences, it is obviously possible to give a characterization of continuity involving sequences. We will see that such a characterization is very useful in many instances.

Definition 4.19

Let $A \subset \mathbb{R}$ and $f : A \to \mathbb{R}$. The function f is said to be **sequentially continuous** at $a \in A$ if for all sequences (a_n) in A, which converge to a, the sequence $(f(a_n))$ converges to $f(a)$.

Theorem 4.20

Let $A \subset \mathbb{R}$, $f : A \to \mathbb{R}$ and $a \in A$. Then f is continuous at a if and only if it is sequentially continuous.

Proof

Except for a few obvious changes, the proof proceeds in exactly the same way as in the proof of Theorem 4.6. We leave it as an exercise. □

Using the definition of continuity and the results of Theorem 4.5, one easily proves the following properties.

Theorem 4.21

Let f and g be real-valued functions on a set A. Suppose that both f and g are continuous at $a \in A$. Then

(1) $f + g$ is continuous at a;

(2) $f \cdot g$ is continuous at a;

(3) if in addition $g(a) \neq 0$, then f/g is continuous at a.

Theorem 4.22

If f is continuous at a and g is continuous at $f(a)$, then $g \circ f$ is continuous at a.

Proof

It is given that $a \in \mathrm{dom}\,(f) = A$, and $f(a) \in \mathrm{ran}\,(f) \subset \mathrm{dom}\,(g)$. Let (a_n) be a sequence in A converging to a. Since f is continuous at a, $f(a_n) \to f(a)$. Since g is continuous at $f(a)$, $g(f(a_n)) \to g(f(a))$, that is $g \circ f(a_n) \to g \circ f(a)$. Hence $g \circ f$ is continuous at a. □

Obvious changes of the argument above can be used to prove the following useful theorem.

Theorem 4.23

If $\lim_{x \to a} g(a) = b$ and f is continuous at b, then

$$\lim_{x \to a} f(g(x)) = f(b) = f\left(\lim_{x \to b} g(x)\right).$$

For example, $f(x) = \cos x$ is continuous everywhere; if g is a real-valued function defined and continuous on a set A containing a number a, then we can evaluate

$$\lim_{x \to a} \cos(g(x)) = \cos\left(\lim_{x \to \infty} g(x)\right).$$

Similarly, we have $\lim_{x \to a} e^{g(x)} = e^{\lim_{x \to \infty} g(x)}$, $\lim_{x \to a} |g(x)| = |\lim_{x \to \infty} g(x)|$, $\lim_{x \to a} \sqrt{|g(x)|} = \sqrt{|\lim_{x \to \infty} g(x)|}$, and so on.

The results of the last three theorems can be combined to obtain further examples of continuous functions.

Example 4.24

Let f and g be real-valued functions on a set A. Suppose that both f and g are continuous at $a \in A$. Then

(1) $\max\{f, g\}$ is continuous at a;

(2) $\min\{f, g\}$ is continuous at a;

(3) $f^+ = \max\{f, 0\}$ and $f^- = \max\{-f, 0\}$ are continuous at a.

Solution

It suffices to show (1). The rest can be proved by similar arguments. First we notice that $\max\{f, g\} = \frac{1}{2}[(f + g) + |f - g|]$. By Theorem 4.21, $f + g$ and $f - g$ are continuous at a. Then Example 4.17 and Theorem 4.22 imply that $|f - g|$ is continuous at a, and thus Theorem 4.21 again ensures that $\max\{f, g\} = \frac{1}{2}[(f + g) + |f - g|]$ is continuous at a. $\square$

4.3 Properties of Continuous Functions

Two very useful properties of continuous functions are the so-called *extreme value theorem* (also known as the *boundedness property*) and the so-called *intermediate value theorem*.

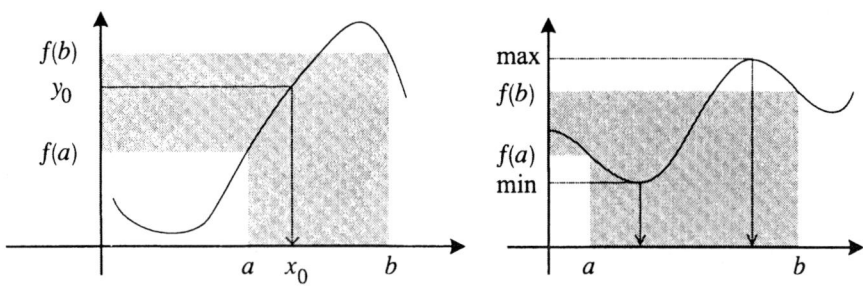

Intermediate value theorem   Extreme value theorem

We say that a function **has a maximum** (resp. minimum) or **attains its maximum** at the point x_0 on a given interval I if

(1) $f(x_0)$ is defined, and

(2) $f(x) \le f(x_0)$ (resp. $f(x) \ge f(x_0)$) for all $x \in I$.

Theorem 4.25 (Extreme Value Theorem)

A continuous real-valued function on a closed interval $[a, b]$ attains its maximum and its minimum.

Proof

Suppose that $f : [a, b] \to \mathbb{R}$ is continuous. Consider

$$y_0 = \sup \{f(x) : x \in [a, b]\}.$$

Suppose that $y_0 = +\infty$. Then for each $n \in \mathbb{N}$, there exists $a_n \in [a, b]$ such that $f(a_n) > n$. Since (a_n) is bounded, it has a convergent subsequence (Bolzano–Weierstrass Theorem 2.30), say $(a_{n_k}) \to c$. Since $a \le a_{n_k} \le b$, we have $a \le c \le b$. Now since f is continuous at on $[a, b]$, $f(a_{n_k}) \to f(c)$; but this is impossible since $f(a_{n_k}) > n_k \to \infty$. Thus $y_0 < +\infty$.

For each $n \in \mathbb{N}$, choose $b_n \in [a, b]$ such that $y_0 - 1/n < f(b_n) \le y_0$. Then we have $f(b_n) \to y_0$. By the Bolzano–Weierstrass theorem, the sequence (b_n) has a convergent subsequence (b_{n_k}) which converges to a limit $b_0 \in [a, b]$. We have on the one hand, since $(f(b_{n_k}))$ is a subsequence of $(f(b_n))$, $\lim f(b_{n_k}) = y_0$. On the other hand, since f is continuous at b_0, $\lim f(b_{n_k}) = f(b_0)$. Thus $f(b_0) = y_0$.

A similar argument shows that f attains its minimum value as well.   □

In particular, if a function is continuous on a closed bounded interval, then it is bounded. It must be noted that the continuity condition is essential in the extreme value theorem: indeed it is easily checked that the function $f(x) = $ fra (x) is bounded, but does not have a maximum value, on the closed interval $[0, 2]$.

Theorem 4.26 (Intermediate Value Theorem)

If f is a real-valued continuous function on a closed bounded interval $[a, b]$, and if y_0 is any number between $f(a)$ and $f(b)$, then there exists $x_0 \in (a, b)$ such that $f(x_0) = y_0$.

Proof

Let us assume that $f(a) < y_0 < f(b)$. Consider the set

$$A = \{x \in [a, b] : f(x) < y_0\}.$$

Then A is a nonempty subset of $[a, b]$ ($a \in A$). Let $x_0 = \sup A$. Then $x_0 \in [a, b]$. For each n there exists $x_n \in A$ such that $x_0 - 1/n < x_n \le x_0$. Thus $f(x_n) < y_0$ for each n. Since $x_n \to x_0$ and since f is continuous at x_0, we have

$$f(x_0) = \lim f(x_n) \le y_0. \tag{4.7}$$

On the other hand, for each n, consider $b_n = \min\{b, x_0 + 1/n\}$. Then since $b_n \notin A$, $f(b_n) \ge y_0$ for each n. Thus, since f is continuous at b,

$$f(x_0) = \lim f(b_n) \ge y_0. \tag{4.8}$$

Equations (4.7) and (4.8) imply that $f(x_0) = y_0$.   □

Example 4.27

Show that a polynomial of odd degree with real coefficients has at least one real zero.

Solution

Let $P(x) = a_n x^n + a_{n-1} x^{n-1} + \cdots + a_0$ be a polynomial with n odd and $a_n \ne 0$, say $a_n > 0$ (if $a_n < 0$, consider $-P(x)$). Then we can write

$$P(x) = x^n \left[a_n + \frac{a_{n-1}}{x} + \cdots + \frac{a_0}{x^n} \right].$$

Then it is clear that

$$\lim_{x \to \infty} P(x) = \infty \quad \text{and} \quad \lim_{x \to -\infty} P(x) = -\infty.$$

Let $a, b \in \mathbb{R}$ such that $a < 0$, $b > 0$, and $P(a) < 0 < P(b)$. Since P is continuous on $\mathbb{R}$, the intermediate value theorem applies on the interval $[a, b]$ and implies that $p(c) = 0$ for some c between a and b. $\qquad\qquad\qquad\qquad\qquad\square$

Another way to express the intermediate value theorem is: *a continuous function f on an interval $[a, b]$ takes on every value between $f(a)$ and $f(b)$.*

Example 4.28

Show that if a real-valued function f is continuous on an interval I, then $f(I) = \{f(x) : x \in I\}$ is an interval.

Solution

We are done if we show that any point between $\inf f(I)$ and $\sup f(I)$ is in $f(I)$. Let y be between $\inf f(I)$ and $\sup f(I)$. Then by the extreme value theorem, there exist $y_1 = f(x_1)$ and $y_2 = f(x_2)$ in $f(I)$ such that $y_1 < y < y_2$. We may assume for simplicity that $x_1 \leq x_2$. Since f is continuous on the closed interval $[x_1, x_2]$, the intermediate value theorem applies and so $y = f(x)$ for some x between x_1 and x_2. Hence $y \in f(I)$. $\qquad\qquad\qquad\square$

The above example claims that continuous functions map intervals onto intervals. In particular, if a real-valued function f is continuous on an interval I and attains its maximum and its minimum on I, then f takes on, at least once, any values lying between its maximum and its minimum. A combination of this result with the extreme value theorem clearly implies the following.

Corollary 4.29

If f is a real-valued continuous function on a closed bounded interval $I = [a, b]$, then $f(I)$ is a closed and bounded interval.

The inverse of this theorem need not be true: the function defined by

$$f(t) = \begin{cases} 2x & \text{if } 0 \leq x \leq 1/2, \\ 0 & \text{if } 1/2 < x \leq 1 \end{cases}$$

satisfies $f([0, 1]) = [0, 1]$, yet it is not continuous on $[0, 1]$.

4.4 Uniform Continuity

In the definition of the continuity of a function f at a point $a \in A$, we emphasize that the number $\delta_a > 0$ depends not only on the given $\varepsilon > 0$, but also on the point a. A stronger property is stated in the following definition.

Definition 4.30

Let $A \subset \mathbb{R}$ and $f : A \to \mathbb{R}$. The function f is said to be **uniformly continuous** if

$$\text{for every } \varepsilon > 0 \text{ there exists } \delta > 0 \text{ such that}$$
$$s,t \in A \text{ and } |s - t| < \delta \text{ implies } |f(s) - f(t)| < \varepsilon. \tag{4.9}$$

Here, the number $\delta > 0$ in (4.9) depends solely on the given $\varepsilon > 0$, and does not depend on any elements in A. Comparing (4.9) with (4.1), we easily see that uniform continuity is a stronger property than continuity: *if f is uniformly continuous on A, then it is continuous on A.* However, the converse need not be true, as the following example shows.

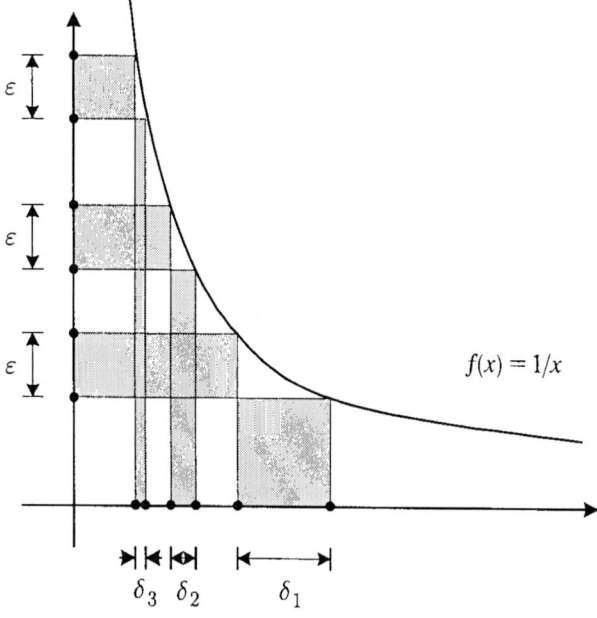

$f(x) = 1/x$

Figure 4.3   Nonuniform continuity

The function $f(x) = 1/x$ is continuous on $(0,1)$. However, for each $n \in \mathbb{N}$, consider $s_n = 1/n$ and $t_n = 1/2n$ in $(0,1)$. Then

$$|s_n - t_n| < \frac{1}{2n} \quad \text{and} \quad |f(s_n) - f(t_n)| = n.$$

For any $\delta > 0$, one can choose n large enough so that $1/2n < \delta$. Then for such n, $|f(s_n) - f(t_n)| = n$. Figure 4.3 shows that for a fixed value of $\varepsilon > 0$, the value of δ changes as the value of x changes.

So if we have chosen $\varepsilon = 1$, there will be no $\delta > 0$ fulfilling the requirement described in (4.9). Hence f is not uniformly continuous on $(0,1)$.

Example 4.31

Show that $f(x) = 1/x$ is uniformly continuous on $(1,2)$.

Solution

Let $\varepsilon > 0$. For every s and t in $(1,2)$, we have

$$\left| \frac{1}{s} - \frac{1}{t} \right| = \frac{|t - s|}{|ts|} \leq |s - t|.$$

Consider $\delta = \varepsilon$. Then for all s and t in $(1,2)$ we have

$$|s - t| < \delta \quad \text{implies} \quad \left| \frac{1}{s} - \frac{1}{t} \right| < \varepsilon.$$

Hence, $f(x) = 1/x$ is uniformly continuous on $(1,2)$. $\square$

Example 4.32

Let A be a nonempty subset of $\mathbb{R}$. Show that the function $f_A : \mathbb{R} \to \mathbb{R}$ defined by

$$f_A(x) = \inf\{|x - a| : a \in A\}$$

is uniformly continuous.

Solution

For arbitrary $x, y \in \mathbb{R}$, and for each $z \in A$, we have

$$|x - z| \leq |x - y| + |y - z|.$$

Therefore,

$$f_A(x) \le |x - y| + |y - z| \text{ or } f_A(x) - |y - z| \le |x - y|$$

for any $z \in A$. Taking the infimum over all $z \in A$, we get

$$f_A(x) - f_A(y) \le |x - y|.$$

Interchanging x and y, we obtain

$$f_A(y) - f_A(x) \le |x - y|.$$

Hence,

$$|f_A(x) - f_A(y)| \le |x - y|.$$

Taking $\delta = \varepsilon$ in the definition of uniform continuity, this inequality shows that f_A is indeed uniformly continuous.   □

We have noticed that continuity does not imply uniform continuity. However, if the domain of the function is a closed and bounded interval, then we have the following result.

Theorem 4.33

If f is a real-valued function continuous on a closed bounded interval $[a, b]$, then f is uniformly continuous on $[a, b]$.

Proof

Suppose that f is not uniformly continuous on $[a, b]$. Then there exists $\varepsilon > 0$ such that for every $\delta > 0$ there exist s and t in $[a, b]$ such that

$$|s - t| < \delta \text{ implies } |f(s) - f(t)| \ge \varepsilon.$$

In particular, for each $n \in \mathbb{N}$ there exist s_n and t_n in $[a, b]$ such that

$$|s_n - t_n| < \frac{1}{n} \text{ implies } |f(s_n) - f(t_n)| \ge \varepsilon.$$

By the Bolzano–Weierstrass theorem (Theorem 2.30), there exists a convergent subsequence (s_{n_k}) of (s_n). Let $s = \lim s_{n_k}$. Since

$$|t_{n_k} - s| \le |t_{n_k} - s_{n_k}| + |s_n - t_n| < \frac{1}{n_k} + |s_n - t_n|,$$

we infer that $s = \lim t_{n_k}$. Since f is continuous at s we can find k large enough so that

$$|f(s_{n_k}) - f(s)| < \frac{\varepsilon}{2} \text{ and } |f(t_{n_k}) - f(s)| < \frac{\varepsilon}{2}.$$

Thus for such k, using the triangle inequality we have

$$\varepsilon \leq |f(s_{n_k}) - f(t_{n_k})| \leq |f(s_{n_k}) - f(s)| + |f(s) - f(t_{n_k})|$$
$$< \frac{\varepsilon}{2} + \frac{\varepsilon}{2} = \varepsilon,$$

a contradiction. Thus f must be uniformly continuous on $[a, b]$. $\qquad\square$

As in the case of continuity, it is possible to give a sequential characterization of uniform continuity.

Theorem 4.34

Let $A \subset \mathbb{R}$, $f : A \to \mathbb{R}$. Then f is uniformly continuous on A if and only if for any two sequences (t_n) and (s_n) in A such that $t_n - s_n \to 0$, we have $f(t_n) - f(s_n) \to 0$.

Proof

Suppose that f is uniformly continuous on A. Let (t_n) and (s_n) be two sequences in A such that $t_n - s_n \to 0$. Fix $\varepsilon > 0$. There exists $\delta > 0$ such that

$$t, s \in A \text{ and } |t - s| < \delta \text{ implies } |f(t) - f(s)| < \varepsilon. \qquad (4.10)$$

Since $t_n - s_n \to 0$, there exists an N such that

$$n > N \text{ implies } |t_n - s_n| < \delta.$$

It follows from (4.10) that

$$\text{for } n > N, \ |f(t_n) - f(s_n)| < \varepsilon,$$

hence $f(t_n) - f(s_n) \to 0$.

Conversely, suppose that for any two sequences (t_n) and (s_n) in A such that $t_n - s_n \to 0$, we have $f(t_n) - f(s_n) \to 0$. Suppose to the contrary that f is not uniformly continuous. Then there exists an $\varepsilon > 0$ such that for every $\delta > 0$ there exist t and s in A such that

$$|t - s| < \delta \text{ implies } |f(t) - f(s)| \geq \varepsilon.$$

For each n, we choose $\delta = \frac{1}{n}$. Thus there exist t_n and s_n in A such that

$$|t_n - s_n| < \frac{1}{n} \text{ implies } |f(t_n) - f(s_n)| \geq \varepsilon.$$

Thus the sequences (t_n) and (s_n) are such that $t_n - s_n \to 0$ but $f(t_n) - f(s_n) \not\to 0$. Contradiction! This completes the proof. $\qquad\square$

We know that continuous functions map convergent sequences to convergent sequences. The next theorem asserts that a uniformly continuous function takes Cauchy sequences into Cauchy sequences. It provides a method of showing that a given function is not uniformly continuous.

Theorem 4.35

Let f be a uniformly continuous function on a set A, and suppose that (a_n) is a Cauchy sequence in A. Then the sequence $(f(a_n))$ is Cauchy.

Proof

Let (a_n) be a Cauchy sequence in A and let $\varepsilon > 0$. Since f is uniformly continuous on A, there exists $\delta > 0$ such that

$$s, t \in A \text{ and } |t - s| < \delta \text{ implies } |f(t) - f(s)| < \varepsilon. \qquad (4.11)$$

Since (a_n) is Cauchy, there exists $N \in \mathbb{N}$ such that

$$|a_n - a_m| < \delta \text{ whenever } n, m > N.$$

From (4.11), it follows that

$$|f(a_n) - f(a_m)| < \varepsilon \text{ whenever } n, m > N.$$

Hence $(f(a_n))$ is Cauchy. $\qquad \Box$

Example 4.36

Show that the function $f(x) = \sin \frac{1}{x}$ is not uniformly continuous on $(0,1]$.

Solution

Consider the Cauchy sequence $\left(\frac{1}{n}\right)$ in $(0,1]$. $f\left(\frac{1}{n}\right) = \sin n$ is not Cauchy. Therefore f is not uniformly continuous. $\qquad \Box$

We have noticed that the extreme value theorem (Theorem 4.25) implies that if a function is continuous on a closed bounded interval, then it is bounded. The function $f(x) = 1/x$ is unbounded even though it is continuous on $(0, 1]$. Therefore a continuous function on a bounded set A is not necessarily bounded. However, we have the following result.

Theorem 4.37

A uniformly continuous function f on a bounded set A is bounded.

Proof

Suppose that f is uniformly continuous but it is not bounded. Then for each n, there exists a_n in A such that

$$|f(a_n)| > n. \tag{4.12}$$

Since A is bounded, so is the sequence (a_n). By the Bolzano–Weierstrass theorem, (a_n) has a convergent subsequence, say (a_{n_k}). Then (a_{n_k}) is Cauchy and by Theorem 4.35, the sequence $(f(a_{n_k}))$ is also Cauchy. As such, $(f(a_{n_k}))$ is bounded. This is impossible since by (4.12), for all k

$$|f(a_{n_k})| > n_k \geq k.$$

This contradiction proves that f is bounded. □

EXERCISES

4.1 Find each of the indicated limits if it exists.

(a) $\lim\limits_{x \to 2} \frac{x^2+3x+5}{x^2+2}$

(b) $\lim\limits_{x \to 1} \frac{x-1}{|x-1|}$

(c) $\lim\limits_{x \to 1} \frac{x^n-1}{x^p-1}$

(d) $\lim\limits_{x \to 1} \frac{1-x}{1-\sqrt[3]{x}}$

(e) $\lim\limits_{x \to 0} \frac{x}{\sqrt{1-\cos x}}$

(f) $\lim\limits_{x \to 0} \frac{\ln(1+\alpha x)}{x}$

(g) $\lim\limits_{x \to 0} \frac{\sqrt{x^2+p^2}-p}{\sqrt{x^2+q^2}-q}$

(h) $\lim\limits_{x \to 0} \frac{\sqrt{1+x}-1}{x}$

(i) $\lim\limits_{x \to +\infty} \frac{5x^3-x^2+3}{-3x^3+2x-x}$

(j) $\lim\limits_{x \to +\infty} \left(1-\frac{1}{x}\right)^x$

(k) $\lim\limits_{x \to \infty} \left(\frac{2x+3}{2x+1}\right)^{x+1}$

(l) $\lim\limits_{x \to \infty} x\left(\sqrt{x^2+1}-x\right)$

4.2 Find each of the indicated limits if it exists.

(a) $\lim\limits_{x \to 0} \frac{\sin^2 \frac{x}{3}}{x^2}$

(b) $\lim\limits_{x \to 1+} \frac{\sqrt{x-1}}{x-1}$

(c) $\lim\limits_{x \to -\infty} \frac{2^x-1}{x-1}$

(d) $\lim\limits_{x \to 1} \frac{1-x}{1-\sqrt[3]{x}}$

(e) $\lim\limits_{x \to 0} \frac{2\arcsin x}{3x}$

(f) $\lim\limits_{x \to 0} \frac{e^{ax}-e^{bx}}{x}$

(g) $\lim\limits_{x \to 0} \frac{e^{ax}-e^{bx}}{\sin ax-\sin bx}$

(h) $\lim\limits_{x \to a-} \frac{x-a}{|x-a|}$

(i) $\lim\limits_{x \to +\infty} \left(\frac{x}{x+1}\right)^x$

(j) $\lim\limits_{x \to a} \frac{x^2-a^2}{x-a}$

(k) $\lim\limits_{x \to \infty} \frac{\sqrt{x}-\sqrt{a}}{x-a}$

(l) $\lim\limits_{x \to \infty} x\left(\ln(1+x)-\ln x\right)$

4.3 Prove that $\lim_{x \to 0} \frac{x^2 \sin(1/x)}{\sin x} = 0$.

4.4 Show that if $\lim_{x \to a} f(x)$ exists, then it must be unique.

4.5   Write in terms of ε and δ the negation of "f is continuous at a".

4.6   Find the points of discontinuity of each function.

(a) $y = \frac{x+1}{x(x^2-2)}$ (b) $y = \tan\frac{1}{x}$ (c) $y = 2^{\frac{1}{x}}$ (d) $y = \frac{1}{\ln|x-3|}$

4.7   Complete the proof of Example 4.24.

4.8   Let f be a real-valued function. Suppose that $a \in \operatorname{dom} f$

$$\lim_{x \to a^{\operatorname{dom} f}} \frac{f(x) - f(a)}{x - a}$$

exists, where $a \in \operatorname{dom} f$. Show that f is continuous at a.

4.9   Show that if a continuous function f satisfies $f(x) = 0$ whenever x is rational, then f is identically equal to 0.

4.10 Let f be a nondecreasing function on (a, b).

(1)   Show that for each $x \in (a, b)$, $f(x^+)$ and $f(x^-)$ exist and $f(x^-) \leq f(x) \leq f(x^+)$.

(2)   Show that if $x < y$ in (a, b), then $f(x^+) \leq f(y^-)$.

4.11 Show that the set of all points of discontinuity of a nondecreasing function is at most countable. (Hint: Use the result of the previous exercise together with the density theorem.)

4.12 Let (a_n) and (b_n) be two sequences of real numbers. Suppose that the a_n are nonnegative and that the series $\sum a_n$ converges with sum a. Define for each n the function

$$f_n(x) = \begin{cases} 0 & \text{if } x < b_n \\ a_n & \text{if } x \geq b_n. \end{cases}$$

Let $f(x) = \sum f_n(x)$ for all x. Show that

(1) f is nondecreasing;

(2) f is discontinuous on $A = \{b_n : n \in \mathbb{N}\}$;

(3) f is continuous on $\mathbb{R} \setminus A$.

4.13 Give an example of a continuous function $f : A \to \mathbb{R}$ such that

(1) f is bounded above but does not attain its maximum;

(2) f is unbounded;

(3) f is bounded and attains its maximum but not its minimum.

4.14   Complete the proof of Theorem 4.26, by dealing with the case $f(a) > f(b)$.

4.15   Write out the proof of Theorem 4.21.

4.16   Write out the proof of Theorem 4.23.

4.17   Let $f : \mathbb{R} \to \mathbb{R}$ be nonidentically zero and satisfy $f(x+y) = f(x) \cdot f(y)$ for all $x,y \in \mathbb{R}$. Show that f is continuous at 0 if and only if it is continuous everywhere.

4.18   Let $f : \mathbb{R} \to \mathbb{R}$. Then f is said be **homogeneous** if $f(\lambda x) = \lambda f(x)$ for all λ, $x \in \mathbb{R}$. Show that f is continuous.

4.19   Let $f : \mathbb{R} \to \mathbb{R}$ satisfy $f(x+y) = f(x) + f(y)$ for all $x, y \in \mathbb{R}$.

   (1)   Show that $f(0) = 0$, $f(-x) = -f(x)$ for all $x \in \mathbb{R}$ and $f(n) = nf(1)$ for all $n \in \mathbb{N}$.

   (2)   Show that $nf(1/n) = f(1)$ for all $n \in \mathbb{N}$.

   (3)   Show that $f(q) = qf(1)$ for all $q \in \mathbb{Q}$.

   (4)   Show that if f is continuous, then $f(\lambda x) = \lambda f(x)$ for all $\lambda, x \in \mathbb{R}$.

4.20   **Dirichlet's discontinuous function.** Let $f : \mathbb{R} \to \mathbb{R}$ be defined by
$$f(t) = \begin{cases} 1 & \text{if } t \text{ is rational} \\ 0 & \text{if } t \text{ is irrational.} \end{cases}$$
Show that f is continuous at no point.

4.21   Let $f : \mathbb{R} \to \mathbb{R}$ be defined by
$$f(t) = \begin{cases} t & \text{if } t \text{ is rational} \\ 0 & \text{if } t \text{ is irrational.} \end{cases}$$
Show that f is continuous at 0 but nowhere else.

4.22   Let $f : \mathbb{R} \to \mathbb{R}$ be defined by
$$f(t) = \begin{cases} t - t^3 & \text{if } t \text{ is rational} \\ 0 & \text{if } t \text{ is irrational.} \end{cases}$$
Show that f is continuous at $t = 0, \pm 1$ but at no other point.

4.23   Let $f : [0,1] \to \mathbb{R}$ be defined by
$$f(t) = \begin{cases} t & \text{if } t \text{ is rational} \\ 1 - t & \text{if } t \text{ is irrational.} \end{cases}$$
Show that f satisfies the conclusion of the intermediate value theorem although it is not continuous.

4.24 Show that the set of all the points of discontinuity of a nondecreasing function f on an interval $[a, b]$ is countable.

4.25 We write any rational number t in the form $t = \frac{m}{n}$ where m and n are integers with no common factors and where $n > 0$. Let $f : \mathbb{R} \to \mathbb{R}$ be defined by

$$f(t) = \begin{cases} \frac{1}{n} & \text{if } t = \frac{m}{n} \text{ is rational} \\ 0 & \text{if } t \text{ is irrational.} \end{cases}$$

Show that f is continuous at every irrational but at no rational point.

4.26 Show that if f and g are real-valued functions both continuous at a, then $\max\{f, g\}$ (resp. $f^+ = \max\{f, 0\}$; $f^- = \max\{-f, 0\}$) is continuous at a.

4.27 Let f be a real-valued function defined and continuous on $[a, b]$. Suppose that $f(a)$ and $f(b)$ have opposite signs. Show that there is a number c between a and b such that $f(c) = 0$.

4.28 Show that the equation $x = \cos x$ has a solution in $(0, \pi/2)$.

4.29 **Fixed point**. Let f be a real-valued function defined and continuous on $[a, b]$. Suppose that ran $f \subset (a, b)$. Show that there exists a number c in $[a, b]$ such that $f(c) = c$. Note: such a point c is called a fixed point for f.

4.30 Show that the function $f(x) = 2x(1 - x)$ admits a fixed point on $[0,1]$.

4.31 Write in terms of ε and δ the negation of "f is uniformly continuous on a set A".

4.32 Determine whether the given function is uniformly continuous on the given interval.

(a) $f(x) = x^2$ on $(-\infty, \infty)$ (e) $f(x) = \sqrt{x}$ on $[0, \infty)$
(b) $f(x) = \sin x$ on $(-\infty, \infty)$ (f) $f(x) = e^x$ on $(-\infty, \infty)$
(c) $f(x) = x \sin x$ on $(-\infty, \infty)$ (g) $f(x) = \ln x$ on $(0, \infty)$
(d) $f(x) = x \sin x^2$ on $(-\infty, \infty)$ (h) $f(x) = x \ln x$ on $(0, \infty)$

4.33 Show that the function

$$f(x) = \begin{cases} x \sin(1/x) & \text{if } 0 < x \leq 1 \\ 0 & \text{if } x = 0 \end{cases}$$

is uniformly continuous on $[0, 1]$.

4.34   Show that $f(x) = \sin(1/x)$ is continuous but not uniformly continuous on $(0, 2\pi]$.

4.35   Let $f(x) = 1/x^2$. Show that

   (1) f is continuous on (a, ∞) if $a \geq 0$;

   (2) f is uniformly continuous on (a, ∞) if $a > 0$;

   (3) f is not uniformly continuous on $(0, \infty)$.

4.36   Suppose that f is uniformly continuous on the closed intervals I_1 and I_2. Show that f is uniformly continuous on $S = I_1 \cup I_2$.

4.37   Let f and g be real-valued functions uniformly continuous on A.

   (1)   Show that $f + g$ is uniformly continuous on A.

   (2)   Show that $f \circ g$ is uniformly continuous on $g(A) \cap A$.

   (3)   Show that f^+ is uniformly continuous on A.

   (4)   Show that if $A = [a, b]$, then fg is uniformly continuous on A. Give a counterexample if A is not a closed interval.

4.38   Let A be a nonempty subset of R. Define a function $f_A : \mathbb{R} \to \mathbb{R}$ by

$$f_A(x) = \inf \{|x - a| : a \in A\}.$$

Show that f_A is uniformly continuous on A. ($f_A(x)$ is called the **distance** from x to A.)

4.39   Show that if f is continuous on $(0, 1)$ but unbounded, then f cannot be uniformly continuous.

4.40   Show that a function is uniformly continuous on (a, b) if it can be extended to a continuous function on $[a, b]$.

4.41   Explain why there could not exist a continuous mapping from $[0, 1]$ onto $(0, 1)$.

5
Differentiation

5.1 Derivatives

Let $y = f(x)$ be a real function defined in a certain interval (a, b). Suppose that the value of the argument changes from x to $x + h$ in the interval. Then the value of the function will change from $f(x)$ to $f(x + h)$. Thus a change $\Delta x = (x + h) - x$ of the argument brings about a change

$$\Delta f(x) = f(x + h) - f(x)$$

of the value of the function. See Figure 5.1. Using such notation, we can rewrite the definition of a continuous function as follows.

A function f is continuous at a point $x \in \operatorname{dom} f$ if and only if

$$\lim_{\Delta x \to 0} \Delta f(x) = \lim_{h \to 0} [f(x + h) - f(x)] = 0.$$

The ratio

$$\frac{\Delta f(x)}{\Delta x} = \frac{f(x + h) - f(x)}{h}$$

is called the **difference quotient** of f at x.

Definition 5.1

Let $y = f(x)$ be a real function and let $x \in \operatorname{dom}(f)$. If

$$f'(x) = \lim_{\Delta x \to 0} \frac{\Delta f(x)}{\Delta x} = \lim_{h \to 0} \frac{f(x + h) - f(x)}{h}$$

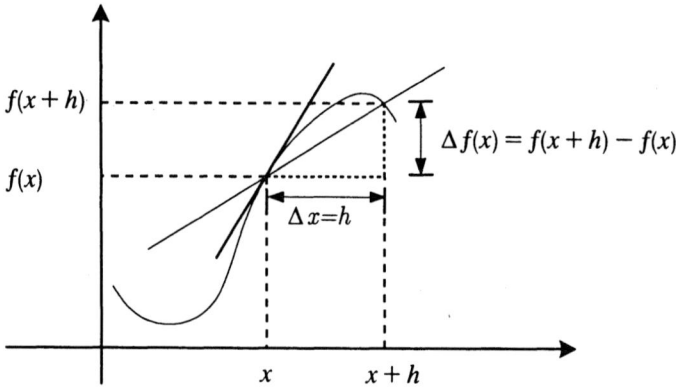

Figure 5.1   Derivative

exists, then we say that f is **differentiable** at x, or that f has a **derivative** at x.

The subset of $\text{dom}\,(f)$ defined by

$$\left\{ x \in \text{dom}\,(f) : \lim_{h \to 0} \frac{f\,(x+h) - f\,(x)}{h} \text{ exists} \right\}$$

is called **the domain of differentiability** of f and is denoted by $\text{dom}\,(f')$. Then a function f is said to be **differentiable on an open interval** (a, b) inside its domain if $(a, b) \subset \text{dom}\,(f')$. We say that f is **differentiable on a closed interval** $[a, b]$ if $(a, b) \subset \text{dom}\,(f')$ and if both of the limits

$$\lim_{h \to 0^+} \frac{f\,(a+h) - f\,(a)}{h} \quad \text{and} \quad \lim_{h \to 0^-} \frac{f\,(b+h) - f\,(b)}{h}$$

exist. Such limits are respectively called **right hand derivative** at a and **left hand derivative** at b and are respectively denoted by $f'_+\,(a)$ and $f'_-\,(b)$. The extension of the definition of differentiablity of a function to any other type of interval should be clear. Note that a function f has a derivative at x if and only if both $f'_+\,(x)$ and $f'_-\,(x)$ exist and have equal values.

Definition 5.2

Let f be a real-valued function defined on an open interval (a, b). The derivative of the function f is the function

$$f' : \text{dom}\,(f') \to \mathbb{R} : x \mapsto \lim_{h \to 0} \frac{f\,(x+h) - f\,(x)}{h}.$$

Example 5.3

Find the derivative of $f(x) = x^2$. Evaluate the derivative at $x = 5$.

Solution

For a corresponding change $\Delta x = h$ of the variable x, the change in the values of f is

$$\Delta f(x) = (x+h)^2 - x^2 = 2xh + (h)^2.$$

Forming the ratio with $\Delta x = h$, we have

$$\frac{\Delta f(x)}{\Delta x} = \frac{f(x+h) - f(x)}{h} = \frac{2xh + (h)^2}{h} = 2x + h.$$

Passing to the limit, we get that the derivative of f at x is

$$f'(x) = \lim_{h \to 0} \frac{f(x+h) - f(x)}{h} = \lim_{h \to 0} (2x + h) = 2x,$$

defined at every point $x \in \mathbb{R}$. Thus the domain of differentiability of f is $\text{dom}(f') = \mathbb{R}$.

The value of the derivative at $x = 5$ is $f'(5) = 2 \cdot 5 = 10$. $\square$

Example 5.4

Find the derivative of $f(x) = \frac{1}{x}$ at every point where it exists.

Solution

For a corresponding change $\Delta x = h$ of the variable x, the change in the values of f is

$$\Delta f(x) = \frac{1}{x+h} - \frac{1}{x} = -\frac{h}{x(x+h)}.$$

Forming the difference quotient, we have

$$\frac{\Delta f(x)}{\Delta x} = -\frac{1}{x(x+h)}.$$

Passing to the limit, we get that the derivative of f at x is

$$f'(x) = \lim_{\Delta x \to 0} \frac{\Delta f}{\Delta x} = \lim_{h \to 0} -\frac{1}{x(x+h)} = -\frac{1}{x^2},$$

defined at every point $x \in \mathbb{R} \setminus \{0\}$. Thus the domain of differentiability of f is $\text{dom}(f') = \mathbb{R} \setminus \{0\}$. We note in passing that, also, $\text{dom}(f) = \mathbb{R} \setminus \{0\}$. $\square$

It is worth noticing that a function cannot have a derivative at points of discontinuity:

Theorem 5.5

If a function f is differentiable at some point x, then f is continuous at x.

Proof

Suppose that f is differentiable at a point x. Then $f'(x) = \lim_{h \to 0} \frac{f(x+h) - f(x)}{h}$ exists. We write, with $\Delta x = h$,

$$\Delta f(x) = f(x+h) - f(x) = \Delta x \frac{f(x+h) - f(x)}{h}.$$

Therefore

$$\lim_{\Delta x \to 0} \Delta f(x) = \lim_{\Delta x \to 0} \Delta x \lim_{h \to 0} \frac{f(x_0 + h) - f(x_0)}{h} = 0 \cdot f'(x_0) = 0.$$

Hence f is continuous at x. $\qquad\square$

The converse of this theorem is not true. The domain of differentiability of a function may be strictly contained in the set of all points of continuity of f.

Example 5.6

The function f defined by

$$f(x) = \begin{cases} 1 - x^2 & \text{for } x > 0 \\ x + 1 & \text{for } x \le 0 \end{cases}$$

is continuous but not differentiable at 0.

Solution

If $h > 0$ we have

$$\lim_{h \to 0^+} [f(0 + h) - f(0)] = \lim_{h \to 0^+} \left(\left[1 - (0 + h)^2 \right] - 1 \right) = 0 \qquad (5.1)$$

$$\lim_{h \to 0^+} \frac{f(0 + h) - f(0)}{h} = \lim_{h \to 0^+} \frac{\left[1 - (0 + h)^2 \right] - 1}{h} = \lim_{h \to 0^+} h = 0. \qquad (5.2)$$

If $h < 0$, we get

$$\lim_{h \to 0^-} [f(0 + h) - f(0)] = \lim_{h \to 0^-} \left([(0 + h) + 1] - 1 \right) = 0 \qquad (5.3)$$

$$\lim_{h \to 0^-} \frac{f(0+h) - f(0)}{h} = \lim_{h \to 0^-} \frac{[(0+h)+1]-1}{h} = \lim_{h \to 0^-} \frac{h}{h} = 1. \qquad (5.4)$$

While (5.1) and (5.3) ensure the continuity of f at 0, (5.2) and (5.4) imply that $f'(0) = \lim_{h \to 0} \frac{f(0+h)-f(0)}{h}$ does not exist. $\qquad \square$

Theorem 5.7

Let f and g be real-valued functions defined on (a, b). Suppose that both f and g are differentiable at a point $x \in (a, b)$. Then $f + g$, and $f \cdot g$ are differentiable at x, and

(1) $(f + g)'(x) = f'(x) + g'(x)$;

(2) $(f \cdot g)'(x) = f'(x)g(x) + f(x)g'(x)$.

If, in addition, $g(x) \neq 0$, $x \in (a, b)$, then f/g is differentiable at x, and

(3) $\left(\frac{f}{g}\right)'(x) = \frac{f'(x)g(x)+f(x)g'(x)}{g^2(x)}$.

Proof

(1) follows from the identity

$$\frac{\Delta(f+g)(x)}{\Delta x} = \frac{f(x+h) - f(x)}{h} + \frac{g(x+h) - g(x)}{h}.$$

(2) Let $F = f \cdot g$. Then

$$\frac{\Delta F(x)}{\Delta x} = f(x+h) \frac{g(x+h) - g(x)}{h} + g(x) \frac{f(x+h) - f(x)}{h}.$$

Letting $h \to 0$ and noting that $\lim_{h \to 0} f(x+h) = f(x)$, we obtain

$$(f \cdot g)'(x) = f(x)g'(x) + f'(x)g(x).$$

(3) Let $F = f/g$ and suppose $g(x) \neq 0$. Since by Theorem 5.5 g is continuous at x, it follows that $g(x+h) \neq 0$ for all h sufficiently small. Now,

$$\Delta F(x) = \frac{f(x+h)g(x) - f(x)g(x+h)}{g(x+h)g(x)}$$
$$= \frac{f(x+h)g(x) - f(x)g(x) + f(x)g(x) - f(x)g(x+h)}{g(x+h)g(x)}.$$

If we divide by $\Delta x = h$, we get

$$\frac{\Delta F(x)}{\Delta x} = \frac{1}{g(x+h)g(x)} \left[\frac{f(x+h) - f(x)}{h} g(x) - f(x) \frac{g(x+h) - g(x)}{h} \right].$$

The result follows on passing to the limit as $h \to 0$. $\qquad \square$

The following is a useful result.

Theorem 5.8 (Differentiation Lemma)

Suppose that f is differentiable at x_0. Then there exists a function l defined on a neighborhood of 0 with $\lim_{h \to 0} l(0) = 0$ such that

$$f(x_0 + h) - f(x_0) = [f'(x_0) + l(h)] \cdot h. \tag{5.5}$$

Proof

We define l by

$$l(h) = \begin{cases} \frac{1}{h}[f(x_0 + h) - f(x_0)] - f'(x_0) & \text{if } h \neq 0 \\ 0 & \text{if } h = 0. \end{cases}$$

Since f is differentiable at x_0, we have $\lim_{h \to 0} l(h) = 0$. Equation (5.5) is a restatement of the first line in the definition of l.   □

We use the above result to prove the following

Example 5.9

Suppose that u is differentiable at $x \in \text{dom}(u)$ and g is differentiable at $u(x) \in \text{dom}(g)$. Show that the composite function $g \circ u$ is differentiable at x and

$$(g \circ u)'(x) = g'(u(x)) \cdot g'(x).$$

Solution

Clearly $x \in \text{dom}(g \circ u)$. Let $\Delta x = h$. Then $\Delta u(x) = u(x + h) - u(x)$ which we write $u(x + h) = u(x) + \Delta u(x)$. Then

$$\begin{aligned} \Delta(g \circ u)(x) &= g \circ u(x + h) - g \circ u(x) \\ &= g[u(x + h)] - g[u(x)] \\ &= g[u(x) + \Delta u(x)] - g[u(x)]. \end{aligned}$$

Since g is differentiable at $u(x)$, by the differentiation lemma,

$$g(u(x) + k) - g(u(x)) = [g'(u(x)) + l(k)] \cdot k,$$

where $l(k) \to 0$ as $k \to 0$. We notice that since u is continuous at x, $\lim_{h \to 0} \Delta u(x) = 0$. Thus we can write

$$g(u(x) + \Delta u(x)) - g(u(x)) = [g'(u(x)) + l(\Delta u(x))] \cdot \Delta u(x).$$

Dividing by h and taking the limit as $h \to 0$, we obtain

$$\lim_{\Delta x \to 0} \frac{\Delta (g \circ u)(x)}{\Delta x} = \lim_{h \to 0} [g'(u(x)) + l(\Delta u(x))] \cdot \lim_{h \to 0} \frac{\Delta u}{h} = g'[u(x)] u'(x),$$

as desired. $\square$

5.2 Mean Value Theorem

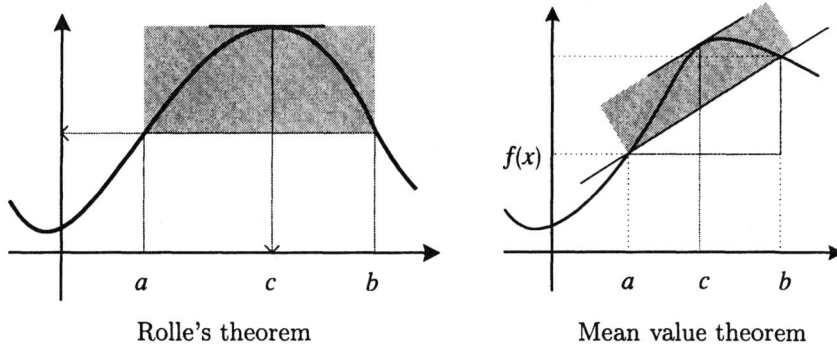

Rolle's theorem   Mean value theorem

Theorem 5.10 (Rolle)

Suppose that f is a continuous function on a closed interval $[a, b]$ differentiable on (a, b), and that it satisfies $f(a) = f(b)$. Then there exists $c \in (a, b)$ such that $f'(c) = 0$.

Proof

Since f is continuous on $[a, b]$, there exist x_0 and y_0 such that

$$f(x_0) \leq f(x) \leq f(y_0)$$

for every $x \in [a, b]$.

If x_0 and y_0 are both endpoints of $[a, b]$, then $f(x) = f(a)$ for all $x \in [a, b]$. It follows that $\Delta y = f(x + h) - f(x) = 0$ for all $x \in [a, b]$ such that $x + h \in [a, b]$. But then $f'(x) = 0$ for all $x \in [a, b]$.

Now suppose that $x_0, y_0 \in (a, b)$. Thus for every h such that $y_0 + h \in [a, b]$, we have $f(y_0 + h) - f(y_0) \leq 0$, and so

$$\lim_{h \to 0^+} \frac{f(y_0 + h) - f(y_0)}{h} \leq 0;$$

$$\lim_{h \to 0^-} \frac{f(y_0 + h) - f(y_0)}{h} \geq 0.$$

But since by our hypothesis f is differentiable at y_0, i.e. since we have

$$\lim_{h \to 0^+} \frac{f(y_0 + h) - f(y_0)}{h} = \lim_{h \to 0^-} \frac{f(y_0 + h) - f(y_0)}{h} = f'(y_0),$$

it follows that $f'(y_0) = 0$. The proof is analogous if only one of x_0 or y_0 belongs to (a, b).   □

We can use Rolle's theorem to prove the following result.

Theorem 5.11

If f and g are continuous functions on $[a, b]$ and differentiable on (a, b), then there exists at least one point $x \in (a, b)$ such that

$$[f(b) - f(a)] g'(x) = [g(b) - g(a)] f'(x). \tag{5.6}$$

Proof

Consider

$$F(x) = [f(b) - f(a)] g(x) - [g(b) - g(a)] f(x).$$

Then F is continuous on $[a, b]$, differentiable on (a, b) and for all $x \in (a, b)$,

$$F'(x) = [f(b) - f(a)] g'(x) - [g(b) - g(a)] f'(x).$$

Since $F(a) = f(b) g(a) - f(a) g(b) = F(b)$, Rolle's theorem applies and implies that there exists $x \in (a, b)$ such that $F'(x) = 0$, i.e.

$$[f(b) - f(a)] g'(x) = [g(b) - g(a)] f'(x)$$

as desired.   □

The above theorem is called the **generalized mean value theorem** and Equation (5.6) is known as Cauchy's formula. The next result is a special case $(g(x) = x)$ and is simply called the **mean value theorem**, also known as the **law of the mean**.

Theorem 5.12 (Mean Value Theorem)

Let f be a continuous function on the closed interval $[a, b]$ and have a derivative at every x in the open interval (a, b). Then there is at least one number x in the open interval (a, b) such that

$$f(b) - f(a) = (b - a) f'(x).$$

The mean value theorem has many applications in analysis. It is for example used to determine the variation of a function on a given interval.

Corollary 5.13

Let f be differentiable on (a, b).

(1) If $f'(x) > 0$ for all x on (a, b), then f is increasing on (a, b).

(2) If $f'(x) = 0$ for all x on (a, b), then f is constant (a, b).

(3) If $f'(x) < 0$ for all x on (a, b), then f is decreasing (a, b).

Proof

Let $x_1, x_2 \in (a, b)$ be such that

$$a < x_1 < x_2 < b.$$

Then f is continuous on $[x_1, x_2]$ and differentiable on (x_1, x_2). By the mean value theorem

$$f(x_2) - f(x_1) = (x_2 - x_1) f'(x)$$

for some $x \in (x_1, x_2)$. (1), (2), and (3) clearly follow.   □

The mean value theorem is also useful when proving inequalities.

Example 5.14

Show that $\frac{b-a}{1+b^2} < \tan^{-1} b - \tan^{-1} a < \frac{b-a}{1+a^2}$ if $a < b$.

Solution

Consider $f(x) = \tan^{-1} x$ defined on the interval $[a, b]$. Then f is continuous $f'(x) = 1/(1 + x^2)$ on $[a, b]$. By the mean value theorem, there exists $x \in (a, b)$ such that

$$\tan^{-1} b - \tan^{-1} a = (b - a) / (1 + x^2).$$

Since $a < x < b$, we have

$$\frac{b-a}{1+b^2} < \tan^{-1} b - \tan^{-1} a < \frac{b-a}{1+a^2}.$$

□

5.3 L'Hôspital's Rule

Cauchy's formula (Theorem 5.11) is useful in finding limits of certain functions.

Theorem 5.15

Let f and g be two real-valued functions both of which are differentiable on (a, b). Suppose that $g'(x) \neq 0$ for all $x \in (a, b)$. If

$$\lim_{x \to a^+} f(x) = \lim_{x \to a^+} g(x) = 0 \quad \text{and} \quad \lim_{x \to a^+} \frac{f'(x)}{g'(x)} = l, \qquad (5.7)$$

then

$$\lim_{x \to a^+} \frac{f(x)}{g(x)} = \lim_{x \to a^+} \frac{f'(x)}{g'(x)} = l.$$

Proof

Since $\lim_{x \to a^+} \frac{f(x)}{g(x)}$ does not depend on the values of $f(a)$ and $g(a)$, if $f(a)$ and $g(a)$ are not defined or are not equal to 0, we may redefine f and g so that

$$\lim_{x \to a^+} f(x) = f(a) = 0 = g(a) = \lim_{x \to a^+} g(x).$$

Take any point $x \in (a, b)$. Then f is continuous on $[a, x]$ and differentiable on (a, x). Applying Cauchy's formula (Theorem 5.11), there is a $c \in (a, x)$ such that

$$\frac{f(x) - f(a)}{g(x) - g(a)} = \frac{f'(c)}{g'(c)}.$$

Since we assumed that $f(a) = g(a) = 0$, we have

$$\frac{f(x)}{g(x)} = \frac{f'(c)}{g'(c)}.$$

Taking the limit as $x \to a^+$, and noting that then $c \to a^+$ too, we get

$$\lim_{x \to a^+} \frac{f(x)}{g(x)} = \lim_{c \to a^+} \frac{f'(c)}{g'(c)} = \lim_{x \to a^+} \frac{f'(x)}{g'(x)}$$

which completes our proof.   □

Note

L'Hôspital's rule also applies for the case

$$\lim_{x \to b^-} f(x) = \lim_{x \to b^-} g(x) = 0, \quad \text{and} \quad \lim_{x \to b^-} \frac{f'(x)}{g'(x)} = l. \qquad (5.8)$$

For example,

$$\lim_{x \to 0} \frac{\sin 2x}{3x} = \lim_{x \to 0} \frac{(\sin 2x)'}{(3x)'} = \lim_{x \to 0} \frac{2 \cos 2x}{3} = \frac{2}{3}$$

$$\lim_{x \to 0} \frac{\ln(1 + x)}{x} = \lim_{x \to 0} \frac{(\ln(1 + x))'}{x'} = \lim_{x \to 0} \frac{\frac{1}{1+x}}{1} = 1.$$

Corollary 5.16

L'Hôspital's rule holds if a^+ is replaced by $-\infty$ in (5.7) and if b^- is replaced by $+\infty$ in (5.8).

Proof

We assume that

$$\lim_{x \to +\infty} f(x) = \lim_{x \to +\infty} g(x) = 0 \quad \text{and} \quad \lim_{x \to +\infty} \frac{f'(x)}{g'(x)} = l.$$

If we apply L'Hôspital's rule to the functions $f(1/x)$ and $g(1/x)$, letting $z = 1/x$ we have

$$\lim_{x \to +\infty} \frac{f(x)}{g(x)} = \lim_{z \to 0} \frac{f(1/z)}{g(1/z)}$$
$$= \lim_{z \to 0} \frac{f'(1/z)\left(-1/z^2\right)}{g'(1/z)\left(-1/z^2\right)}$$
$$= \lim_{z \to 0} \frac{f'(1/z)}{g'(1/z)} = \lim_{x \to +\infty} \frac{f'(x)}{g'(x)},$$

as desired. The $-\infty$ case is dealt with in a similar fashion. □

Example 5.17

Find $\lim_{x \to -\infty} x \tan^{-1} x$ if it exists.

Solution

We let $f(x) = \tan^{-1} x$ and $g(x) = 1/x$. Then

$$\lim_{x \to -\infty} f(x) = \lim_{x \to -\infty} g(x) = 0$$

and L'Hôspital's rule applies. We have

$$\lim_{x \to -\infty} \frac{f(x)}{g(x)} = \lim_{x \to -\infty} \frac{f'(x)}{g'(x)} = \lim_{x \to -\infty} \frac{1/(1+x^2)}{-1/x^2}$$

$$= \lim_{x \to -\infty} \frac{-x^2}{1+x^2} = -1.$$

Thus $\lim_{x \to -\infty} x \tan^{-1} x = -1$.   □

Example 5.18

Find $\lim_{x \to 0+} x \ln x$ if it exists.

Solution

We set $f(x) = \ln x$ and $g(x) = 1/x$. Then $\lim_{x \to 0+} f(x) = -\infty$ and $\lim_{x \to 0+} g(x) = +\infty$. We apply L'Hôspital's rule:

$$\lim_{x \to 0+} \frac{f(x)}{g(x)} = \lim_{x \to 0+} \frac{f'(x)}{g'(x)} = \lim_{x \to 0+} \frac{1/x}{-1/x^2}$$

$$= \lim_{x \to 0+} -x = 0.$$

Hence $\lim_{x \to 0+} x \ln x = 0$.   □

Example 5.19

Find $\lim_{x \to 0+} x^x$ if it exists.

Solution

Let $f(x) = x^x$. Then $\ln(f(x)) = x \ln x$. By the previous example we have $\lim_{x \to 0+} \ln f(x) = 0$. Since $y = e^x$ is a continuous function, we have

$$\lim_{x \to 0+} f(x) = \lim_{x \to 0+} e^{\ln f(x)} = e^{\lim_{x \to 0+} \ln f(x)} = e^0 = 1.$$

□

5.4 Inverse Function Theorems

Recall that a real function f with domain D and range R is said to be **injective** or **one-to-one** if whenever $a \neq b$ in D, then $f(a) \neq f(b)$ in R. This is logically equivalent to: if $f(a) = f(b)$ in R, then $a = b$ in D.

For example, to show that $f(x) = 2x$ is one-to-one on $\mathbb{R}$, we start by considering a and $b \in \mathbb{R}$ so that $f(a) = f(b)$, that is $2a = 2b$. It follows that $a = b$, so f is one-to-one. On the other hand, the function $g(x) = x^2$ is not one-to-one on $\mathbb{R}$ because $g(1) = g(-1)$ but $1 \neq -1$.

If $f : D \to \mathbb{R}$ is one-to-one, then a mapping $f^{-1} : f(D) \to \mathbb{R}$ can be defined by the rule

$$f^{-1}(y) = x \text{ if and only if } y = f(x).$$

Such a mapping f^{-1} is called the **inverse function** of f.

Theorem 5.20 (Inverse Function Theorem)

Suppose that f is a real function continuous and increasing on an interval I. Then f has an inverse f^{-1} on $f(I)$ which is increasing and continuous. A similar result holds if f is decreasing on I.

Proof

Suppose that f is increasing. Then f is one-to-one and so f^{-1} exists and is defined on $f(I)$. To see that f^{-1} is increasing, consider $y_1 < y_2$ in $f(I)$, and suppose to the contrary that $f^{-1}(y_1) \geq f^{-1}(y_2)$. Since f is increasing, we would have $f(f^{-1}(y_1)) \geq f(f^{-1}(y_2))$ and hence $y_1 \geq y_2$, a contradiction. So f^{-1} must also be increasing. The proof of the decreasing case is analogous.

We now prove that f^{-1} is continuous. Let $y_0 \in f(I)$. Then there exists $x_0 \in I$ such that $y_0 = f(x_0)$ and $x_0 = f^{-1}(y_0)$. We first assume that x_0 is not an endpoint of $f(I)$. Then $f^{-1}(y_0)$ is not an endpoint of I so there is $\varepsilon_0 > 0$ such that $\left(f^{-1}(y_0) - \varepsilon_0, f^{-1}(y_0) + \varepsilon_0\right) \subset I$. Let $\varepsilon > 0$. We assume $\varepsilon_0 > \varepsilon$. Then there exists y_1 and y_2 in $f(I)$ such that

$$f^{-1}(y_1) = f^{-1}(y_0) - \varepsilon \text{ and } f^{-1}(y_2) = f^{-1}(y_0) + \varepsilon.$$

Since f is increasing, we have $y_1 < y_0 < y_2$. Also since f^{-1} is increasing, for all y such that $y_1 < y < y_2$,

$$f^{-1}(y_0) - \varepsilon < f^{-1}(y) < f^{-1}(y_0) + \varepsilon,$$

or equivalently,

$$\left|f^{-1}(y) - f^{-1}(y_0)\right| < \varepsilon.$$

Choosing $\delta = \min\{y_2 - y_0, y_0 - y_1\}$, we have

$$\left|f^{-1}(y) - f^{-1}(y_0)\right| < \varepsilon \text{ whenever } |y - y_0| < \delta.$$

A slight change in the above argument shows that if y_0 is a left endpoint (resp. right endpoint), then f^{-1} is continuous on the right (resp. on the left) at x_0. $\square$

For example, the function $f(x) = e^x$ is continuous on the interval $(-\infty, \infty)$, with range $(0, \infty)$. Since its derivative $f'(x) = e^x > 0$ for all $x \in (-\infty, \infty)$, f is also an increasing function. Therefore it is one-to-one. Its inverse is the function $f^{-1}(x) = \ln x$ defined on the interval $(0, \infty)$, and with range $(-\infty, \infty)$.

It is worth noticing that the graphs of a one-to-one function and of its inverse are symmetric about the bisector $y = x$.

The next theorem provides a method for finding the derivative of the inverse of a function.

Theorem 5.21

Suppose that f is continuous and one-to-one on an open interval I. If $f'(a)$ exists and is nonzero for $a \in I$, then f^{-1} is differentiable at $f(a)$ and

$$f^{-1'}(f(a)) = \frac{1}{f'(a)}. \tag{5.9}$$

Proof

We first notice that by Example 4.28 (page 111), $f(I)$ is an interval. Clearly, $f : I \to f(I)$ is one-to-one and onto. We claim that f is either increasing or decreasing on I. Indeed, suppose that there are numbers a_1, a_2 and a_3 in I such that $a_1 < a_2 < a_3$ and $f(a_1) < f(a_3) < f(a_2)$. Applying the intermediate value theorem on $[a_1, a_2]$, we have for some $c \in (a_1, a_2)$, $f(c) = f(a_3)$. This contradicts the fact that f is one-to-one, which proves our claim.

By the inverse function theorem, f^{-1} is either increasing or decreasing and continuous on $f(I)$. It follows that, for $x \neq a$ in I, we have $f(x) \neq f(a)$, and thus

$$\frac{f(x) - f(a)}{x - a} \neq 0.$$

Whence, using the continuity of f^{-1} on $f(I)$,

$$\lim_{y \to f(a)} \frac{f^{-1}(y) - f^{-1}(f(a))}{y - f(a)} = \lim_{f(x) \to f(a)} \frac{f^{-1}(f(x)) - f^{-1}(f(a))}{f(x) - f(a)}$$

$$= \lim_{x \to a} \frac{x - a}{f(x) - f(a)}$$

$$= \frac{1}{f'(a)}.$$

This proves the theorem. □

Example 5.22

Show that the derivative of $\sin^{-1} x$ (also denoted by $\arcsin x$) is given by $1/\sqrt{1-x^2}$, $x \in (-1,1)$. The function $\sin^{-1} x$ is also denoted by $\arcsin x$.

Solution

We first notice that the function $f(x) = \sin x$ is defined everywhere. Then $f'(x) = \cos x > 0$ for $x \in (-\pi/2 + 2k\pi, \pi/2 + 2k\pi)$, $k \in \mathbb{Z}$. In particular, the function $f(x) = \sin x$ is continuous and increasing when restricted to $(-\pi/2, \pi/2)$. Its inverse $y = \sin^{-1} x$ is defined on $(-1,1)$. It follows from (5.9) that

$$f^{-1'}(\sin y) = \frac{1}{\cos y} = \frac{1}{\sqrt{1 - \sin^2 y}}.$$

Therefore,

$$\left(\sin^{-1} x\right)' = \frac{1}{\sqrt{1-x^2}}.$$

$\square$

5.5 Taylor's Theorem

Suppose that f is differentiable on an interval I, and suppose that f' is also differentiable on I. Then the derivative of f' is denoted by f'' and called the **second derivative** of f. In a similar manner, we define the third derivative of f to be the derivative of its second derivative (whenever it exists), and so on. Inductively, we define the n-**th derivative** of f, or **derivative of order** n of f to be the derivative of its $(n-1)$-th derivative whenever it exists. The n-th derivative of f will be denoted by $f^{(n)}$ whenever it exists.

Theorem 5.23 (Taylor's Theorem)

Let f be defined on (a,b), and suppose that the $(n+1)$-th derivative $f^{(n+1)}$ exists on (a,b). Let $x, x_0 \in (a,b)$ and consider

$$P_n(f, x_0)(x) = \sum_{k=0}^{n} \frac{f^{(k)}(x_0)}{k!} (x - x_0)^k. \tag{5.10}$$

Then there exists a point ξ between x and x_0 such that

$$f(x) = P_n(f, x_0)(x) + \frac{f^{(n+1)}(\xi)}{(n+1)!} (x - x_0)^{n+1}. \tag{5.11}$$

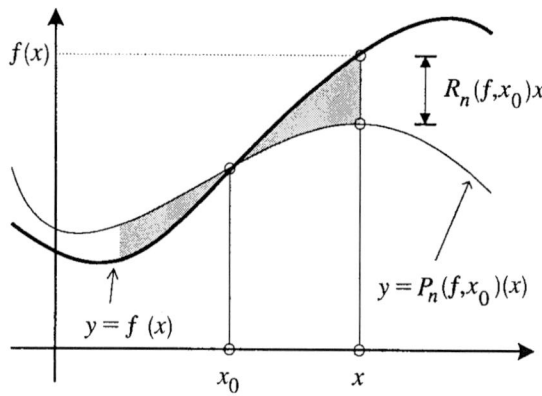

Figure 5.2   Taylor approximation

Proof

Let M be the number defined by

$$f(x) = P_n(f, x_0)(x) + M(x - x_0)^{n+1} \qquad (5.12)$$

and define

$$F(t) = f(t) - P_n(f, x_0)(t) - M(t - x_0)^{n+1}$$

on (a, b). We need to show that $(n + 1)!M = f^{(n+1)}(\xi)$ for some ξ between x and x_0. A direct calculation shows that

$$F^{(n+1)}(t) = f^{(n+1)}(t) - (n + 1)!M$$

for $t \in (a, b)$. Thus we are done if we show that $F^{(n+1)}(\xi) = 0$ for some ξ between x and x_0.

Since $P_n^{(k)}(f, x_0)(x_0) = f^{(k)}(x_0)$ for $k = 0, 1, \cdots, n$, we have

$$F(x_0) = F'(x_0) = \cdots = F^{(n)}(x_0) = 0.$$

By the choice of the number M in (5.12), we also have $F(x) = 0$. Hence by Rolle's theorem, $F'(c_1) = 0$ for some c_1 between x_0 and x. Since $F'(x_0) = 0$, a second application of Rolle's theorem implies that $F''(c_2) = 0$ for some c_2 between x_0 and c_1. Continuing in this way, we obtain $F^{(n+1)}(c_{n+1}) = 0$ for some c_{n+1} between x_0 and c_n. Hence, it suffices to take $\xi = c_{n+1}$. The proof is complete. $\square$

Theorem 5.23 states that the Taylor polynomial is an approximation of f in the neighborhood of x_0 whose accuracy is measured by the **remainder** (see Figure 5.2)

$$R_n \left(f, x_0 \right)(x) = \frac{f^{(n+1)}(\xi)}{(n+1)!} \left(x - x_0 \right)^{n+1}.$$

Example 5.24

Use Taylor's formula to give e correct to three decimal places.

Solution

For some ξ between 0 and x, we have

$$R_n \left(e^x, 0 \right)(x) = \frac{e^\xi}{(n+1)!} x^{n+1}.$$

If $x = 1$, then $0 < \xi < 1$ and

$$R_n \left(e^x, 0 \right)(1) = \frac{e^\xi}{(n+1)!} < \frac{3}{(n+1)!}.$$

We wish to have $R_n \left(e^x, 0 \right)(1) < 0.0005$. It is enough to choose n such that $\frac{3}{(n+1)!} < 0.0005$ or $(n+1)! > 6000$. Since $7! = 5040$ and $8! = 40\,320$, $n = 7$ will do. Hence, we have

$$e \approx 1 + \frac{1}{1!} + \frac{1}{2!} + \frac{1}{3!} + \frac{1}{4!} + \frac{1}{5!} + \frac{1}{6!} + \frac{1}{7!} \approx 2.7183$$

□

Example 5.25

Let f be defined on (a, b), and suppose that the $(n+1)$-th derivative $f^{(n+1)}$ exists on (a, b). Show that the remainder of the Taylor formula of f at a point $x_0 \in (a, b)$ is also given by

$$R_n \left(f, x_0 \right)(x) = \frac{f^{(n+1)}(\eta)}{n!} \left(x - \eta \right)^n \left(x - x_0 \right),$$

for some η between x_0 and x.

Solution

With x_0 and x fixed, and let t be any number between x_0 and x, we put

$$F(t) = f(x) - \sum_{k=0}^{n} \frac{f^{(k)}(t)}{k!} (x - t)^k.$$

Then,

$$f(x) = \sum_{k=0}^{n} \frac{f^{(k)}(t)}{k!} (x - t)^k + F(t)$$

where $F(t) = R_n(f, t)(x)$. Differentiating with respect to t, we have

$$0 = f'(t) + \left[-f'(t) + \frac{f''(t)}{1!} (x - t) \right]$$
$$+ \left[-\frac{f''(t)}{1!} (x - t) + \frac{f'''(t)}{2!} (x - t)^2 \right] + \cdots$$
$$+ \left[-\frac{f^{(n)}(t)}{(n-1)!} (x - t)^{n-1} + \frac{f^{(n+1)}(t)}{n!} (x - t)^n \right] + F'(t).$$

Hence, after cancellation, we get

$$F'(t) = -\frac{f^{(n+1)}(t)}{n!} (x - t)^n.$$

We then apply the mean value theorem to the function F on the interval with endpoints x_0 and x, and we obtain a number ξ between x_0 and x such that

$$F'(\xi) = -\frac{f^{(n+1)}(\xi)}{n!} (x - \xi)^n = \frac{F(x) - F(x_0)}{x - x_0}.$$

Since $F(x) = R_n(f, x)(x) = 0$, we have

$$F(x_0) = R_n(f, x_0)(x) = \frac{f^{(n+1)}(\xi)}{n!} (x - \xi)^n (x - x_0)$$

as desired.   □

EXERCISES

5.1 Consider the function defined by $f(x) = x^2 \sin(1/x)$ if $x \neq 0$ and $f(0) = 0$. Show that the sets of its points of continuity are strictly contained in dom(f').

5.2 Show that the function defined in Exercise 4.23 (page 120) is nowhere differentiable.

5.3 Use mathematical induction to prove the **Leibnitz rule** of differentiation:

$$(fg)^{(n)} = f^{(n)}g + \frac{n}{1!}f^{(n-1)}g' + \frac{n(n-1)}{2!}f^{(n-2)}g'' + \cdots + fg^{(n)}.$$

5.4 State the hypotheses which assure the validity of the formula

$$[f \circ u \circ v]'(x_0) = f'(u(v(x_0))) \cdot u'(v(x_0)) \cdot v'(x_0),$$

and prove the formula.

5.5 Let $\Delta f(x) = f(x + \Delta x) - f(x)$.

(1) Show that

$$\Delta(\Delta f(x)) = \Delta^2 f(x) = f(x + 2\Delta x) - 2f(x + \Delta x) + f(x).$$

(2) Derive an expression for $\Delta^n f$.

(3) Show that $\lim_{n \to \infty} \frac{\Delta^n f(x)}{\Delta x}^n = f(x)\}$ if this limit exists.

5.6 Show that if f is differentiable on (a, b) and if $f' > 0$ (resp. $f' < 0$), then f is one-to-one.

5.7 Suppose that f is differentiable at x_0. Show that
•
$$\lim_{n \to \infty} n\left[f\left(x_0 + \frac{\alpha}{n}\right) - f\left(x_0 - \frac{\beta}{n}\right)\right] = (\alpha + \beta) f'(x_0).$$

Give an example to show that the existence of this limit does not imply the existence of $f'(x_0)$.

5.8 Suppose that f is differentiable at x_0. Let (h_n) and (k_n) be two nonincreasing sequences both converging to 0. Show that

$$\lim_{n \to \infty} \frac{f(x_0 + h_n) - f(x_0 - k_n)}{h_n + k_n} = f'(x_0).$$

Give an example to show that the existence of this limit does not imply the existence of $f''(x_0)$.

5.9 Suppose that f'' exists at x_0. Show that

$$\lim_{h \to 0} \frac{f(x_0 + h) + f(x_0 - h) - 2f(x_0)}{h^2} = f''(x_0).$$

5.10 A real-valued function f defined on (a, b) is said to be **convex** if

$$f(\lambda x + (1 - \lambda) y) \leq \lambda f(x) + (1 - \lambda) f(y)$$

whenever $x, y \in (a, b)$ and $0 < \lambda < 1$.

(1) Show that if f is convex on (a, b), then it is continuous on (a, b).

(2) Show that if $w < x < y < z$ in (a, b), then
$$\frac{f(x) - f(w)}{x - w} \leq \frac{f(z) - f(y)}{z - y}.$$

(3) Show that a differentiable function f is convex on (a, b) if and only if f' is nondecreasing.

(4) Show that a twice-differentiable function f is convex on (a, b) if and only if $f''(x) \geq 0$ for all $x \in (a, b)$.

5.11 Evaluate each of the indicated limits.

(a) $\lim\limits_{x \to 0} \frac{\sin^2 \frac{x}{3}}{x^2}$ (e) $\lim\limits_{x \to +\infty} \left(\frac{x}{x+1}\right)^x$ (i) $\lim\limits_{x \to 1} \frac{x-1}{x^n - 1}$

(b) $\lim\limits_{x \to 0} (\cos x)^{1/x^2}$ (f) $\lim\limits_{x \to \infty} x^{\sin(1/x)}$ (j) $\lim\limits_{x \to 1} x^{\frac{1}{1-x}}$

(c) $\lim\limits_{x \to 0^-} \frac{e^{1/x} - 1}{e^{1/x} + 1}$ (g) $\lim\limits_{x \to +\infty} \frac{x^3}{e^{2x}}$ (k) $\lim\limits_{x \to \pi/2} \frac{(2x - \pi)^2}{\ln \sin x}$

(d) $\lim\limits_{x \to 0^+} \frac{e^{1/x} - 1}{e^{1/x} + 1}$ (h) $\lim\limits_{x \to +\infty} (e^x + x)^{1/x}$ (l) $\lim\limits_{x \to \infty} x (\ln(1+x) - \ln x)$

5.12 Prove that $\lim_{x \to 0} \frac{x^2 \sin(1/x)}{\sin x} = 0$. Explain why one cannot use L'Hôspital's rule.

5.13 Show that the function $f(x) = (1 + 1/x)^x$ is increasing for $x > 0$.

5.14 Prove that $|\cos x - \cos y| \leq |x - y|$ for all $x, y \in \mathbb{R}$.

5.15 Let f be defined on $\mathbb{R}$ and suppose that $|f(x) - f(y)| \leq (x - y)^2$ for all $x, y \in \mathbb{R}$. Prove that f is a constant function.

5.16 Verify that the roots of the derivative of $f(x) = \sqrt[3]{x^2 - 5x + 6}$ lie between the roots of f.

5.17 Consider the function $f(x) = 1 - \sqrt[5]{x^4}$. Then $f(-1) = f(1) = 0$. Explain why Rolle's theorem is not applicable here.

5.18 Show that $b^n - a^n < nb^{n-1}(b - a)$ for $b > a$ and for $n \in \mathbb{N}$.

5.19 Write Taylor's formula for $\cos x, \ln(1 + x), (1 - x^{-1})$ with $x_0 = 0$.

5.20 Use Taylor's formula for e^{-x} to calculate $1/e$ to 5 decimal places.

5.21 Suppose f' is continuous on $[a, b]$. Show that for every $\varepsilon > 0$, there is $\delta > 0$ such that
$$\left| \frac{f(t) - f(x)}{t - x} - f'(x) \right| < \varepsilon$$
whenever $0 < |t - x| < \delta$, and $x, t \in [a, b]$.

5.22 Use the mean value theorem to show that $1/6 < \ln 1.2 < 1/5$.

5.23 Let $a_1, a_2, \ldots, a_n \in \mathbb{R}$. Determine x so that $\sum_{i=1}^{n} (a_i - x)^2$ is minimum.

5.24 Show that $\frac{\sin b - \sin a}{\cos a - \cos b} = \cot x$, where x is some point in (a, b).

5.25 Prove the following inequalities:

$$
\begin{array}{lll}
(a) & 1 + x < e^x < 1 + xe^x & \text{for } x \neq 0; \\
(b) & \frac{x}{x+1} < \ln(1 + x) < x & \text{for } x > -1, x \neq 0; \\
(c) & \frac{2}{\pi} \leq \frac{\sin x}{x} < 1 & \text{for } 0 < |x| < \frac{\pi}{2}; \\
(d) & x > \arctan x & \text{for all } x; \\
(e) & e^x \leq 1/(1 - x) & \text{for } x < 1.
\end{array}
$$

5.26 Prove that if $x > 0$, $\ln\left(1 + \frac{1}{x}\right) < 1/\sqrt{x^2 + x}$.

5.27 Prove that if $x > 0$, then

$$1 + \frac{x}{2} - \frac{x^2}{8} \leq \sqrt{1 + x} \leq 1 + \frac{x}{2}.$$

5.28 Show that if $0 < r < 1$ and $-1 < x$, then $(1 + x)^r \leq 1 + rx$. Moreover, show that equality holds if and only if $x = 0$.

5.29 Prove that if f and g are differentiable on $\mathbb{R}$, $f'(x) \leq g'(x)$ for all x, and $f(0) = g(0)$, then $f(x) \leq g(x)$ for all $x \geq 0$.

5.30 Suppose that f is continuous on $[a, b]$ and differentiable on (a, b) and that $f'(a)$ and $f'(b)$ have opposite signs. Prove that there is a point $c \in [a, b]$ for which $f'(c) = 0$.

5.31 Prove that $\ln' x = 1/x$ for $x > 0$.

5.32 Suppose that f is continuous on $[a, b]$, twice differentiable on (a, b) and that $f'(x) > 0$ and $f''(x) > 0$ for $x \in (a, b)$.

(1) Find an expression for the second derivative of f^{-1} the inverse of f.

(2) Show that $(f^{-1})''(x) < 0$ on its domain. What can you say about the convexity of f^{-1}?

5.33 Let $f(x) = x^\alpha$, where $\alpha \in \mathbb{R}$. Prove that $f'(x) = \alpha x^{\alpha-1}$.

5.34 Prove that if f is differentiable on an open interval I, with $f'(x) \neq 0$ for all $x \in I$, then f is one-to-one and either $f'(x) > 0$ on I or $f'(x) < 0$ on I.

5.35 Prove that if f is differentiable on an open interval I, then $f'(I)$ is an interval.

5.36 Let $0 < \alpha < 1$ and define $f(t) = \alpha t - t^\alpha$, $t \geq 0$.

(1) Show that $f(t) \geq f(1)$ for all $t \geq 0$. Deduce that if $a, b \geq 0$, then
$$a^\alpha b^{1-\alpha} \leq \alpha a + (1-\alpha) b.$$

(2) Let p and q be such that $1 < p < \infty$ and $1/p + 1/q = 1$. If $A, B \geq 0$, then show that
$$AB \leq \frac{A^p}{p} + \frac{B^q}{q}.$$

5.37 Show that if $f^{(n+1)}$ exists and is continuous on $[a, b]$, then for each $x \in [a, b]$,
$$f(x) = \sum_{k=0}^{n} \frac{f^{(k)}(a)}{k!} (x-a)^k + \frac{1}{n!} \int_a^x (x-t)^n f^{(n+1)}(t)\, dt.$$

5.38 A root x_0 of a polynomial p is said to be **simple** if $p'(x_0) \neq 0$ and to have **multiplicity** n if $p(x_0) = p'(x_0) = \cdots = p^{(n-1)}(x_0) = 0$, but $p^{(n)}(x_0) \neq 0$. Suppose that a and b are consecutive roots of a polynomial p and that $a < b$. Show that there is an odd number (counting multiplicity) of roots of p'.

5.39 Show that if the roots of a polynomial p are all real, then the roots of p' are all real whenever they exist.

5.40 Show that if the real roots of a polynomial p are all simple, then the roots of p' are all simple.

5.41 Let $f(x) = (x^2 - 1)^n$. Show that the n-th derivative of f is a polynomial whose roots are all simple and lie in the open interval $(-1, 1)$.

5.42 Let f be differentiable on $[a, b]$. Suppose that $0 < m \leq f'(x) \leq M < \infty$ for $x \in [a, b]$ and that $f(a) < 0 < f(b)$. Given $x_1 \in [a, b]$, define the sequence (x_n) by
$$x_{n+1} = x_n - \frac{1}{M} f(x_n) \quad \text{for } n \in \mathbb{N}.$$

Show that (x_n) converges to the number x_0 such that $f(x_0) = 0$. Moreover show that
$$|x_{n+1} - x_0| \leq \frac{f(x_1)}{m} \left(1 - \frac{m}{M}\right)^n.$$

6
Elements of Integration

Integration theory has always been at the center of any analysis-related courses. This explains the many different approaches to this theory. In this chapter, we discuss the somewhat traditional approach of the Riemann integral and some other closely related topics.

6.1 Step Functions

Let E be a subset of $\mathbb{R}$. The **characteristic function** of E is the function χ_E defined by

$$\chi_E(x) = \left\{ \begin{array}{ll} 1 & \text{if } x \in E, \\ 0 & \text{if } x \notin E. \end{array} \right.$$

For example, if $E = \mathbb{Q} \cap [0,1]$, then the characteristic function

$$\chi_{\mathbb{Q} \cap [0,1]}(x) = \left\{ \begin{array}{ll} 1 & \text{if } x \text{ is rational}, \\ 0 & \text{if } x \text{ is irrational} \end{array} \right.$$

is nothing else but Dirichlet's discontinuous function of Example 4.4.

For subsets A and B of $\mathbb{R}$, the following relations are immediate. We leave the proofs as an exercise.

- $\chi_\varnothing = 0$ and $\chi_\mathbb{R} = 1$.

- If $A \subset B$, then $\chi_A \leq \chi_B$.

- $\chi_{A \cap B} = \chi_A \cdot \chi_B = \chi_A \wedge \chi_B$.

- $\chi_{A \cup B} = \chi_A + \chi_B - \chi_{A \cap B} = \chi_A \vee \chi_B$.

- $\chi_{A \setminus B} = \chi_A - \chi_{A \cap B}$.

- If $A = \bigcup_{n=1}^{\infty} A_n$ and (A_n) is a disjoint sequence of subsets of $\mathbb{R}$, then $\chi_A = \sum_{n=1}^{\infty} \chi_{A_n}$.

Definition 6.1

A function $\varphi : \mathbb{R} \to \mathbb{R}$ is called a **step function** if it can be expressed as a linear combination of characteristic functions of bounded intervals.

Thus a step function $\varphi : \mathbb{R} \to \mathbb{R}$ can be represented in the form

$$\varphi(x) = \alpha_1 \chi_{I_1}(x) + \alpha_2 \chi_{I_2}(x) + \cdots + \alpha_n \chi_{I_n}(x) = \sum_{i=1}^{n} \alpha_i \chi_{I_i}(x)$$

where the α_i are real numbers and the I_i are bounded intervals. We denote by S the set of all step functions.

The following properties are immediate from the definition.

- The step functions form a linear space, that is, if $\varphi = \sum_{i=1}^{n} \alpha_i \chi_{I_i}$ and $\psi = \sum_{j=1}^{m} \beta_j \chi_{J_j}$ belong to S, then for every real pair of real numbers a and b, we have

$$a\varphi + b\psi = \sum_{i=1}^{n} a\alpha_i \chi_{I_i} + \sum_{j=1}^{m} b\beta_j \chi_{J_j} \in S.$$

- If $\varphi = \sum_{i=1}^{n} \alpha_i \chi_{I_i} \in S$, then φ takes only a finite number of values: 0 and/or finite sums of the α_i.

If $\varphi = \sum_{i=1}^{n} \alpha_i \chi_{I_i} \in S$, then φ has at most a finite number of discontinuities. Indeed, let $a_0 < a_1 < \cdots < a_m$ be the distinct endpoints of all the I_i arranged in increasing order. For each j and for each i, the interval (a_{j-1}, a_j) satisfies either $(a_{j-1}, a_j) \cap I_i = \varnothing$ or $(a_{j-1}, a_j) \subset I_i$. Thus for $x \in (a_{j-1}, a_j)$, $\varphi(x)$ is the sum of those α_i for which the corresponding interval I_i contains (a_{j-1}, a_j). Hence

- φ is constant on each (a_{j-1}, a_j);

- φ is continuous except possibly at the points $a_0, a_1, \ldots, a_m$; and

- φ vanishes outside the bounded interval $[a_0, a_m]$.

Functions such as $\chi_{[0,1]}$, $2\chi_{[0,1]} - \pi\chi_{(1,3]}$, $-\chi_{(-3,1)} + 3\chi_{(-1,3]} + 2\chi_{[3,5)}$ are examples of step functions. On the other hand, since an increasing continuous function on an open interval (a, b) takes on infinitely many values it cannot be a simple function. Also, since Dirichlet's discontinuous function is discontinuous everywhere, it cannot be a step function.

Example 6.2

Find the points of discontinuity of the function $\varphi = 3\chi_{[0,3]} - \chi_{(2,4]}$.

Solution

The distinct endpoints of the intervals $[0,3]$ and $(2,4]$ are $0, 2, 3$, and 4. We have

$$\varphi(x) = \begin{cases} 3 & \text{if } x \in [0,2] \\ 2 & \text{if } x \in (2,3] \\ -1 & \text{if } x \in (3,4] \\ 0 & \text{otherwise.} \end{cases}$$

Thus $0, 2, 3$, and 4 are all points of discontinuity.   □

In general, a step function may be represented in more than one way as a linear combination of characteristic functions of bounded intervals. For example, for the step function φ of Example 6.2, we have

$$\begin{aligned} \varphi &= 3\chi_{[0,3]} - \chi_{(2,4]} \\ &= 2\chi_{[0,2]} + \chi_{[0,3]} + \chi_{(2,4]} - 2\chi_{(3,4]} \\ &= 3\chi_{[0,1]} + 3\chi_{(1,2]} + 2\chi_{(2,3]} - \chi_{(3,4]} \\ &= 3\chi_{[0,2]} + 2\chi_{(2,3]} - \chi_{(3,4]}. \end{aligned}$$

In the last two representations of φ, we notice that the intervals are disjoint or nonoverlapping. A representation with such property is called a **disjoint representation**.

Proposition 6.3

A step function can be expressed as a linear combination of characteristic functions of disjoint bounded intervals.

Proof

Let $\varphi = \sum_{i=1}^{n} \alpha_i \chi_{I_i}$ be a step function. Let $a_0 < a_1 < \cdots < a_r$ be the distinct endpoints of all the I_i, arranged in increasing order. Then φ is constant on each (a_{j-1}, a_j). Let β_j be the different values taken by φ on each (a_{j-1}, a_j). Then $\varphi = \sum_{j=1}^{r} \beta_j \chi_{J_j}$ where the J_j are disjoint bounded intervals with endpoints $a_0, a_1, \ldots, a_r$.   □

In the disjoint representation $\varphi = \sum_{j=1}^{r} \beta_j \chi_{J_j}$ obtained in the above proposition, the coefficients β_j may not be all distinct. By gathering terms involving the same coefficient, we obtain a representation $\sum_{k=1}^{s} \beta_k \chi_{K_k}$ of φ where the β_k are the distinct nonzero values of φ, and $K_k = \{x \in \mathbb{R} : \varphi(x) = \beta_k\}$. (Notice that K_k is a finite disjoint union of some of the J_j.) Such a representation is called the **standard representation** of φ. The standard representation of the function φ of Example 6.2 is $3\chi_{[0,2]} + 2\chi_{(2,3]} - \chi_{(3,4]}$.

Our next example states a very important fact about standard representations.

Example 6.4

Show that a standard representation is unique for a given step function.

Solution

Let $\varphi \in S$. Suppose that $\sum_{i=1}^{r} \alpha_i \chi_{I_i}$ and $\sum_{k=1}^{s} \beta_k \chi_{K_k}$ be two standard representations of φ. Fix $1 \leq i \leq r$. Then there is $1 \leq k \leq s$ such that $\alpha_i = \beta_k$. Thus $I_i \cap K_k \neq \varnothing$. Since $K_k = \{x : \varphi(x) = \beta_k\}$ and $I_i = \{x : \varphi(x) = \alpha_i\}$, it also follows that $I_i = K_k$ and $r \leq s$. Reversing the roles of the intervals I_i and K_k, we also have $s \leq r$. The uniqueness property follows.

Before giving the definition of the integral of step functions, we will discuss the notion of the length of intervals. If the endpoints of a bounded interval I are a and b, we define the **length** of I to be the unique number $\ell(I) = b - a$. We note that the length of I is the same whether or not I contains one or both of its endpoints. Also, we observe that $\ell(I) \geq 0$ and $\ell(I) = 0$ if and only if $a = b$. If $a = -\infty$ and/or $b = \infty$, we let $\ell(I) = \infty$. The length of $I \cup J$, where I and J are disjoint intervals, is defined to be the sum of lengths of the two intervals; i.e.

$$\ell(I \cup J) = \ell(I) + \ell(J) \text{ if } I \cap J = \varnothing.$$

The following facts about the length of bounded intervals are intuitively clear.

Lemma 6.5

Let I and J be two bounded intervals. Then

(1) $\ell(I) \leq \ell(J)$ if $I \subset J$;

(2) $\ell(I \cup J) = \ell(I) + \ell(J) - \ell(I \cap J)$.

Proof

Let a_I and b_I, $a_I \leq b_I$, be the endpoints of I and let a_I and b_J, $a_J \leq b_J$, be the endpoints of J.

If $I \subset J$, then $a_J \leq a_I \leq b_I \leq b_I$ and hence $b_I - a_I \leq b_J - a_J$. This proves (1).

If $I \cap J = \varnothing$, then $\ell(I \cap J) = 0$ and we have by definition

$$\ell(I \cup J) = \ell(I) + \ell(J) = \ell(I) + \ell(J) - \ell(\varnothing) = \ell(I) + \ell(J) - \ell(I \cap J).$$

If $a_I \leq b_I = a_J \leq b_J$ then either $I \cap J = \varnothing$ or $I \cap J = \{a_J\}$. If $I \cap J = \varnothing$ we have the previous case. If $I \cap J = \{a_J\}$

$$\ell(I \cup J) = b_J - a_I = (b_I - a_I) + (b_J - a_J) + 0 = \ell(I) + \ell(J) - \ell(I \cap J).$$

Finally, if $a_I \leq a_J < b_I \leq b_J$, then $I \cap J$ is an interval with endpoints a_J and b_I. Therefore $\ell(I \cap J) = b_I - a_J$ and

$$\begin{aligned}
\ell(I \cup J) = b_J - a_I &= (b_I - a_I) + (b_J - a_J) - (b_I - a_J) \\
&= \ell(I) + \ell(J) - \ell(I \cap J).
\end{aligned}$$

This completes the proof.

$\square$

It clearly follows that if $I \cup J$ is a disjoint union, then $\ell(I \cup J) = \ell(I) + \ell(J)$. Such an observation sets up the following

Example 6.6

Let $\sum_{i=1}^{n} \alpha_i \chi_{I_i}$ and $\sum_{j=1}^{m} \beta_j \chi_{J_j}$ be two disjoint representations of the same step function φ. Show that

$$\sum_{i=1}^{n} \alpha_i \ell(I_i) = \sum_{j=1}^{m} \beta_j \ell(J_j).$$

Solution

Let $\alpha_0 = \beta_0 = 0$, $I_0 = \bigcup_{i=1}^n I_i \setminus \bigcup_{j=1}^m J_j$ and $J_0 = \bigcup_{j=1}^m J_j \setminus \bigcup_{i=1}^n I_i$. Then clearly, $\sum_{i=0}^n \alpha_i \chi_{I_i}$ and $\sum_{j=0}^m \beta_j \chi_{J_j}$ are two disjoint representations of φ, and we have

$$\sum_{i=1}^n \alpha_i \ell(I_i) = \sum_{i=0}^n \alpha_i \ell(I_i) \text{ and } \sum_{j=1}^m \beta_i \ell(J_j) = \sum_{j=0}^m \beta_i \ell(J_j).$$

We notice that for each i and j, the intersection $I_i \cap J_j$ is either empty, in which case $\ell(I_i \cap J_j) = 0$, or $I_i \cap J_j$ is not empty, in which case $\alpha_i = \beta_j$. Finally, for each i, $I_i = \bigcup_{j=0}^m I_i \cap J_j$ is a disjoint union and $\ell(I_i) = \sum_{j=0}^m \ell(I_i \cap J_j)$. Similarly for each j, $J_j = \bigcup_{i=0}^n I_i \cap J_j$ is a disjoint union and $\ell(J_j) = \sum_{i=0}^n \ell(I_i \cap J_j)$. Putting all of these together, we get

$$\sum_{i=0}^n \alpha_i \ell(I_i) = \sum_{i=0}^n \alpha_i \sum_{j=0}^m \ell(I_i \cap J_j) = \sum_{i=0}^n \sum_{j=0}^m \alpha_i \ell(I_i \cap J_j)$$

$$= \sum_{j=0}^m \sum_{i=0}^n \beta_j \ell(I_i \cap J_j) = \sum_{j=0}^m \beta_i \sum_{i=0}^n \ell(I_i \cap J_j) = \sum_{j=0}^m \beta_i \ell(J_j).$$

The proof is finished.   □

Thus to each step function φ, we can associate the unique number given by $\sum_{i=1}^n \alpha_i \ell(I_i)$ if $\sum_{i=1}^n \alpha_i \chi_{I_i}$ is a disjoint representation of φ. This fact justifies the introduction of the following definition.

Definition 6.7

Let $\varphi \in \mathcal{S}$. If $\sum_{i=1}^n \alpha_i \chi_{I_i}$ is a disjoint representation of φ, then the **integral** of φ is defined to be the real number $\sum_{i=1}^n \alpha_i \ell(I_i)$ and is denoted by $\int \varphi$.

For example, a disjoint representation of the step function $3\chi_{[0,3]} - \chi_{(2,4]}$ is $3\chi_{[0,2]} + 2\chi_{(2,3]} - \chi_{(3,4]}$. Therefore

$$\int \left(3\chi_{[0,3]} - \chi_{(2,4]} \right) = 3(2 - 0) + 2(3 - 2) - (4 - 3) = 7.$$

Example 6.8

Let $\sum_{k=1}^p \gamma_k \chi_{K_k}$ be a standard representation of a step function φ. Show that $\int \varphi = \sum_{k=1}^p \gamma_k \ell(K_k)$.

Solution

It suffices to notice that for each k, K_k is a disjoint union of bounded intervals, say $K_k = \bigcup_{i=1}^{n_k} I_{k,i}$. It then follows that $\ell(K_k) = \sum_{i=1}^{n_k} \ell(I_{k,i})$ and that $\sum_{k=1}^{p} \sum_{i=1}^{n_k} \gamma_k \chi_{I_{k,i}}$ is a disjoint representation of φ. Hence

$$\int \varphi = \sum_{k=1}^{p} \sum_{i=1}^{n_k} \gamma_k \ell(I_{k,i}) = \sum_{k=1}^{p} \gamma_k \sum_{i=1}^{n_k} \ell(I_{k,i}) = \sum_{k=1}^{p} \gamma_k \ell(K_k),$$

as required. $\square$

Notice that the value of a step function can be changed at its points of discontinuity without affecting $\int \varphi$.

We previously mentioned that $\mathcal{S}$ a linear space. The next result shows that the operator $\int$ is linear on $\mathcal{S}$.

Theorem 6.9

Let φ, $\psi \in \mathcal{S}$ and let $a, b \in \mathbb{R}$. Then

$$\int (a\varphi + b\psi) = a \int \varphi + b \int \psi.$$

Proof

We first notice that if $\sum_{i=1}^{n} \alpha_i \chi_{I_i}$ is a disjoint representation of φ, then $\sum_{i=1}^{n} a\alpha_i \chi_{I_i}$ is a disjoint representation of $a\varphi$. Therefore,

$$\int a\varphi = \sum_{i=1}^{n} a\alpha_i \ell(I_i) = a \sum_{i=1}^{n} \alpha_i \ell(I_i) = a \int \varphi. \tag{6.1}$$

Let $\sum_{i=1}^{n} \alpha_i \chi_{I_i}$ (resp. $\sum_{j=1}^{m} \beta_j \chi_{J_j}$ (resp. $\sum_{k=1}^{p} \gamma_k \chi_{K_k}$)) be the standard representation of the step function φ (resp. ψ (resp. $\varphi + \psi$)). Consider the sets $A = \bigcup_{i=1}^{n} I_i$, $B = \bigcup_{j=1}^{m} J_j$, and $C = \bigcup_{k=1}^{p} K_k$. If $x \in C$, then $(\varphi + \psi)(x) \neq 0$, and thus $\varphi(x) \neq 0$ or $\psi(x) \neq 0$. Therefore $C \subset A \cup B$. Let $I_0 = (A \cup B) \setminus A$, $J_0 = (A \cup B) \setminus B$, and $K_0 = (A \cup B) \setminus C$. If we set $a_0 = b_0 = c_0$, then we see that $\sum_{i=0}^{n} \alpha_i \chi_{I_i}$ (resp. $\sum_{j=0}^{m} \beta_j \chi_{J_j}$ (resp. $\sum_{k=0}^{p} \gamma_k \chi_{K_k}$)) is a disjoint representation of φ (resp. ψ (resp. $\varphi + \psi$)).

Now the family $\left((I_i \cap J_j \cap K_k)_{i=0}^{n} \right)_{j=0}^{m}$ is a family of disjoint bounded intervals and for $0 \leq k \leq p$,

$$K_k = \bigcup_{i=0}^{n} \bigcup_{j=0}^{m} (I_i \cap J_j \cap K_k).$$

Hence $\ell(K_k) = \sum_{i=0}^{n} \sum_{j=0}^{m} \ell(I_i \cap J_j \cap K_k)$.

Similarly, for each i, $I_i = \bigcup_{j=0}^{m} I_i \cap J_j$ is a disjoint union and $\ell(I_i) = \sum_{j=0}^{m} \ell(I_i \cap J_j)$; and for each j, $J_j = \bigcup_{i=0}^{n} I_i \cap J_j$ is a disjoint union and $\ell(J_j) = \sum_{i=0}^{n} \ell(I_i \cap J_j)$.

Finally, noticing that on the one hand if $x \in I_i \cap J_j \cap K_k$, then $\gamma_k = (\varphi + \psi)(x) = \varphi(x) + \psi(x) = a_i + b_j$, and on the other hand if $I_i \cap J_j \cap K_k = \varnothing$, then $\ell(I_i \cap J_j \cap K_k) = 0$, we have

$$\int (\varphi + \psi) = \sum_{k=1}^{p} \gamma_k \ell(K_k) = \sum_{k=0}^{p} \gamma_k \ell(K_k) = \sum_{k=0}^{p} \gamma_k \sum_{i=0}^{n} \sum_{j=0}^{m} \ell(I_i \cap J_j \cap K_k)$$

$$(6.2)$$

$$= \sum_{k=0}^{p} \sum_{i=0}^{n} \sum_{j=0}^{m} (a_i + b_j) \ell(I_i \cap J_j \cap K_k)$$

$$= \sum_{i=0}^{n} \sum_{j=0}^{m} (a_i + b_j) \sum_{k=0}^{p} \ell(I_i \cap J_j \cap K_k) = \sum_{i=0}^{n} \sum_{j=0}^{m} (a_i + b_j) \ell(I_i \cap J_j)$$

$$= \sum_{i=0}^{n} \sum_{j=0}^{m} a_i \ell(I_i \cap J_j) + \sum_{j=0}^{m} \sum_{i=0}^{n} b_j \ell(I_i \cap J_j)$$

$$= \sum_{i=0}^{n} a_i \ell(I_i) + \sum_{j=0}^{m} b_j \ell(J_j) = \sum_{i=1}^{n} a_i \ell(I_i) + \sum_{j=1}^{m} b_j \ell(J_j)$$

$$= \int \varphi + \int \psi.$$

The combination of (6.1) and (6.2) now completes the proof. $\qquad\square$

It follows immediately from the definition that the integral of a nonnegative step function is nonnegative. More precisely we have

Theorem 6.10

Let $\varphi, \psi \in \mathcal{S}$. Then the following statements hold.

(1) If $\varphi \geq \psi$, then $\int \varphi \geq \int \psi$.

(2) If $\varphi = \psi$, then $\int \varphi = \int \psi$.

Proof

Let φ and ψ be step functions on $[a, b]$

(1) Assume first that $\psi = 0$, and let $\sum_{i=1}^{n} \alpha_i \chi_{I_i}$ be the standard representation of φ. Since $\varphi \geq 0$, $\alpha_i \geq 0$ for all i, and $\int \varphi = \sum_{i=1}^{n} \alpha_i \ell(I_i) \geq 0$. Now

if $\varphi \geq \psi$, then $\varphi - \psi \geq 0$ and by the previous argument $\int (\varphi - \psi) \geq 0$. By linearity, $\int \varphi - b \int \psi \geq 0$; hence $\int \varphi \geq \int \psi$.

(2) It suffices to notice that $\varphi = \psi$ if and only if $\varphi \geq \psi$ and $\varphi \leq \psi$. Applying the result of part (1) to each of these inequalities, we obtain $\int \varphi \geq \int \psi$ and $\int \varphi \leq \int \psi$ and thus $\int \varphi = \int \psi$. □

We notice that for any interval I, $\chi_I (x - d) = 1$ if and only if $x - d \in I$ if and only if $x \in I + d = \{t + d : t \in I\}$. The set $I + d$ is called the d-**translate** of A. In such a case, it is easily seen that $\ell (I) = \ell (I + d)$. This observation sets up the following basic but very important property of the integral of step functions.

Theorem 6.11 (Translation Invariance)

Let $\varphi \in \mathcal{S}$. For $d \in \mathbb{R}$, let $\tau_d \varphi$ be defined by $\tau_d \varphi (x) = \varphi (x + d)$. Then $\tau_d \varphi \in \mathcal{S}$ and $\int \varphi = \int \tau_d \varphi$.

Proof

Let $\varphi = \sum_{i=1}^n \alpha_i \chi_{I_i}$ be in its standard representation. Then

$$\tau_d \varphi (x) = \sum_{i=1}^n \alpha_i \chi_{I_i} (x + d) = \sum_{i=1}^n \alpha_i \chi_{I_i - d} (x),$$

so

$$\int \tau_d \varphi = \sum_{i=1}^n \alpha_i \ell (I_i - d) = \sum_{i=1}^n \alpha_i \ell (I_i) = \int \varphi.$$

□

6.2 Riemann Integral

Let $[a, b]$ be given. A **partition** P of $[a, b]$ is a finite set of points $t_0, t_1, \ldots, t_n$ where

$$a = t_0 < t_1 < \cdots < t_n = b.$$

The length of the largest subinterval is called the **norm** of the partition P and denoted $\|P\|$, i.e.

$$\|P\| = \max \{t_i - t_{i-1} : i = 1, 2, \ldots, n\}.$$

A partition Q is called a **refinement** of P if $P \subset Q$.

Suppose now that f is a real function defined on $[a, b]$. For each partition $P_n = \{t_0, t_1, \ldots, t_n\}$ of $[a, b]$, we can associate to f a step function of the form

$$f^P = \sum_{i=1}^n f(x_i) \chi_{E_i},$$

where $E_n = [t_{n-1}, t_n]$, $E_i = [t_{i-1}, t_i)$ for $i = 1, 2, \ldots, n-1$, and $t_{i-1} \le x_i \le t_i$ for $i = 1, 2, \ldots, n$. It is readily seen (prove it!) that if (P_n) is a sequence of partitions of $[a, b]$ ordered by refinement, that is to say $P_n \subset P_{n+1}$ for all $n \in \mathbb{N}$, then the sequence of step functions (f^{*P_n}) converges pointwise to f.

For each partition $P_n = \{t_0, t_1, \ldots, t_n\}$ of the interval $[a, b]$, we also define

$$M_i(f) = \sup\{f(x) : t_{i-1} \le x \le t_i\} \quad \text{and} \quad m_i(f) = \inf\{f(x) : t_{i-1} \le x \le t_i\}.$$

Then the step functions

$$\overline{f}^P = \sum_{i=1}^n M_i(f) \chi_{E_i} \quad \text{and} \quad \underline{f}^P = \sum_{i=1}^n m_i(f) \chi_{E_i}$$

are called respectively the **upper** and the **lower step function** associated to f relative to the partition P.

Example 6.12

Let f be a function defined on $[a, b]$. Let $M = \sup\{f(x) : a \le x \le b\}$ and $m = \inf\{f(x) : a \le x \le b\}$. Show that for every partition P of $[a, b]$,

$$m(b-a) \le \int \underline{f}^P \le \int f^P \le \int \overline{f}^P \le M(b-a). \tag{6.3}$$

Solution

Let $P = \{t_0, t_1, \ldots, t_n\}$. For each i, let $t_{i-1} \le x_i \le t_i$. Then we clearly have

$$m \le m_i(f) \le f(x_i) \le M_i(f) \le M$$

for each i. Hence

$$m(b-a) \le \sum_{i=1}^n m(t_i - t_{i-1})$$

$$\le \sum_{i=1}^n m_i(f)(t_i - t_{i-1}) = \int \underline{f}^P$$

$$\le \sum_{i=1}^n f(x_i)(t_i - t_{i-1}) = \int f^P$$

$$\le \sum_{i=1}^n M_i(f)(t_i - t_{i-1}) = \int \overline{f}^P$$

$$= \sum_{i=1}^n M_i(f)(t_i - t_{i-1})$$

$$\le \sum_{i=1}^n M(t_i - t_{i-1}) = M(b-a).$$

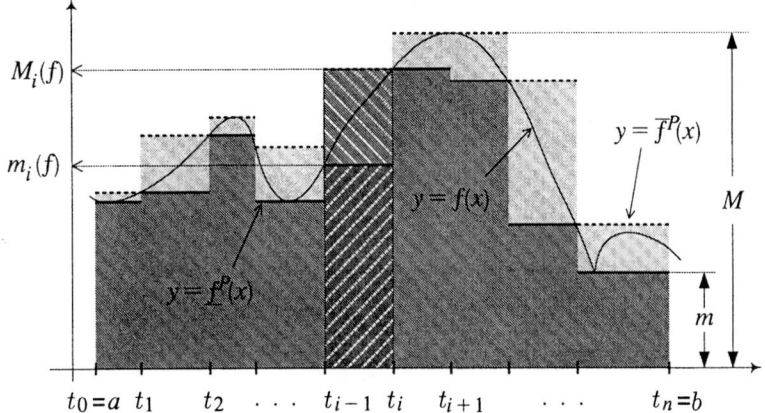

Figure 6.1 Upper and lower Riemann integrals

$\square$

Definition 6.13

If f is a function defined on $[a, b]$, we define its **upper** and **lower Riemann integrals** over the interval $[a, b]$, respectively by

$$\overline{\int}_a^b f(t)\, dt = \inf\left\{\int \overline{f}^P : P \text{ partition of } [a, b]\right\},$$

$$\underline{\int}_a^b f(t)\, dt = \sup\left\{\int \underline{f}^P : P \text{ partition of } [a, b]\right\}.$$

(See Figure 6.1.)

Lemma 6.14

Let f be a real function defined on $[a, b]$, and let P, Q be partitions of $[a, b]$. Then $\int \underline{f}^P \leq \int \overline{f}^Q$.

Proof

Suppose that $P = \{t_0, t_1, \cdots t_n\}$ and let $R = P \cup Q$. Then R is a refinement of both P and Q. Suppose first that $R = P \cup \{s\}$ where $t_{i-1} < s < t_i$. Set $m^* = \inf\{f(x) : t_{i-1} \leq x \leq s\}$, and $m^{**} = \inf\{f(x) : s \leq x \leq t_i\}$. Then

$m_i(f) = \min\{m^*, m^{**}\}$, and thus

$$\int \underline{f}^R - \int \underline{f}^P = m^*[s - t_{i-1}]$$
$$+ m^{**}[t_i - s] - m_i(t_i - t_{i-1})$$
$$= (m^* - m_i(f))[s - t_{i-1}] + (m^{**} - m_i(f))[t_i - s]$$
$$\geq 0.$$

Hence $\int \underline{f}^P \leq \int \underline{f}^{P \cup \{s\}}$.

Now, if $R = P \cup \{s_1, s_2, \cdots, s_k\}$ where $s_k \notin P$, then

$$\int \underline{f}^P \leq \int \underline{f}^{P \cup \{s_1\}} \leq \int \underline{f}^{P \cup \{s_1, s_2\}} \leq \cdots \leq \int \underline{f}^R.$$

Similar arguments show that $\int \overline{f}^R \leq \int \overline{f}^Q$. It then follows that

$$\int \underline{f}^P \leq \int \underline{f}^R \leq \int \overline{f}^R \leq \int \overline{f}^Q.$$

$\square$

The following theorem immediately follows.

Theorem 6.15

Let f be a function defined on $[a, b]$. Then $\underline{\int}_a^b f(t)\, dt \leq \overline{\int}_a^b f(t)\, dt$.

The upper and lower Riemann integrals of a bounded function f on $[a, b]$ always exist as real numbers although they are not necessarily equal.

Example 6.16

Show that the Dirichlet discontinuous function $\chi_{Q \cap [0,1]}$ satisfies

$$\underline{\int}_0^1 \chi_{Q \cap [0,1]}(t)\, dt < \overline{\int}_0^1 \chi_{Q \cap [0,1]}(t)\, dt.$$

Solution

For any partition $P = \{t_0, t_1, \ldots, t_n\}$ of $[0, 1]$, we have

$$\int \left(\overline{\chi}_{Q \cap [0,1]}^P\right) = \sum_{k=1}^n M_i\left(\chi_{Q \cap [0,1]}\right)(t_k - t_{k-1}) = \sum_{k=1}^n 1 \cdot (t_k - t_{k-1}) = 1$$

and

$$\int \left(\chi^P_{\underline{Q \cap [0,1]}} \right) = \sum_{k=1}^{n} m_i \left(\chi_{Q \cap [0,1]} \right) (t_k - t_{k-1}) = \sum_{k=1}^{n} 0 \cdot (t_k - t_{k-1}) = 0.$$

Thus $\overline{\int}_0^1 \chi_Q (t)\, dt > \int_0^1 \chi_Q (t)\, dt$.   □

We now introduce our definition of Riemann integrable functions.

Definition 6.17

Let f be a real function. We say that f is **Riemann integrable over an interval** $[a, b]$ if f is defined on $[a, b]$ and $\int_{\underline{a}}^{b} f(t)\, dt = \overline{\int}_a^b f(t) dt$. The common value of the upper and lower integrals is then called the **Riemann integral of f over the interval** $[a, b]$ and is denoted by $\int_a^b f(t)\, dt$.

We will use the notation $f \in \mathcal{R}([a, b])$ to indicate that f is Riemann integrable over the interval $[a, b]$. It is worth noticing that a step function is always Riemann integrable over any closed bounded interval.

Definition 6.18

We say that a function $f : \mathbb{R} \to \mathbb{R}$ is **Riemann integrable** if there exists a closed bounded interval $[a, b]$ such that

(1) f vanishes outside $[a, b]$ and

(2) $f \in \mathcal{R}([a, b])$.

We denote simply by $\mathcal{R}$ the sets of all the Riemann integrable functions. Since a step function vanishes outside a closed bounded interval and is Riemann integrable over any interval, it is readily seen that $\mathcal{S} \subset \mathcal{R}$. Also if $\mathcal{V}$ denotes the class of functions which vanish outside closed bounded intervals, then it is plain that $\mathcal{R} \subset \mathcal{V}$.

A closed bounded interval satisfying condition (1) of the above definition will be called a **support interval** of f. We should convince ourselves that if $f \in \mathcal{R}$ and if $[a, b]$ and $[c, d]$ are both support intervals of f, then $\int_c^d f(t)\, dt = \int_a^b f(t)\, dt$. We define the integral of a Riemann integrable function to be $\int_a^b f(t)\, dt$ for any support interval $[a, b]$ of f and we simply denote

$$\int f = \int_a^b f(t)\, dt.$$

A function $f : [a, b] \to \mathbb{R}$ is always identified to the function $\tilde{f} : \mathbb{R} \to \mathbb{R}$ defined by

$$\tilde{f}(x) = \begin{cases} f(x) & \text{if } x \in [a, b], \\ 0 & \text{otherwise.} \end{cases}$$

It is readily seen that $f \in \mathcal{R}([a, b])$ if and only if $\tilde{f} \in \mathcal{R}$. On the other hand, we also notice that

Remark 6.19

If $f \in \mathcal{R}$, then for any interval $[a, b]$, the function $\chi_{[a,b]} f \in \mathcal{R}$, and $\int \chi_{[a,b]} f = \int_a^b f(t) \, dt$.

Example 6.16 shows that the Dirichlet function is not Riemann integrable. The next theorem gives a convenient criterion for Riemann integrability.

Theorem 6.20

Let $f \in \mathcal{V}$. Then $f \in \mathcal{R}$ if and only if for every support interval $[a, b]$ of f and for every $\varepsilon > 0$ there exists a partition P of $[a, b]$ such that

$$\int \overline{f}^P - \int \underline{f}^P < \varepsilon. \tag{6.4}$$

Proof

Suppose that $f \in \mathcal{R}$ and let $[a, b]$ be a support interval of f. Then $f \in \mathcal{R}([a, b])$. Fix $\varepsilon > 0$. Then there exist two partitions P and Q such that

$$\int \overline{f}^P < \int_a^b f(t) \, dt + \frac{\varepsilon}{2} \text{ and } \int_a^b f dt - \frac{\varepsilon}{2} < \int \underline{f}^Q.$$

Let $R = P \cup Q$. By Lemma 6.14, we have

$$\int \overline{f}^R \le \int \overline{f}^P < \int_a^b f(t) \, dt + \frac{\varepsilon}{2} < \int \underline{f}^Q + \varepsilon \le \int \underline{f}^R + \varepsilon.$$

Hence (6.4) holds for the partition R.

Conversely, suppose that for every support interval $[a, b]$ of f for any given $\varepsilon > 0$, there exists P such that (6.4) holds. Then

$$\int \underline{f}^P \le \underline{\int_a^b} f(t) \, dt \le \overline{\int_a^b} f(t) \, dt \le \int \overline{f}^P.$$

It follows that $0 \le \overline{\int_a^b} f(t) \, dt - \underline{\int_a^b} f(t) \, dt < \varepsilon$. Since ε is arbitrary, we have $f \in \mathcal{R}([a, b])$, and hence $f \in \mathcal{R}$. $\square$

The following corollary is easily obtained from the above theorem and inequalities (6.3).

Corollary 6.21

Let $f \in V$. Then $f \in R$ if and only if for every support interval $[a, b]$ of f and for every $\varepsilon > 0$ there exists a $\delta > 0$ such that

$$\left| \int f^P - \int_a^b f(t)dt \right| < \varepsilon$$

for each partition $P = \{t_0, t_1, \ldots, t_n\}$, with $\|P\| < \delta$.

It follows from the above corollary that if $f \in R$ and if $[a, b]$ is a support interval of f, then

$$\int f = \int_a^b f(t) \, dt = \lim_{\|P\| \to 0} \int f^P. \tag{6.5}$$

The figures show an approximation of the Riemann integral of $f(x) = x \sin x$ over the interval $[0, 3]$.

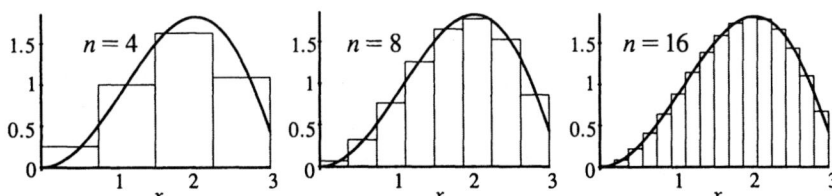

Theorem 6.22

Let $f \in R$. Then f is bounded.

Proof

Suppose that $f \in R$. Let $[a, b]$ be a support interval of f. Then $f \in R([a, b])$. Let P be a partition of $[a, b]$ such that

$$\left| \sum_{i=1}^n f(x_i)(t_i - t_{i-1}) - \int_a^b f(t) \, dt \right| < 1$$

and

$$\left| \sum_{i=1}^n f(x_i')(t_i - t_{i-1}) - \int_a^b f(t) \, dt \right| < 1$$

where x_i, $x_i' \in [t_{i-1}, t_i]$ for $i = 1, 2, \ldots, n$. Then

$$\left| \sum_{i=1}^{n} [f(x_i) - f(x_i')] (t_i - t_{i-1}) \right| < 2.$$

Fix x_1. Then for each $t \in [t_0, t_1]$, we have $|f(x_1) - f(t)| (t_1 - t_0) < 2$, and hence

$$|f(t)| \leq \frac{2}{t_1 - t_0} + |f(x_1)|.$$

This shows that f is bounded on $[t_0, t_1]$. The same argument can be repeated for each $[t_{i-1}, t_i]$, $i = 2, 3, \ldots, n$. We conclude that f is bounded on $[a, b]$. □

We note that not every bounded function is Riemann integrable, as evidenced by Example 6.16. The following example is worth noting.

Example 6.23

Show that $\mathcal{C}([a, b]) \subset \mathcal{R}([a, b])$.

Solution

Suppose that $f \in \mathcal{C}([a, b])$, and let $\varepsilon > 0$. Choose $N \in \mathbb{N}$ large enough so that $b - a < N\varepsilon$. Since f is uniformly continuous on $[a, b]$, there exists $\delta > 0$ such that

$$|f(x) - f(y)| < \frac{1}{N} \text{ for all } x, y \in [a, b] \text{ such that } |x - y| < \delta. \qquad (6.6)$$

Let $P = \{t_0, t_1, \ldots, t_n\}$ be a partition of $[a, b]$ such that $\|P\| < \delta$. Then by (6.6) $M_i - m_i < 1/N$ for each $i = 1, 2, \ldots, n$. Therefore

$$\int \overline{f}^P - \int \underline{f}^P = \sum_{i=1}^{n} (M_i(f) - m_i(f)) (t_i - t_{i-1})$$

$$\leq \sum_{i=1}^{n} \frac{1}{N} (t_i - t_{i-1}) = \frac{1}{N} (b - a) < \varepsilon.$$

By Theorem 6.20, $f \in \mathcal{R}([a, b])$. □

The next result establishes the linearity and the monotonicity of the Riemann integral.

Theorem 6.24

Let $f, g \in \mathcal{R}$ and let k be a constant. Then

(1) $f + g \in \mathcal{R}$, and $\int (f + g) = \int f + \int g$;

(2) $cf \in \mathcal{R}$, and $\int kf = k \int f$;

(3) if $f(x) \le g(x)$ on (a, b), then $\int_a^b f(t)\, dt \le \int_a^b g(t)\, dt$.

Proof

The proofs of these assertions use similar arguments. We only give the proof of (1) and leave the rest as exercises.

To prove (1), let $f, g \in \mathcal{R}$. Let $[a_f, b_f]$ and $[a_g, b_g]$ be support intervals respectively for f and g. Then $[a, d]$, where $a = \min\{a_f, a_g\}$ and $b = \max\{b_f, b_g\}$ is a support interval for both f and g, as well as for $f + g$. Then there exist two partitions P_1 and P_2 of $[a, b]$ such that

$$\int \overline{f}^{P_1} - \int \underline{f}^{P_1} < \frac{\varepsilon}{2} \text{ and } \int \overline{g}^{P_2} - \int \underline{g}^{P_2} < \frac{\varepsilon}{2}.$$

Let $P = P_1 \cup P_2$. By Lemma 6.14, we have

$$\int \overline{f}^{P} - \int \underline{f}^{P} < \frac{\varepsilon}{2} \text{ and } \int \overline{g}^{P} - \int \underline{g}^{P} < \frac{\varepsilon}{2}.$$

On the other hand, since

$$\int \underline{f}^{P} + \int \underline{g}^{P} \le \int \underline{(f + g)}^{P} \tag{6.7}$$

$$\le \int \overline{(f + g)}^{P} \le \int \overline{f}^{P} + \int \overline{g}^{P},$$

we have $\int \overline{(f + g)}^{P} - \int \underline{(f + g)}^{P} < \varepsilon$. Thus $f + g \in \mathcal{R}$.

The inequalities (6.7) also imply that

$$\int_a^b (f + g)(t)\, dt \le \int \left(\overline{f}^{P}\right) + \int \left(\overline{g}^{P}\right)$$

$$< \left(\int_a^b f(t)\, dt + \frac{\varepsilon}{2}\right) + \left(\int_a^b g(t)\, dt + \frac{\varepsilon}{2}\right)$$

$$= \int_a^b f(t)\, dt + \int_a^b g(t)\, dt + \varepsilon$$

and that

$$\int_a^b f(t)\, dt + \int_a^b g(t)\, dt \le \left(\int \underline{f}^{P} + \frac{\varepsilon}{2}\right) + \left(\int \underline{g}^{P} + \frac{\varepsilon}{2}\right)$$

$$= \int \underline{f}^{P} + \int \underline{g}^{P} + \varepsilon$$

$$\le \int \underline{(f + g)}^{P} + \varepsilon \le \int_a^b (f + g)(t)\, dt + \varepsilon.$$

Since ε was arbitrary, we conclude that

$$\int_a^b (f + g)(t)\, dt = \int_a^b f(t)\, dt + \int_a^b g(t)\, dt.$$

$\square$

The following example can be considered as a corollary of the above theorem.

Example 6.25

Suppose that $f \in \mathcal{R}([a, b])$. Show that if $a < c < b$, then $f \in \mathcal{R}([a, c])$ and $f \in \mathcal{R}([c, b])$, and

$$\int_a^b f(t)\, dt = \int_a^c f(t)\, dt + \int_c^b f(t)\, dt.$$

Solution

Let $\tilde{f}$ be the function which vanishes off the interval $[a, b]$, and the values of which coincide with that of f on $[a, b]$. Then for every $a < c < b$, we easily check that

$$f = \chi_{[a,b]}\tilde{f} = \chi_{[a,c]}\tilde{f} + \chi_{[c,b]}\tilde{f}.$$

Now Remark 6.19 and condition (1) of Theorem 6.24 imply

$$\int_a^b f(t)\, dt = \int \chi_{[a,b]}\tilde{f} = \int \left(\chi_{[a,c]}\tilde{f} + \chi_{[c,b]}\tilde{f}\right)$$

$$= \int \chi_{[a,c]}\tilde{f} + \int \chi_{[c,b]}\tilde{f} = \int_a^c f(t)\, dt + \int_c^b f(t)\, dt,$$

as desired.

$\square$

Lemma 6.26

Let $f \in \mathcal{R}([a, b])$ and let $g \in \mathcal{C}([m, M])$ where

$$M = \sup\{f(x) : a \le x \le b\} \text{ and } m = \inf\{f(x) : a \le x \le b\}.$$

Then $g \circ f \in \mathcal{R}([a, b])$.

Proof

Let $\varepsilon > 0$. Since g is uniformly continuous on $[m, M]$, there exists $\delta > 0$ such that

$$|g(s) - g(t)| < \varepsilon \text{ whenever } s, t \in S \text{ with } |s - t| < \delta. \tag{6.8}$$

Since $f \in \mathcal{R}([a, b])$, there is a partition $P = \{t_0, t_1, \ldots, t_n\}$ of $[a, b]$ such that $\int \overline{f}^P - \int \underline{f}^P < \varepsilon \delta$. We consider the sets $K = \{i : M_i(f) - m_i(f) < \delta\}$ and $K^c = \{i : M_i(f) - m_i(f) \geq \delta\}$. Then we have

$$\sum_{i \in K^c} (t_i - t_{i-1}) = \frac{1}{\delta} \sum_{i \in K^c} \delta (t_i - t_{i-1})$$

$$\leq \frac{1}{\delta} \sum_{i=1}^{n} [M_i(f) - m_i(f)] (t_i - t_{i-1})$$

$$= \frac{1}{\delta} \left(\int \overline{f}^P - \int \left(\underline{f}^P \right) \right) < \varepsilon.$$

According to (6.8), we have $M_i(g \circ f) - m_i(g \circ f) < \varepsilon$ for $i \in K$. Hence

$$\int \overline{(f \circ g)}^P - \int \underline{(f \circ g)}^P = \sum_{i=1}^{n} [M_i(g \circ f) - m_i(g \circ f)] (t_i - t_{i-1})$$

$$= \sum_{i \in K} [M_i(g \circ f) - m_i(g \circ f)] (t_i - t_{i-1})$$

$$+ \sum_{i \in K^c} [M_i(g \circ f) - m_i(g \circ f)] (t_i - t_{i-1})$$

$$\leq \varepsilon \sum_{i \in K} (t_i - t_{i-1}) + C \sum_{i \in K^c} (t_i - t_{i-1})$$

$$\leq \varepsilon (b - a) + C\varepsilon = (C + b - a) \varepsilon,$$

where $C = \max \{g(u) : m \leq u \leq M\} - \min \{g(u) : m \leq u \leq M\}$. Since $\varepsilon > 0$ is arbitrary, this shows that $f \circ g \in \mathcal{R}([a, b])$. $\square$

Further properties of $\mathcal{R}$ are given in the next theorem.

Theorem 6.27

If $f, g \in \mathcal{R}$, then $|f|, f^2, fg, f \vee g$, and $f \wedge g \in \mathcal{R}$ and $\left| \int f \right| \leq \int |f|$.

Proof

We first notice that it is possible to consider one common support interval to all the functions in consideration in the theorem. So we assume that $[a, b]$ is such interval. To see that $|f|$ (resp. f^2) $\in \mathcal{R}$, we apply the result of the previous theorem to f and g where $g(x) = |x|$ (resp. $g(x) = x^2$) on the interval $[a, b]$. To prove the inequality, we apply (3) of Theorem 6.24 to f, $-f$ and $|f|$ and note that $\pm f(x) \leq |f|(x)$ for all $x \in \mathbb{R}$. To see that (3) and (4) hold, it suffices to notice that $fg = \frac{1}{2}\left[(f+g)^2 - (f^2 + g^2)\right]$, $f \vee g = \frac{1}{2}[(f+g) + |f-g|]$, and $f \wedge g = \frac{1}{2}[(f+g) - |f-g|]$. $\qquad \square$

We finish this section by another feature of the Riemann integral.

Theorem 6.28

Let $f \in \mathcal{R}$. For $d \in \mathbb{R}$, let $\tau_d f$ be defined by $\tau_d f(x) = f(x+d)$. Then $\tau_d f \in \mathcal{R}$ and $\int f = \int \tau_d f$.

The proof of the above theorem is obtained by combining the definition of Riemann integral with the result of Theorem 6.11 and is left as an exercise.

6.3 Functions of Bounded Variation

For this section, we will only consider real-valued functions defined on a closed bounded interval.

Definition 6.29

Let $f : [a, b] \to \mathbb{R}$. The **variation** of f over $[a, b]$ is defined to be

$$V_f[a, b] = \sup\left\{\sum_{i=1}^{n} |f(t_i) - f(t_{i-1})| : \{t_0, t_1, \ldots, t_n\} \text{ is a partition of } [a, b]\right\}.$$

If $V_f[a, b]$ exists as a real number, then f is said to be of **bounded variation**.

Notation

The collection of all functions of bounded variation on $[a, b]$ is denoted by $\mathcal{BV}([a, b])$. For every $x < y$ in $[a, b]$, we also denote by $V_f[x, y]$ the variation of f on the interval $[x, y]$.

Lemma 6.30

Let $f \in \mathcal{BV}([a, b])$. Then the function $\varphi : [a, b] \to \mathbb{R}$ defined by $\varphi(x) = V_f[a, x]$ is nondecreasing.

Proof

Let $x_1 < x_2$ in $[a, b]$. Let $P = \{t_0, t_1, \ldots, t_n\}$ be a partition of $[a, x_1]$. Then $P \cup \{x_2\}$ is a partition of $[a, x_2]$ and

$$\sum_{i=1}^{n} |f(t_i) - f(t_{i-1})| \leq \sum_{i=1}^{n} |f(t_i) - f(t_{i-1})| + |f(x_2) - f(x_1)| \leq V_f[a, x_2].$$

Taking the supremum over all partitions of $[a, x_1]$, we have $V_f[a, x_1] \leq V_f[a, x_2]$. This proves the lemma. $\square$

An immediate consequence of this lemma is the following

Example 6.31

Show that a function of bounded variation is necessarily bounded.

Solution

We notice that if $x \in [a, b]$, then

$$V_f[a, b] \geq V_f[a, x] \geq |f(x) - f(a)| \geq ||f(x)| - |f(a)|| \geq |f(x)| - |f(a)|.$$

Hence $\sup\{|f(x)| : x \in [a, b]\} \leq V_f[a, b] + |f(a)|$. $\square$

Example 6.32

Show that if $f \in \mathcal{R}([a, b])$, then $F \in \mathcal{BV}([a, b]) \cap C[a, b]$ where $F(x) = \int_a^x f(t)\, dt$.

Solution

Since $f \in \mathcal{R}([a, b])$, f is bounded on $[a, b]$. Let $M > 0$ such that $|f(t)| \leq M$ for every $t \in [a, b]$. It follows that for every x, y in $[a, b]$, we have

$$|F(x) - F(y)| = \left| \int_a^x f(t)\, dt - \int_a^y f(t)\, dt \right| \leq M |x - y|.$$

This establishes the continuity of F.

Now let $P = \{t_0, t_1, \ldots, t_n\}$ be a partition of $[a, b]$. Then

$$\sum_{i=1}^{n} |F(t_i) - F(t_{i-1})| = \sum_{i=1}^{n} \left| \int_{t_{i-1}}^{t_i} f(t)\, dt \right|$$

$$\leq \sum_{i=1}^{n} \int_{t_{i-1}}^{t_i} |f(t)|\, dt = \int_{a}^{b} |f(t)|\, dt.$$

Thus $V_F[a, b] \leq \int_a^b |f(t)|\, dt < \infty$. The proof is complete. $\qquad\square$

The proofs of the following properties are routine and are left as exercises.

Proposition 6.33

Let f and $g \in BV([a, b])$ and $\alpha \in \mathbb{R}$. Then $f + g$, αf, fg and $|f| \in BV([a, b])$.

We notice that if f is a nondecreasing function on $[a, b]$, and if $\{t_0, t_1, \ldots, t_n\}$ is a partition of $[a, b]$, then

$$\sum_{i=1}^{n} |f(t_i) - f(t_{i-1})| = \sum_{i=1}^{n} [f(t_i) - f(t_{i-1})] = f(b) - f(a).$$

Therefore $f \in BV([a, b])$. In fact, functions of bounded variation can be characterized in terms of nondecreasing functions.

Theorem 6.34

A function $f \in BV([a, b])$ if and only if there exist nondecreasing functions g and h such that $f = g - h$.

Proof

Suppose that $f = g - h$ where g and h are nondecreasing. Then $g, h \in BV([a, b])$, and by Proposition 6.33, $f = g - h \in BV([a, b])$.

Conversely, suppose $f \in BV([a, b])$. We write for every $x \in [a, b]$

$$f(x) = V_f[a, x] - [V_f[a, x] - f(x)].$$

By Lemma 6.30, the function $g(x) = V_f[a, x]$ is nondecreasing on $[a, b]$. From the proof of the same lemma, if $x_1 < x_2$ in $[a, b]$, then

$$f(x_2) - f(x_1) \leq V_f[a, x_2] - V_f[a, x_1],$$

from which we obtain $V_f[a, x_1] - f(x_1) \leq V_f[a, x_2] - f(x_2)$ whenever $x_1 < x_2$ in $[a, b]$. Thus the function $h(x) = V_f[a, x] - f(x)$ is also nondecreasing on $[a, b]$. This completes the proof. $\qquad\square$

Recall that the **fundamental theorem of calculus** states that if f is a real-valued continuous function on $[a, b]$, then the function $F(x) = \int_a^x f(t)\, dt$ is differentiable on $[a, b]$ and $F'(x) = f(x)$ for all $x \in [a, b]$. The following result generalizes this result to the case of functions of bounded variation.

Theorem 6.35

Let $f \in \mathcal{BV}([a, b])$ and let $F(x) = \int_a^x f(t)\, dt$ for $x \in [a, b]$. Then there exists a subset D of $[a, b]$, at most countable, such that $F'(x) = f(x)$ for all $x \in [a, b] \setminus D$.

Proof

Combining the results of Theorem 6.34 and Example 4.18 (page 106), we see that the set D of all the points of discontinuity of f is at most countable. Let $x_0 \in [a, b] \setminus D$. Then we observe that if $x \neq x_0$ in $[a, b]$, then

$$\frac{F(x) - F(x_0)}{x - x_0} = \frac{1}{x - x_0} \int_{x_0}^x f(t)\, dt$$

and therefore

$$\frac{F(x) - F(x_0)}{x - x_0} - f(x_0) = \frac{1}{x - x_0} \int_{x_0}^x [f(t) - f(x_0)]\, dt.$$

Now since f is continuous at x_0, for every $\varepsilon > 0$, there exists $\delta > 0$ such that

$$|t - x_0| < \delta \quad \text{implies} \quad |f(t) - f(x_0)| < \varepsilon.$$

It follows that

$$|x - x_0| < \delta \quad \text{implies} \quad \left| \frac{F(x) - F(x_0)}{x - x_0} - f(x_0) \right| < \varepsilon.$$

This shows that

$$F'(x_0) = \lim_{x \to x_0} \frac{F(x) - F(x_0)}{x - x_0} = f(x_0).$$

$\qquad\square$

Definition 6.36

A function $f : [a, b] \to \mathbb{R}$ is said to be **absolutely continuous** if

for every $\varepsilon > 0$ there exists $\delta > 0$ such that
whenever $(a_1, b_1), \ldots, (a_n, b_n)$ are disjoint subintervals of $[a, b]$
and $\sum_{i=1}^{n} |b_i - a_i| < \delta$, then $\sum_{i=1}^{n} |f(b_i) - f(a_i)| < \varepsilon$.

Notation

The collection of all absolutely continuous functions on $[a, b]$ is denoted by $\mathcal{AC}([a, b])$.

An absolutely continuous function is necessarily continuous. Again the proofs of the following properties are routine and are left as exercises.

Proposition 6.37

Let f and $g \in \mathcal{AC}([a, b])$ and $\alpha \in \mathbb{R}$. Then $f + g, \alpha f, fg$, and $|f| \in \mathcal{AC}([a, b])$.

Example 6.38

Show that $\mathcal{AC}([a, b]) \subset \mathcal{BV}([a, b])$.

Solution

Let $f \in \mathcal{AC}([a, b])$, and fix $\varepsilon = 1$. There is $\delta > 0$ such that

$$\sum_{i=1}^{n} |f(b_i) - f(a_i)| < 1 \tag{6.9}$$

for every $(a_1, b_1), \ldots, (a_n, b_n)$ disjoint subintervals of $[a, b]$ with $\sum_{i=1}^{n} |b_i - a_i| < \delta$. Now choose N large enough so that $(b - a)/N < \delta$. Let $P = \{t_0, t_1, \ldots, t_N\}$ be the partition of $[a, b]$ into subintervals of length $(b - a)/N$. Then for each $i = 1, 2, \ldots, n$, $V_f[t_{i-1}, t_i] < 1$. Hence since $V_f[a, b] = \sum_{i=1}^{N} V_f[t_{i-1}, t_i]$ (verify!), it follows that $V_f[a, b] < N < \infty$. □

6.4 Riemann–Stieltjes Integral

For the sake of completeness, we end this chapter by sketching a generalization of the Riemann integral. Throughout the rest of this section we fix a function of bounded variation $\nu : [a, b] \to \mathbb{R}$.

Let $f : [a, b] \to \mathbb{R}$ be a bounded function and let $P = \{t_0, t_1, \ldots, t_n\}$ be a partition of $[a, b]$. An expression of the form

$$\sum_{i=1}^{n} f(x_i) \left[\nu(t_i) - \nu(t_{i-1}) \right],$$

where $x_i \in [t_i, t_{i-1}]$ for each i, is called a **Riemann–Stieltjes sum** of f with respect to ν.

Definition 6.39

A function $f : [a, b] \to \mathbb{R}$ is said to be **Riemann–Stieltjes integrable** with respect to ν if there exists a real number r with the property

for every $\varepsilon > 0$ there exists $\delta > 0$ such that
whenever $P = \{t_0, t_1, \ldots, t_n\}$ is a partition of $[a, b]$,
$\|P\| < \delta$ implies $\left| \sum_{i=1}^{n} f(x_i) \left[\nu(t_i) - \nu(t_{i-1}) \right] - r \right| < \varepsilon.$

The number r in the above definition is uniquely determined if it exists and is denoted by

$$r = \int_a^b f(t) \, d\nu(t) \text{ or } \int_a^b f \, d\nu.$$

It is called the **Riemann–Stieltjes integral of** f with respect to ν over $[a, b]$. Sometimes we write "f is ν-integrable" to indicate that f is integrable with respect to ν. The set of all Riemann–Stieltjes integrable functions with respect to ν over $[a, b]$ will be denoted by $\mathcal{R}([a, b], \nu)$.

In the special case where $\nu(x) = x$, it is a good exercise to show that the Riemann–Stieltjes integral reduces to the ordinary Riemann integral, i.e. $\mathcal{R}([a, b], \nu) = \mathcal{R}([a, b])$.

Many of the results in the section on Riemann integrable functions can be extended to the Riemann–Stieltjes integral. For example, the following theorem is easily seen to be the counterpart of Theorem 6.20.

Theorem 6.40

A function $f \in \mathcal{R}([a, b], \nu)$ if and only if

for every $\varepsilon > 0$ there exists a partition $P = \{t_0, t_1, \ldots, t_n\}$ such that
$\sum_{i=1}^{n} \left[M_i(f) - m_i(f) \right] |\nu(t_i) - \nu(t_{i-1})| < \varepsilon,$

where $M_i = \sup \{ \nu(x) : t_{i-1} \leq x \leq t_i \}$, and $m_i = \inf \{ \nu(x) : t_{i-1} \leq x \leq t_i \}$ for $i = 1, 2, \ldots, n$.

The proofs of the following results use similar arguments as in the ordinary Riemann integral case and are therefore left as exercises.

Proposition 6.41

Let f and $g \in \mathcal{R}\left([a,b],\nu\right)$ and $\alpha \in \mathbb{R}$. Then $f+g$, αf, fg, and $|f| \in \mathcal{R}\left([a,b],\nu\right)$, and

(1) $\int_a^b (f+g)\, d\nu = \int_a^b f d\nu + \int_a^b g d\nu$;

(2) $\int_a^b \alpha f d\nu = \alpha \int_a^b g d\nu$;

(3) $\left|\int_a^b f d\nu\right| \leq \int_a^b |f|\, d\nu$;

(4) $\int_a^b f d\nu \leq \int_a^b g d\nu$, whenever $f \leq g$ on $[a,b]$.

The case where ν has a derivative presents a special feature. In such a case it is possible to reduce the Riemann–Stieltjes integral to the ordinary Riemann integral.

Example 6.42

Suppose that ν is differentiable on $[a,b]$ and $\nu' \in \mathcal{R}\left([a,b]\right)$. Show that if $f \in \mathcal{R}\left([a,b]\right)$, then $f \in \mathcal{R}\left([a,b],\nu\right)$, and $\int_a^b f(t)\, d\nu(t) = \int_a^b f(t)\, \nu'(t)\, dt$.

Solution

Suppose that $f \in \mathcal{R}\left([a,b]\right)$. Then f is bounded and let

$$M = \sup\left\{|f(x)| : x \in [a,b]\right\}.$$

It follows that $f\nu' \in \mathcal{R}\left([a,b]\right)$ (Theorem 6.27). Let $r = \int_a^b f(t)\, \nu'(t)\, dt$. Therefore there exists a partition P_ε such that for any refinement $P = \{t_0, t_1, \ldots, t_n\}$ of P_ε we have

$$\left|\sum_{i=1}^n f(x_i)\, \nu'(x_i)\, (t_i - t_{i-1}) - r\right| < \frac{\varepsilon}{2}$$

where $x_i \in [t_{i-1}, t_i]$ for $i = 1, 2, \ldots, n$ and

$$\sum_{i=1}^n \left[M_i(f) - m_i(f)\right] |t_i - t_{i-1}| < \frac{\varepsilon}{2M}$$

where $M_i = \sup\{\nu'(x) : t_{i-1} \leq x \leq t_i\}$, and $m_i = \inf\{\nu'(x) : t_{i-1} \leq x \leq t_i\}$ for $i = 1, 2, \ldots, n$. Applying the mean value theorem to the function ν on each interval (t_{i-1}, t_i), we obtain points $y_i \in (t_{i-1}, t_i)$ such that $\nu(t_i) - \nu(t_{i-1}) = \nu'(y_i)(t_i - t_{i-1})$. It follows that

$$
\left| \sum_{i=1}^{n} f(x_i)\left[\nu(t_i) - \nu(t_{i-1})\right] - r \right| = \left| \sum_{i=1}^{n} f(x_i)\nu'(y_i)(t_i - t_{i-1}) - r \right|
$$

$$
\leq \left| \sum_{i=1}^{n} f(x_i)\left[\nu'(y_i) - \nu'(x_i)\right](t_i - t_{i-1}) \right|
$$

$$
+ \left| \sum_{i=1}^{n} f(x_i)\nu'(x_i)(t_i - t_{i-1}) - r \right|
$$

$$
\leq \sum_{i=1}^{n} M\left[M_i(f) - m_i(f)\right]|t_i - t_{i-1}|
$$

$$
+ \left| \sum_{i=1}^{n} f(x_i)\nu'(x_i)(t_i - t_{i-1}) - r \right|
$$

$$
< \frac{M\varepsilon}{2M} + \frac{\varepsilon}{2} = \varepsilon.
$$

Hence we have established that $f \in \mathcal{R}([a,b], \nu)$ and that

$$
\int_a^b f(t)\, d\nu(t) = \int_a^b f(t)\nu'(t)\, dt.
$$

The proof is complete. $\qquad\qquad\qquad\qquad\qquad\qquad\qquad\qquad\qquad\qquad$ □

EXERCISES

6.1   Write each of the following step functions in standard form:

$$
\chi_{[0,4)} + 3\chi_{(1,5)}; \quad \chi_{[0,4)} + 2\chi_{(2,3)} - 3\chi_{[3,5)}; \quad 2\chi_{[-2,2)} - \chi_{(0,4)} + \chi_{[4,5)}.
$$

6.2   Decide whether f is a step function or not.

      (a) $f = \operatorname{sgn}(\sin x)$ (d) $f = \chi_{\mathbb{Q} \cap [0,1]}$

      (b) $f = \sum_{i=1}^{\infty} n\chi_{\left(\frac{1}{n+1}, \frac{1}{n}\right]}$ (e) $f = \operatorname{int}(x)$

      (c) $f = \chi_{(-\infty,2]} - \chi_{(-\infty,3]}$ (f) $f = \operatorname{int}(x)\chi_{[-100,100]}(x)$.

6.3   Let $\varphi \in \mathcal{S}$. Show that $|\varphi| \in \mathcal{S}$ and that $\left|\int \varphi\right| \leq \int |\varphi|$.

6.4   Let $\varphi \in \mathcal{S}$. Let $m_k\varphi$ be defined by $m_k\varphi(x) = \varphi(kx)$. Show that $m_k\varphi \in \mathcal{S}$ and $\int m_k\varphi = \frac{1}{|k|}\int \varphi$.

6.5   Let φ be a nonnegative step function. Show that $\sqrt{\varphi} \in \mathcal{S}$.

6.6   If $\varphi, \psi \in \mathcal{S}$, show that $\max\{\varphi, \psi\}, \min\{\varphi, \psi\}, \varphi \circ \psi \in \mathcal{S}$.

6.7   If $\varphi, \psi \in \mathcal{S}$, show that $\varphi\psi \in \mathcal{S}$ and that $\left(\int (\varphi\psi)\right)^2 \leq \int \varphi^2 \int \psi^2$.

6.8   Let $f(x) = x^2$ be defined on $[0, 1]$. For each n, let

$$P_n = \left\{0, \frac{1}{n}, \frac{2}{n}, \ldots, \frac{n}{n}\right\}$$

be a partition of $[0, 1]$. Write out $\overline{f}^{P_n}$ and $\underline{f}^{P_n}$. Show that f is Riemann integrable and $\int_0^1 f(x)\, dx = 1/3$.

6.9   Let $f \in \mathcal{R}([a, b])$. Show that

$$\int_a^b f(x)\, dx = \lim_{n \to \infty} \frac{b-a}{n} \sum_{i=1}^n f\left(a + i\frac{b-a}{n}\right).$$

6.10 Use the Riemann integral to evaluate each of the following limits.

   (1) $\lim_{n \to \infty} \frac{1^p + 2^p + \ldots + n^p}{n^{p+1}}$ for $p = \frac{1}{2}$; for $p \in \mathbb{N}$.

   (2) $\lim_{n \to \infty} \frac{1}{n}(1 + \cos \frac{a}{n} + \cos \frac{2a}{n} + \ldots + \cos \frac{(n-1)a}{n})$ where $a > 0$.

6.11 Prove that an increasing function $f : [a, b] \to \mathbb{R}$ is Riemann integrable.

6.12 Show that the function $\chi_{(\mathbb{R}\backslash\mathbb{Q})\cap[0,1]}$ is not Riemann integrable.

6.13 Show that for any $n \in \mathbb{N}$, the function $f : x \longmapsto x^n$ is in $\mathcal{R}([a, b])$ for any interval $[a, b]$. Does $f \in \mathcal{R}$?

6.14 If $f, g \in \mathcal{R}$, let $\langle f, g \rangle = \int fg$. Show that $\langle \cdot, \cdot \rangle$ satisfies the following properties:

   (1) $\langle f, f \rangle \geq 0$ for all $f \in \mathcal{R}$.

   (2) $\langle f + g, h \rangle = \langle f, h \rangle + \langle g, h \rangle$ for all $f, g, h \in \mathcal{R}$.

   (3) $\langle cf, g \rangle = c\langle f, g \rangle$ for all $f, g \in \mathcal{R}$ and for all $c \in \mathbb{R}$.

   (4) $\langle f, g \rangle = \langle f, g \rangle$ for all $f, g \in \mathcal{R}$.

6.15 Let $f \in \mathcal{R}$. Suppose that $f(x) = g(x)$ except for a finite number of points. Show that $g \in \mathcal{R}$ and that $\int g = \int f$.

6.16 Let $f \in \mathcal{C}([a, b])$. Suppose that f is nonnegative and that $\int_a^b f(x)\, dx = 0$. Show that $f(x) = 0$ for all $x \in [a, b]$.

6.17 Let (f_n) be a sequence in $\mathcal{R}([0, 1])$ defined by $f_n(x) = \frac{nx^{n-1}}{1+x}$ for all $x \in [0, 1]$. Show that $f_n(x) \to 0$ for all $x \in (0, 1)$, but $\lim \int_0^1 f_n(x)\, dx \neq 0$.

6.18   Discuss the limit as $n \to \infty$ of the following integrals.

(a) $\int_0^1 \frac{\sin nx}{x^2+n^2} dx$ (c) $\int_1^\pi \frac{\sin nx}{nx} dx$

(b) $\int_0^1 nx (1-x)^n \, dx$ (d) $\int_0^1 e^{-nx} dx$.

6.19   Let (f_n) be a sequence in $\mathcal{R}([0,1])$ defined by $f_n(x) = nxe^{-nx^2}$. Show that $f_n(x) \to 0$ for all $x \in [0,1]$, but $\lim \int_0^1 f_n(x) \, dx \neq 0$.

6.20   Repeat the previous exercise with
$$f_n(x) = \begin{cases} nx^2 & \text{if} \quad 0 \le x < 1/n, \\ -n^2(x-2/n) & \text{if} \quad 1/n \le x < 2/n, \\ 0 & \text{if} \quad 2/n \le x \le 1. \end{cases}$$

6.21   **Arzelà.** Let (f_n) be a sequence in $\mathcal{R}([a,b])$ which converges point-wise to a function $f \in \mathcal{R}([a,b])$ and which is uniformly bounded on $[a,b]$. Show that $\lim \int_a^b f_n(x) \, dx = \int_a^b f(x) \, dx$.

6.22   Write down the proofs of the results of Proposition 6.33.

6.23   Write down the proofs of the results of Proposition 6.37.

6.24   Write down the proofs of the results of Proposition 6.41.

6.25   Let ν be a nondecreasing function on $[a,b]$. Show that if $f, g \in \mathcal{R}([a,b], \nu)$, then $f \vee g$, $f \wedge g \in \mathcal{R}([a,b], \nu)$.

6.26   Let ν and μ be nondecreasing functions on $[a,b]$. Let $f \in \mathcal{R}([a,b], \nu)$. Suppose that $\mu(x) = \nu(x)$ except for a finite number of points in (a,b) at which f is continuous. Show that $f \in \mathcal{R}([a,b], \mu)$ and that $\int_a^b f d\mu = \int_a^b f d\nu$.

6.27   Let ν be a nondecreasing function on $[a,b]$. Suppose that (f_n) is a sequence in $\mathcal{R}([a,b], \nu)$ which converges uniformly to a function f on $[a,b]$. Show that $f \in \mathcal{R}([a,b], \nu)$ and that $\int_a^b f d\nu = \lim \int_a^b f d\nu$.

6.28   Let ν be a nondecreasing function on $[a,b]$ and let (f_n) be a sequence in $\mathcal{R}([a,b], \nu)$. Suppose that the function defined by $f(x) = \lim f_n(x)$ belongs to $\mathcal{R}([a,b], \nu)$. Then $\int_a^b f d\nu = \lim \int_a^b f d\nu$.

6.29   Let ν be a nondecreasing function on $[a,b]$ and let (f_n) be a monotone sequence in $\mathcal{R}([a,b], \nu)$. Suppose that there exists $M > 0$ such that $|f_n(x)| \le M$ for all $n \in \mathbb{N}$, and $x \in [a,b]$. Then (f_n) converges pointwise to a function $f \in \mathcal{R}([a,b], \nu)$ and $\int_a^b f d\nu = \lim \int_a^b f d\nu$.

6.30   **Change of variable** Let ν be a nondecreasing continuous function on $[a,b]$ and let $f \in \mathcal{R}([a,b], \nu)$. Suppose that φ is an increasing continuous function that maps $[A, B]$ onto $[a,b]$. Define $\beta = \nu \circ \varphi$ and $g = f \circ \varphi$. Show that $g \in \mathcal{R}([A, B], \beta)$ and that $\int_A^B g d\beta = \int_a^b f d\nu$.

6.31   Let $f \in \mathcal{R}([a, b])$. Show that

(1)   the function $F(x) = \int_a^x f(t)\, dt$ is continuous on $[a, b]$.

(2)   if f is continuous at $x \in [a, b]$, then F is differentiable at x, and $F'(x) = f(x)$.

(3)   Suppose that f is continuous and nonnegative and ν is increasing on $[a, b]$. Prove that if $\int_a^b f\, d\nu = 0$, then $f(x) = 0$ for all $x \in [a, b]$.

6.32   Consider $f : [a, b] \to \mathbb{R}$.

(1)   Suppose that f is continuous and nonnegative. Prove that if $\int_a^b f\, dx = 0$, then $f(x) = 0$ for all $x \in [a, b]$.

(2)   Suppose that f is continuous and nonnegative and ν is increasing on $[a, b]$. Prove that if $\int_a^b f\, d\nu = 0$, then $f(x) = 0$ for all $x \in [a, b]$.

6.33   Suppose that $f \in \mathcal{R}([a, b])$ and $\int_a^b f(x) h(x)\, dx = 0$ for all continuous functions h. Show that $f(x) = 0$ for all points of continuity of f.

6.34   Evaluate the following integrals.

(a) $\int_0^1 x\, d\left(x^3\right)$ (d) $\int_1^2 x^3\, d\left(\sqrt{x}\right)$

(b) $\int_{-1}^1 x\, d\left(|x|\right)$ (e) $\int_2^4 x\, d\left(\ln x\right)$

(c) $\int_0^2 x^2\, d\left(\mathrm{int}\,(x)\right)$ (f) $\int_0^\pi \cos\, d\left(\sin x\right)$

6.35   Let $\{q_1, q_2, \ldots\}$ be an enumeration of the rationals in $[0, 1]$. Define $f : [0, 1] \to \mathbb{R}$ by $f(x) = \sum_{q_n \le x} 2^{-n}$.

(1)   Show that f is discontinuous at each rational point and continuous at each irrational point of $[0, 1]$.

(2)   Deduce that $f \in \mathcal{R}([0, 1])$; show that

$$\int_0^x f(t)\, dt = 1 - \sum_{q_n \le x} q_n 2^{-n}.$$

6.36   Show that $\mathrm{Lip}\,([a, b]) \subset \mathcal{AC}\,([a, b])$.

6.37   Show that $f \in \mathcal{AC}\,([a, b]) \cap \mathrm{Lip}\,([a, b])$ if and only if $|f'|$ is bounded.

6.38   Show that $\mathcal{AC}\,([0, 1]) \subsetneq \mathcal{C}\,([0, 1])$ by considering the function $f : [0, 1] \to \mathbb{R}$ defined by $f(0) = 0$ and $f(x) = x^2 \cos\left(x^{-2}\right)$ for $0 < x \le 1$.

6.39   Show that if $f \in \mathcal{AC}\left([a,b]\right)$, then $V_f\left[a,b\right] = \int_a^b |f'\left(t\right)|\, dt$.

6.40   Show that for any given real function f defined on the closed interval $[a,b]$, $V_f\left[a,b\right] = V_f\left[a,c\right] + V_f\left[c,b\right]$ whenever $a < c < b$.

Sequences and Series of Functions

7.1 Sequences of Functions

Definition 7.1

A sequence (f_n) is called a **sequence of functions** if for each $n \in \mathbb{N}$, its n-th term is a real-valued function.

When a sequence of functions is given, it is understood that all the functions f_n are defined on a single given set. The formula $f_n(x) = x^n$ for $x \in \mathbb{R}$ and for $n \in \mathbb{N}$ defines a sequence of functions. Similarly $g_n(x) = x/(1 + nx^2)$ and $h_n(x) = \sin nx$ are examples of sequences of functions both defined on $\mathbb{R}$.

Given (f_n) a sequence of functions defined on a set A, then there are many ways in which we may understand the statement "(f_n) converges to a function f on A".

First, we notice that for every $x \in A$, $(f_n(x))_n$ defines a sequence of real numbers.

Example 7.2

Consider the sequence of functions $f_n(x) = nx^n(1 + x^n)$ for $x \in [0, 1]$. Determine whether or not the sequences $(f_n(1))$, $(f_n(0))$, and $(f_n(1/2))$ converge.

Solution

For each $n \in \mathbb{N}$, $f_n(0) = 0$. Thus the numerical sequence $(f_n(0))$ converges and $\lim f_n(0) = 0$. For each $n \in \mathbb{N}$, $f_n\left(\frac{1}{2}\right) = \frac{n}{1+2^n}$. Thus $\left(f_n\left(\frac{1}{2}\right)\right)$ converges and $\lim f_n\left(\frac{1}{2}\right) = 0$. For each $n \in \mathbb{N}$, $f_n(1) = n/2$. Thus $(f_n(1))$ diverges. $\square$

Hence assigning different values to x, we get different sequences of real numbers which may be convergent or divergent.

Let (f_n) be the sequence defined by $f_n(x) = x^n$ on $[0, 1]$. The figure shows the graphs of some of the f_n.

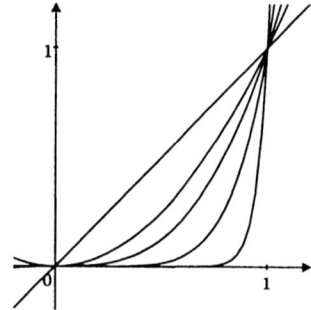

$$f_1(x) = x$$
$$f_2(x) = x^2$$
$$f_3(x) = x^3$$
$$\cdots$$
$$f_6(x) = x^6$$
$$\cdots$$
$$f_{19}(x) = x^{19}$$

The following is easily checked

$$\lim_{n \to \infty} f_n(x) = \begin{cases} 0 & \text{for } 0 \le x < 1 \\ 1 & \text{for } x = 1. \end{cases}$$

Definition 7.3

Let (f_n) be a sequence of real-valued functions on a set A. We say that the sequence (f_n) **converges pointwise** (or **simply**) on A to a function f, and we write $f_n \to f$ pointwise, if $\lim f_n(x) = f(x)$ for every $x \in A$. The function f is then called the **pointwise limit of the sequence** (f_n).

Let us go through a couple of examples.

Example 7.4

Find the pointwise limit of the sequence of functions defined by $f_n(x) = x^n/(1+x^n)$ on the set $[0, +\infty)$.

Solution

We have

- if $x = 0$, then $f_n(0) = 0 \to 0$;

- if $0 < x < 1$, then $0 < f_n(x) = x^n / (1 + x^n) < x^n$, and thus $\lim f_n(x) = 0$;

- if $x = 1$, then $f_n(1) = 1/2 \to 1/2$;

- if $x > 1$, then $f_n(x) = x^n / (1 + x^n) \to 1$.

Therefore we conclude that

$$f_n(x) \to f(x) = \begin{cases} 0 & \text{if } x \in [0,1), \\ 1/2 & \text{if } x = 1, \\ 1 & \text{if } x \in (1,+\infty). \end{cases}$$

□

Example 7.5

For $x \in (-1,1)$, let $f_n(x) = 1 - x + x^2 - x^3 + \cdots + (-1)^n x^n$. Find $f(x) = \lim f_n(x)$ on $(-1,1)$.

Solution

We notice that for each x, $f_n(x)$ is the partial sum of the geometric series $\sum_{n=0}^{\infty} (-x)^n$. Thus if $|x| < 1$, the series converges and

$$\lim f_n(x) = \lim \frac{1 - (-x)^n}{1 + x} = \frac{1}{1 + x}.$$

Hence $f(x) = 1/(1 + x)$ for $x \in (-1,1)$.   □

Using the (ε, δ)-definition of the limit, we have:

Remark 7.6

The sequence (f_n) converges pointwise to f on A if and only if

for each $\varepsilon > 0$ and $x \in A$, there exists N in $\mathbb{N}$ such that $|f_{N+p}(x) - f(x)| < \varepsilon$ for all $p \in \mathbb{N}$.

We observe that the number N depends on the choice of $x \in A$, that is, the same number N may not work for two different values of x. However, certain sequences of functions have the property that for each given $\varepsilon > 0$, one can find a number N which works for all $x \in A$. We say in this situation that the sequence converges uniformly (Figure 7.1). We will see that this type of convergence presents more interesting features than pointwise convergence.

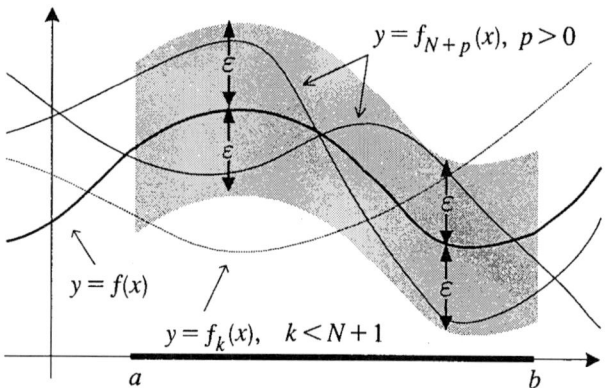

Figure 7.1   Uniform convergence

Definition 7.7

Let (f_n) be a sequence of real-valued functions on a set A. We say that (f_n) **converges uniformly** on A to a function f, and we write $f_n \to f$ uniformly, if

$$\text{for each } \varepsilon > 0, \text{ there exists } N \text{ in } \mathbb{N} \text{ such that}$$
$$|f_{N+p}(x) - f(x)| < \varepsilon \text{ for all } x \in A \text{ and all } p \in \mathbb{N}.$$

To better understand the difference between the two types of convergence, it sometimes helps to write their definitions respectively in the following equivalent forms:

- (f_n) *converges pointwise to f on A if for each $x \in A$, $|f_n(x) - f(x)| \to 0$ as $n \to \infty$;*

- (f_n) *converges uniformly to f on A if $\sup \{|f_n(x) - f(x)| : x \in A\} \to 0$ as $n \to \infty$.*

Example 7.8

Let $f_n(x) = (n \sin x - 1)/n$. Show that (f_n) converges to f uniformly on $\mathbb{R}$.

Solution

Let $\varepsilon > 0$. We notice that for all $x \in \mathbb{R}$

$$|f_n(x) - \sin x| = \left| \frac{(n \sin x - 1)}{n} - \sin x \right| = \frac{1}{n}.$$

Hence, with $f(x) = \sin(x)$ we have $\sup\{|f_n(x) - f(x)| : x \in \mathbb{R}\} = 1/n$. Since $\lim \frac{1}{n} = 0$, we conclude that $f_n \to f$ uniformly on $\mathbb{R}$. $\qquad\qquad$ □

Example 7.9

Let $f_n(x) = x/n$ for $x \in [0, +\infty)$.

(1) Find $f(x) = \lim f_n(x)$.

(2) Show that f_n does not converge uniformly to f on $[0, +\infty)$.

Solution

(1) For each $x \in [0, +\infty)$, we have $|x/n| \to 0$, thus $f_n \to 0$ pointwise on $[0, +\infty)$.

(2) Suppose that $f_n \to 0$ uniformly. Then there would exists an $N \in \mathbb{N}$ such that

$$\left|\frac{x}{n} - 0\right| < 1 \quad \text{for all } x \in [0, +\infty) \text{ whenever } n > N - 1.$$

In particular, we would have $\left|\frac{x}{N}\right| < 1$ for all $x \in [0, +\infty)$. But for example, if $x = 2N$, then $x \in [0, +\infty)$ and

$$1 > \left|\frac{x}{N}\right| = \left|\frac{2N}{N}\right| = 2.$$

A contradiction which proves (2). $\qquad\qquad$ □

We notice from Example 7.4 that the limit of a sequence consisting of continuous functions need not necessarily be continuous. One of the most important features of uniform convergence is the fact that it preserves continuity.

Theorem 7.10

Let (f_n) be a sequence of real-valued continuous functions on a set $A \subset \mathbb{R}$. If $f_n \to f$ uniformly, then f is continuous on A.

Proof

Let $x \in A$ and fix $\varepsilon > 0$. Choose n so that

$$|f_n(a) - f(a)| < \frac{\varepsilon}{3} \quad \text{for every } a \in A.$$

Then since f_n is continuous on A, we can choose $\delta > 0$ so that

$$|f_n(y) - f_n(x)| < \frac{\varepsilon}{3} \text{ for all } y \in A \text{ with } |y - x| < \delta.$$

Then for all $y \in A$ with $|y - x| < \delta$ we have

$$
\begin{aligned}
|f(x) - f(y)| &= |f(x) - f_n(x) + f_n(x) - f_n(y) + f_n(y) - f(y)| \\
&\leq |f(x) - f_n(x)| + |f_n(x) - f_n(y)| + |f_n(y) - f(y)| \\
&< \frac{\varepsilon}{3} + \frac{\varepsilon}{3} + \frac{\varepsilon}{3} = \varepsilon.
\end{aligned}
$$

Thus f is continuous at x, and hence f is continuous on A.   □

This last theorem is mostly used to reveal the lack of uniform convergence for sequences of continuous functions with a discontinuous pointwise limit. For example, the sequence of functions, $f_n(x) = x^n / (1 + x^n)$ (see Example 7.5) is a sequence of continuous functions on $[0, \infty)$. Its pointwise limit f is clearly seen to be discontinuous. Therefore the convergence $f_n \to f$ cannot be uniform.

Definition 7.11

Let (f_n) be a sequence of real-valued functions on a set A. Then (f_n) is said to be **uniformly Cauchy** on A if

for each $\varepsilon > 0$ there exists N in $\mathbb{N}$ such that
$|f_{n+p}(x) - f_n(x)| < \varepsilon$ for all $x \in A$ whenever $n \geq N$ and $p \in \mathbb{N}$.

The figure shows a uniform Cauchy sequence of functions on the closed interval $[a, b]$.

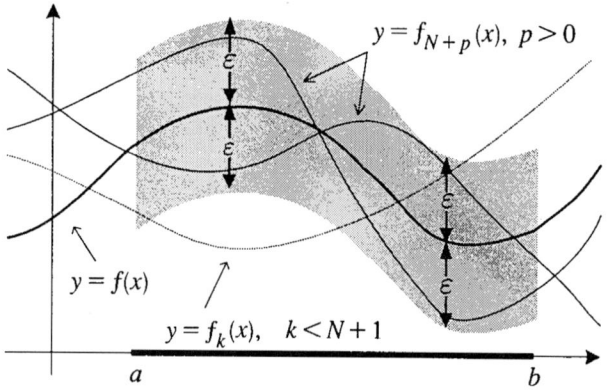

Theorem 7.12

Let (f_n) be a sequence of real-valued functions on a set A. Then the following assertions are equivalent.

(1) (f_n) is uniformly convergent to some function f on A.

(2) (f_n) is uniformly Cauchy on A.

Proof

Suppose that (f_n) is uniformly convergent to some function f on A. Let $\varepsilon > 0$. There exists $N \in \mathbb{N}$ such that

$$|f_n(a) - f(a)| < \frac{\varepsilon}{2} \text{ for all } a \in A \text{ whenever } n > N.$$

Therefore

$$\begin{aligned}
|f_{n+p}(a) - f_n(a)| &= |f_{n+p}(a) - f(a) + f(a) - f_n(a)| \\
&\leq |f_{n+p}(a) - f(a)| + |f(a) - f_n(a)| \\
&< \frac{\varepsilon}{2} + \frac{\varepsilon}{2} = \varepsilon,
\end{aligned}$$

for every $a \in A$ whenever $n > N$ and $p \in \mathbb{N}$. Hence (f_n) is uniformly Cauchy.

Conversely, suppose that (f_n) is uniformly Cauchy on A. For each $\varepsilon > 0$, there exists $N \in \mathbb{N}$ such that

$$|f_{n+p}(a) - f_n(a)| < \varepsilon \text{ for all } a \in A \text{ whenever } n > N \text{ and } p \in \mathbb{N}.$$

In particular, for each $a \in A$ we have

$$|f_{n+p}(a) - f_n(a)| < \varepsilon \text{ whenever } n > N \text{ and } p \in \mathbb{N}.$$

This means that for each $a \in A$, the real-valued sequence $(f_n(a))$ is Cauchy. Hence for each $a \in A$, $\lim f_n(a)$ exists. Define a function f by

$$f(x) = \lim f_n(x) \text{ for each } x \in A.$$

We wish to show that $f_n \to f$ uniformly. Let $\varepsilon > 0$. Choose $N \in \mathbb{N}$ such that

$$|f_{n+p}(x) - f_n(x)| < \frac{\varepsilon}{2} \text{ for all } x \in A \text{ whenever } n > N \text{ and } p \in \mathbb{N}.$$

Then for $n > N$ and for every $x \in A$, we have

$$\begin{aligned}
|f_n(x) - f(x)| &= |f_n(x) - f_{n+p}(x) + f_{n+p}(x) - f(x)| \\
&\leq |f_n(x) - f_{n+p}(x)| + |f_{n+p}(x) - f(x)| \\
&< \frac{\varepsilon}{2} + |f_{n+p}(x) - f(x)|,
\end{aligned}$$

for any $p \in \mathbb{N}$. Since $\lim f_m(x) = f(x)$, we can choose p large enough so that $|f_{n+p}(x) - f(x)| < \frac{\varepsilon}{2}$. Thus

$$|f_n(x) - f(x)| < \varepsilon \text{ for every } x \in A \text{ whenever } n > N,$$

i.e. $f_n \to f$ uniformly. $\qquad\qquad\qquad\qquad\qquad\qquad\qquad\qquad\qquad\quad \square$

The next result is another useful feature of uniform convergence.

Theorem 7.13

Let (f_n) be a sequence of continuous functions on a closed and bounded interval $[a, b]$. Suppose that $f_n \to f$ uniformly on $[a, b]$. Then f is integrable and

$$\lim_{n \to \infty} \int_a^b f_n(x)\, dx = \int_a^b f(x)\, dx.$$

Proof

By Theorem 7.10 the uniform limit function f is continuous on $[a, b]$. Thus for each n, $f_n - f$ is continuous on $[a, b]$, and therefore integrable on $[a, b]$. Let $\varepsilon > 0$. There exists $N \in \mathbb{N}$ such that

$$|f_n(x) - f(x)| < \frac{\varepsilon}{b - a} \text{ for all } x \in [a, b] \text{ whenever } n > N.$$

Then for $n > N$,

$$\left| \int_a^b f_n(x)\, dx - \int_a^b f(x)\, dx \right| = \left| \int_a^b [f_n(x) - f(x)]\, dx \right|$$

$$\leq \int_a^b |f_n(x) - f(x)|\, dx$$

$$\leq \int_a^b \frac{\varepsilon}{b - a}\, dx = \varepsilon.$$

Hence $\lim_{n \to \infty} \int_a^b f_n(x)\, dx = \int_a^b f(x)\, dx$. $\qquad\square$

Example 7.14

Let $f_n(x) = 1 + x + x^2 + \cdots + x^n$ for $x \in (-1, 1)$.

(1) Find $f(x) = \lim f_n(x)$ for $x \in (-1, 1)$.

(2) Find $\lim_{n \to \infty} \int_0^{1/2} f_n(x)\, dx$.

Solution

(1) For each $x \in (-1, 1)$, we have $f_n(x) = \frac{1 - x^{n+1}}{1 - x} \to f(x) = \frac{1}{1 - x}$.

(2) For $x \in \left[0, \frac{1}{2}\right]$, we have

$$|f_n(x) - f(x)| = \left| \frac{1 - x^{n+1}}{1 - x} - \frac{1}{1 - x} \right|$$

$$= \frac{x^{n+1}}{1 - x} \leq \frac{1}{2^n}.$$

Since $\frac{1}{2^n} \to 0$, independently of $x \in \left[0, \frac{1}{2}\right]$, the above inequality implies that $f_n \to f$ uniformly on $\left[0, \frac{1}{2}\right]$. Hence

$$\lim_{n \to \infty} \int_0^{\frac{1}{2}} f_n(x)\, dx = \int_0^{\frac{1}{2}} \frac{1}{1 - x}\, dx = \ln 2.$$

$\square$

Example 7.15

Let (f_n) be a sequence of functions on an open bounded interval (a, b). Assume that

(1) each f_n is differentiable on (a, b) and f_n' is continuous;

(2) (f_n') converges uniformly on each closed subinterval of (a, b);

(3) for some x_0, the sequence $(f_n(x_0))$ converges.

Show that (f_n) converges pointwise on (a, b) to a function f and that $f'(x) = \lim f_n'(x)$ for all $x \in (a, b)$.

Solution

Suppose that $f_n' \to g$ and that $\lim f_n(x_0) = b$. We first notice that by Theorem 7.10 and condition (2), g is continuous at each point $x \in (a, b)$. Fix $x \in (a, b)$. Then Theorem 7.13 applies on the closed interval with endpoints x and x_0. It follows that

$$\int_{x_0}^x g(t)\, dt = \lim \int_{x_0}^x f_n'(t)\, dt$$

$$= \lim (f_n(x) - f_n(x_0))$$

$$= \lim f_n(x) - b.$$

This shows that $f(x) = \lim f_n(x) = b + \int_{x_0}^x g(t)\, dt$ for all $x \in (a, b)$. Moreover, the fundamental theorem of calculus implies that $\lim f_n'(x) = g(x) = f'(x)$ at every $x \in (a, b)$. $\square$

7.2 Series of Functions

Definition 7.16

Let (g_n) be a sequence of real-valued functions on a set A. An **(infinite) series** of functions is an expression of the form $\sum g_n$.

As in the case of series of numbers, we define for each $n \in \mathbb{N}$, the n-th partial sum of the series $\sum g_n$ to be the function $f_n = \sum_{k=1}^{n} g_k$. For each $x \in A$, $f_n(x)$ is the n-th partial sum of the numerical series $\sum g_n(x)$, which may or may not converge.

Definition 7.17

The set D of all those $x \in A$ for which $f_n(x) = \sum_{k=1}^{n} g_k(x)$ converges is called the **domain of convergence** of the series $\sum g_n$; and we say that the series $\sum g_n$ converges pointwise on D.

The sequence of functions studied in Example 7.5 is exactly the sequence of partial sums of the series of functions $\sum g_n$ where $g_n(x) = x^n$ for each n. We saw that $f_n \to f$ where $f(x) = \frac{1}{1-x}$ pointwise on $(-1, 1)$. It is an easy exercise to show that $(f_n(x))$ diverges if $|x| \geq 1$. Thus the domain of convergence of the series $\sum g_n$ is the interval $(-1, 1)$. We then simply write

$$\sum_{n=0}^{\infty} x^n = \frac{1}{1-x} \quad \text{for } x \in (-1, 1).$$

Definition 7.18

A series of functions $\sum g_n$ is said to **converge uniformly** on a set A if the sequence of its partial sums $(f_n = \sum_{k=1}^{n} g_k)$ converges uniformly on A.

Example 7.19

Show that the series $\sum_{n=1}^{\infty} x^n$ converges uniformly on any interval $[-a, a]$ for every $0 < a < 1$.

Solution

We already know that the series converges pointwise to $f(x) = 1/(1-x)$ on

$(-1, 1)$. We notice that for each x, the partial sum of the series $\sum_{n=1}^{\infty} x^n$ is

$$f_n(x) = 1 + x + x^2 + \cdots + x^n = \frac{1 - x^{n+1}}{1 - x}.$$

Now for $0 < a < 1$, and for $|x| < a$, we have

$$|f_n(x) - f(x)| = \frac{|x|^{n+1}}{|1 - x|} \leq \frac{a^{n+1}}{1 - a}. \tag{7.1}$$

Since $0 < a < 1$, $a^{n+1} \to 0$ independently of the choice of $x \in [-a, a]$, we conclude from inequality (7.1) that $f_n \to f$ uniformly on $[-a, a]$, i.e. the series $\sum_{n=1}^{\infty} x^n$ converges uniformly to $f(x) = \frac{1}{1-x}$ on $[-a, a]$. $\square$

Example 7.20

Let (g_n) be a sequence of real-valued continuous functions on a set A. Suppose that $\sum g_n$ converges uniformly on A. Show that the series $\sum g_n$ represents a continuous function on A.

Solution

The sequence $(f_n = \sum_{k=1}^{n} g_k)$ of partial sums of the series $\sum g_n$ is a sequence of continuous functions. Since it converges uniformly on A, then the limit $\sum g_n$ is a continuous function on A by Theorem 7.10. $\square$

Note that if $(f_n = \sum_{k=1}^{n} g_k)$ is the sequence of partial sums of a series $\sum g_n$, then we have for every n and $p \in \mathbb{N}$

$$f_{n+p}(x) - f_n(x) = \sum_{k=n+1}^{n+p} g_k.$$

Hence in view of Theorem 7.12, we have the following

Theorem 7.21

Let (g_n) be a sequence of real-valued continuous functions on a set A. Then the series $\sum g_n$ converges uniformly on A if and only if

for each $\varepsilon > 0$, there exists N in $\mathbb{N}$ such that
$$\left| \sum_{k=n+1}^{n+p} g_k(x) \right| < \varepsilon \text{ for all } x \in A \text{ whenever } n > N \text{ and } p \in \mathbb{N}. \tag{7.2}$$

A series $\sum g_n$ is said to satisfy the **Cauchy criterion** if it satisfies (7.2).

The next result, known as the *Weierstrass M-test*, is a very useful criterion for uniform convergence.

Theorem 7.22 (Weierstrass M-test)

Let (g_n) be a sequence of real-valued functions on an interval I. Suppose that there exists a sequence of positive real numbers (M_n) such that

(1) $|g_n(x)| \le M_n$ for each n and for every $x \in I$, and

(2) $\sum M_n < +\infty$.

Then the series $\sum g_n$ converges uniformly on I.

Under the conditions of Theorem 7.22, we say that the series of functions $\sum g_n$ is **dominated** by the numerical series $\sum M_n$.

Proof

It suffices to show that $\sum g_n$ satisfies the Cauchy criterion (7.2). Let $\varepsilon > 0$, there exists $N \in \mathbb{N}$ such that

$$\sum_{k=n+1}^{n+p} M_k < \varepsilon \text{ whenever } n > N \text{ and } p \in \mathbb{N}.$$

Hence if $n > N$, $p \in \mathbb{N}$, and $x \in I$, we have

$$\left| \sum_{k=n+1}^{n+p} g_k(x) \right| \le \sum_{k=n+1}^{n+p} |g_k(x)| \le \sum_{k=n+1}^{n+p} M_k < \varepsilon.$$

Hence $\sum g_n$ converges uniformly on I. $\square$

Example 7.23

Show that the series $\sum_{n=1}^{\infty} \frac{\text{fra}(nx)}{n^2}$ converges uniformly on $\mathbb{R}$.

Solution

For each $n \in \mathbb{N}$, and for all $x \in \mathbb{R}$, we have $\left| \frac{\text{fra}(nx)}{n^2} \right| \le \frac{1}{n^2}$. Since the series $\sum \frac{1}{n^2}$ is known to converge, the Weierstrass M-test applies and implies that the series $\sum_{n=1}^{\infty} \frac{\text{fra}(nx)}{n^2}$ converges uniformly on $\mathbb{R}$. $\square$

The proof of our next result is reminiscent of the proof of Theorem 3.38 on page 86.

Theorem 7.24 (Dirichlet's Test)

Let (g_n) be a sequence of real-valued functions on an interval I. Suppose that there exists $M > 0$ such that for every $x \in I$,

$$\left| \sum_{k=1}^{n} g_k(x) \right| \leq M \text{ for each } n.$$

Then the series $\sum a_n g_n$ converges uniformly on I whenever (a_n) is a nonincreasing sequence of nonnegative real numbers converging to zero.

Proof

It suffices to show that the series satisfies the Cauchy criterion. Fix $\varepsilon > 0$. Since $\lim a_n = 0$, there exists $N_1 \in \mathbb{N}$ such that

$$|a_n| < \varepsilon / (3M) \text{ for } n > N_1.$$

We also notice that for each n we have

$$0 \leq \sum_{k=1}^{n} (a_k - a_{k+1}) = (a_1 - a_2) + (a_3 - a_2) + \cdots + (a_n - a_{n+1}) < a_1,$$

thus the series $\sum_{k=1}^{\infty} (a_k - a_{k+1})$ converges. Hence there exists N_2 in $\mathbb{N}$ such that

$$\left| \sum_{k=n}^{n+p} (a_k - a_{k+1}) \right| < \varepsilon / (3M) \text{ for } n > N_2 \text{ and for all } p \in \mathbb{N}.$$

Next for each n, consider $s_n = \sum_{k=1}^{n} g_k$ so that for each k, $g_k = s_k - s_{k-1}$. By our hypothesis, $|s_n| < M$ for every n. Now

$$\sum_{k=n}^{n+p} a_k g_k = \sum_{k=n}^{n+p} a_k (s_k - s_{k-1}) = \sum_{k=n}^{n+p} a_k s_k - \sum_{k=n}^{n+p} a_k s_{k-1}$$

$$= \sum_{k=n}^{n+p} a_k s_k - \sum_{k=n-1}^{n+p-1} a_{k+1} s_k$$

$$= \sum_{k=n}^{n+p-1} (a_k - a_{k+1}) s_k + a_{n+p} s_{n+p} - a_n s_{n-1}.$$

Hence for $n > \max\{N_1, N_2\}$ and for all $p \in \mathbb{N}$,

$$\left| \sum_{k=n}^{n+p} a_k g_k \right| \le \sum_{k=n}^{n+p-1} |a_k - a_{k+1}| \, |s_k| + |a_{n+p}| \, |s_{n+p}| + |a_n| \, |s_{n-1}|$$

$$\le M \sum_{k=n}^{n+p-1} |a_k - a_{k+1}| + |a_{n+p}| + |a_n|$$

$$\le M \left(\frac{\varepsilon}{3M} + \frac{\varepsilon}{3M} + \frac{\varepsilon}{3M} \right) = \varepsilon,$$

as desired.   □

Example 7.25

Prove that $\sum_{n=1}^{\infty} \frac{\sin nx}{n}$ is uniformly convergent on any interval not containing $0, \pm\pi, \pm2\pi, \dots$.

Solution

First, we notice that for $x = \pm k\pi$, $k \in \mathbb{N}_0$, then the series is 0. It can also be proved by induction (do it!) that

$$\sin x + \sin 2x + \sin 3x + \cdots + \sin nx = \frac{1}{2} \frac{\cos \frac{1}{2}x - \cos\left(n + \frac{1}{2}\right)x}{\sin \frac{1}{2}x}$$

for $x \ne 0, \pm2\pi, \pm4\pi, \dots$. It follows that if $[a, b]$ is any interval not containing $0, \pm\pi, \pm2\pi, \dots$, and if $x \in [a, b]$, then

$$\left| \sum_{k=1}^{n} \sin kx \right| \le 1/\min\{\sin(x/2) : x \in [a, b]\}.$$

Since the sequence $(1/n)$ is decreasing to 0, the Dirichlet's test applies and shows that $\sum_{n=1}^{\infty} \frac{\sin nx}{n}$ is uniformly convergent on $[a, b]$.   □

The next result is a direct analogue of Theorem 7.13 for series and asserts that it is possible to find the integral of a uniform convergent series by *integrating* the series *term-by-term*.

Theorem 7.26

Let (g_n) be a sequence of real-valued continuous functions on $[a, b]$. Suppose that $\sum_{n=1}^{\infty} g_n$ converges uniformly on $[a, b]$, and let f be its sum. Then

$$\int_{\alpha}^{x} f(t)\, dt = \sum_{n=1}^{\infty} \int_{\alpha}^{x} g_n(t)\, dt \quad \text{for } x, \alpha \in [a, b].$$

Proof

Let $f_n = \sum_{k=1}^{n} g_k$ be the n-th partial sum of the series $\sum_{n=1}^{\infty} g_n$ for each $n \in \mathbb{N}$. By our hypothesis, (f_n) converges uniformly to f on $[a, b]$. Thus by Example 7.20, f is continuous on $[a, b]$ and therefore integrable. Now, let $x, c \in [a, b]$ and assume that $\alpha < x$; the case $x < \alpha$ is similar. Then we have (f_n) converges uniformly to f on $[\alpha, x]$. Applying Theorem 7.13 to the sequence (f_n) on the interval $[\alpha, x]$, we have

$$\int_{\alpha}^{x} f(t)\, dt = \lim \int_{\alpha}^{x} f_n(t)\, dt = \lim \sum_{k=1}^{n} \int_{\alpha}^{x} g_k(t)\, dt = \sum_{n=1}^{\infty} \int_{\alpha}^{x} g_n(t)\, dt,$$

as desired. □

Example 7.27

Consider $s(x) = \sum_{n=1}^{\infty} (-1)^{n+1} \frac{x^{2n-1}}{(2n-1)!}$, $x \in (-\infty, \infty)$. Evaluate $\int_{0}^{x} s(t)\, dt$, where $x \in (-\infty, \infty)$.

Solution

Let $x \in (-\infty, \infty)$. Choose $|x| < r < \infty$, we have for each n

$$\left| (-1)^{n+1} \frac{x^{2n-1}}{(2n-1)!} \right| \leq \frac{r^{2n-1}}{(2n-1)!}.$$

It is quickly seen that the series $\sum \frac{r^{2n-1}}{(2n-1)!}$ converges. (Why?) Hence by the Weierstrass M-test, the series $\sum_{n=1}^{\infty} (-1)^{n+1} \frac{x^{2n-1}}{(2n-1)!}$ converges uniformly on $[-r, r]$. Thus the use of integration term-by-term is legitimate on $[-r, r]$ and we get

$$\int_{0}^{x} s(t)\, dt = \sum_{n=1}^{\infty} \int_{0}^{x} (-1)^{n+1} \frac{t^{2n-1}}{(2n-1)!}\, dt = \sum_{n=1}^{\infty} (-1)^{n+1} \frac{x^{2n}}{2n!}.$$

□

Example 7.28 (Euler's Constant)

Consider the series $\sum_{n=1}^{\infty} \frac{x}{n(x+n)}$.

(1) Show that it converges uniformly on $[0, 1]$.

(2) Deduce that $\lim_{n \to \infty} \sum_{k=1}^{n} \frac{1}{k} - \ln n = \gamma$ exists. (Compare with Exercise 2.35, page 63.)

Solution

(1) It is clear that for $x \in [0,1]$ we have

$$0 \le \frac{x}{n(x+n)} \le \frac{1}{n^2} \text{ for every } n \in \mathbb{N}.$$

Since the series $\sum \frac{1}{n^2}$ converges, by the Weierstrass M-test, $\sum \frac{x}{n(x+n)}$ converges uniformly on $[0,1]$.

(2) From the first part, we see that integration term-by-term applies and so

$$\sum_{n=1}^{\infty} \int_0^1 \frac{x}{n(x+n)} dx = \int_0^1 \sum_{n=1}^{\infty} \frac{x}{n(x+n)} dx = \gamma$$

exists. Since for every n

$$\int_0^1 \frac{x}{n(x+n)} dx = \int_0^1 \left(\frac{1}{n} - \frac{1}{n+t} \right) dx = \frac{1}{n} - \ln \frac{n+1}{n},$$

we have

$$\sum_{n=1}^{N} \int_0^1 \frac{x}{n(x+n)} dx = \sum_{n=1}^{N} \frac{1}{n} - \ln (N+1),$$

and thus

$$\lim_{N \to \infty} \left[\sum_{n=1}^{N} \frac{1}{n} - \ln (N+1) \right] = \gamma.$$

Finally, since $\ln (N+1) - \ln N \to 0$ as $N \to \infty$, it follows that

$$\lim_{N \to \infty} \left[\sum_{n=1}^{N} \frac{1}{n} - \ln N \right] = \lim_{N \to \infty} \left[\sum_{n=1}^{N} \frac{1}{n} - \ln (N+1) \right] = \gamma,$$

as desired. $\square$

The next result gives conditions under which it is also possible to *differentiate term-by-term*.

Theorem 7.29

Let (g_n) be a sequence of real-valued differentiable functions on $[a,b]$. Suppose that

(1) $\sum_{n=1}^{\infty} g_n$ converges uniformly on $[a,b]$, and let F be its sum;

(2) $\sum_{n=1}^{\infty} g_n'$ converges uniformly on $[a,b]$, and let f be its sum.

Then $F'(x) = f(x)$ for all $x \in [a, b]$, i.e.

$$\frac{d}{dx}\left(\sum_{n=1}^{\infty} g_n(x)\right) = \sum_{n=1}^{\infty} \frac{d}{dx} g_n(x).$$

Proof

Since the series $\sum_{n=1}^{\infty} g_n'$ converges uniformly on $[a, b]$ to f, it follows by Theorem 7.26, that for $\alpha, x \in [a, b]$

$$\int_{\alpha}^{x} f(t)\,dt = \sum_{n=1}^{\infty} \int_{\alpha}^{x} g_n'(t)\,dt = \sum_{n=1}^{\infty} [g_n(x) - g_n(\alpha)].$$

On the other hand, since $\sum_{n=1}^{\infty} g_n$ converges uniformly on $[a, b]$ to F, we have

$$\sum_{n=1}^{\infty} [g_n(x) - g_n(\alpha)] = F(x) - F(\alpha) \quad \text{for} \quad \alpha, x \in [a, b].$$

Hence

$$\int_{\alpha}^{x} f(t)\,dt = F(x) - F(\alpha) \quad \text{for} \quad \alpha, x \in [a, b].$$

Differentiating both sides of this equation with respect to x, we obtain

$$F'(x) = f(x) \quad \text{for} \quad x \in [a, b].$$

The proof is complete.   □

Example 7.30

Expand the function $f(x) = \frac{1}{(1-x)^2}$, $|x| < 1$.

Solution

We recognize that f is the derivative of

$$F(x) = \frac{-1}{1-x} = -\sum_{n=0}^{\infty} x^n \quad \text{for} \quad |x| < 1.$$

Now for each $x \in (-1, 1)$, choose r such that $|x| < r < 1$. Since $|x^n| < r^n$, for each n, the series $\sum_{n=0}^{\infty} x^n$ converges uniformly on $|x| < r$ by the Weierstrass M-test. Therefore differentiation term-by-term applies and implies

$$F'(x) = \sum_{n=1}^{\infty} x^{n-1} \quad \text{for} \quad |x| < 1.$$

Hence

$$\frac{1}{(1-x)^2} = -\sum_{n=1}^{\infty} x^{n-1} \text{ for } |x| < 1.$$

$\square$

One should note that the requirement that the series of derivatives converges uniformly is essential. It could be the case that differentiation term-by-term is not possible if this condition on the series of derivatives is not fulfilled. For instance, the series $\sum_{n=1}^{\infty} \frac{\sin 2^n x}{n^2}$ is dominated by the convergent p-series $\sum_{n=1}^{\infty} \frac{1}{n^2}$. Therefore by the Weierstrass M-test, the series $\sum_{n=1}^{\infty} \frac{\sin 2^n x}{n^2}$ converges uniformly on $\mathbb{R}$. However, the series of the derivatives $\sum_{n=1}^{\infty} \frac{2^n \cos 2^n x}{n^2}$ diverges, for example for $x = 0$ or π.

7.3 Power Series

Series of functions which present particularly interesting features are the series of the form $\sum g_n$ where the n-th term is given by $g_n(x) = a_n(x-c)^n$ for each n.

Definition 7.31

A **power series** is a series of functions of the form

$$\sum_{k=0}^{\infty} a_n(x-c)^n$$

where $c, a_0, a_1, a_2, \ldots, a_n, \ldots$ are real numbers.

Example 7.32

Find the domain of convergence of the power series $\sum_{n=1}^{\infty} (nx)^{2n}$.

Solution

The series converges for $x = 0$ and diverges for all $x \neq 0$ because $(nx)^{2n} \to \infty$ for $x \neq 0$.

$\square$

Example 7.33

Find the domain of convergence of the series $\sum_{n=1}^{\infty} \frac{(x-2)^n}{n}$.

Solution

For each x, the n-th term of the given series is $u_n = (1/n)\,(x-2)^n$. Thus

$$\limsup |u_n|^{1/n} = \limsup |(1/n)\,(x-2)^n|^{1/n} = |x-2|.$$

By the root test, the series converges if $|x-2| < 1$ and diverges if $|x-2| > 1$; i.e. the series converges on $(1,3)$. For $x = 1$, the series is the convergent alternating series $\sum_{n=1}^{\infty} (-1)^n /n$. For $x = 3$, the series is the divergent harmonic series $\sum_{n=1}^{\infty} (1/n)$. We conclude that the domain of convergence of the given power series is the interval $[1,3)$.   □

Example 7.34

Find the domain of convergence of the series $\sum_{n=1}^{\infty} \left(\frac{2x}{n}\right)^{2n}$.

Solution

For each x, the n-th term of the given series is $u_n = \left(\frac{2x}{n}\right)^{2n}$. Thus

$$\limsup |u_n|^{1/n} = \limsup \left| \left(\frac{2x}{n}\right)^{2n} \right|^{1/n} = \lim \frac{|4x^2|}{n} = 0 < 1.$$

By the root test, the series converges for all values of x.   □

Theorem 7.35 (Abel)

Let $\sum_{n=0}^{\infty} a_n x^n$ be a power series.

(1) If $\sum_{n=0}^{\infty} a_n x^n$ converges for some nonzero x_0, then it converges absolutely for any x in the open interval $(-|x_0|,|x_0|)$.

(2) If $\sum_{n=0}^{\infty} a_n x^n$ diverges for some nonzero x_0', then it diverges for any x outside the closed interval $[-|x_0'|,|x_0'|]$.

Proof

(1) For all $x \in (-|x_0|,|x_0|)$, we have $|a_n x^n| \le |a_n x_0{}^n| \left| \frac{x}{x_0} \right|^n$ for each n. Since the series $\sum_{n=0}^{\infty} a_n x_0^n$ converges, $\lim a_n x_0^n = 0$ by Proposition 3.7, and so $(a_n x_0^n)$ is bounded, say $|a_n x_0^n| \le M$ for all $n \in \mathbb{N}$. It follows that

$$|a_n x^n| \le |a_n x_0^n| \left| \frac{x}{x_0} \right|^n \le M \left| \frac{x}{x_0} \right|^n.$$

(2) Since $|x| < |x_0|$, the series $\sum \left| \frac{x}{x_0} \right|^n$ is a convergent geometric series. By the comparison test, the series $\sum |a_n x^n|$ converges and hence $\sum a_n x^n$ converges absolutely.

Suppose that $\sum a_n x^n$ converges for some x with $|x| > |x_0'|$. Then by the first part of the theorem, it should converge at x_0'. This contradicts the condition that the series diverges at x_0'.

The proof is complete. $\qquad\square$

It is clear that the power series $\sum_{k=0}^{\infty} a_n (x - c)^n$ always converges for $x = c$. It turns out that the domain of convergence of a power series is always either an interval of one of the following forms

$$(c - R, c + R), \quad (c - R, c + R], \quad [c - R, c + R), \text{ or } [c - R, c + R]$$

or the singleton $\{c\}$. The number R is called the **radius of convergence** of the power series.

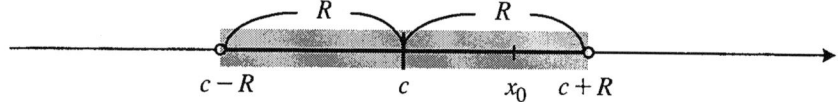

The next theorem gives a method for determining the radius of convergence of a given power series.

Theorem 7.36 (Hadamard)

Let $\sum_{n=0}^{\infty} a_n (x - c)^n$ be a power series. Define

$$R = \begin{cases} \frac{1}{\limsup |a_n|^{1/n}} & \text{if } 0 < \limsup |a_n|^{1/n} < \infty \\ 0 & \text{if } \limsup |a_n|^{1/n} = \infty \\ \infty & \text{if } \limsup |a_n|^{1/n} = 0. \end{cases}$$

Then the power series converges for every $x \in \mathbb{R}$ such that $|x - c| < R$ and diverges for every $x \in \mathbb{R}$ such that $|x - c| > R$. Thus R is the radius of convergence of the power series.

Notice that the condition $|x - c| < \infty$ is satisfied for every $x \in \mathbb{R}$, while $|x - c| > 0$ is satisfied for every $x \neq c$.

Proof

For each $x \in \mathbb{R}$, the n-th term of the series is $u_n = a_n (x - c)^n$. Thus we have $\limsup |u_n|^{1/n} = |x - c| \limsup |a_n|^{1/n}$.

Case 1. Suppose that $0 < \limsup |u_n|^{1/n} < 1$. Then

$$|x - c| < \frac{1}{\limsup |a_n|^{1/n}} = R.$$

Hence by the root test, $\sum_{n=0}^{\infty} a_n (x - c)^n$ converges for $|x - c| < R$ and diverges for $|x - c| > R$.

Case 2. Suppose that $R = \infty$. Then $\limsup |u_n|^{1/n} = 0$, so $\sum_{n=0}^{\infty} a_n (x - c)^n$ converges for every $x \in \mathbb{R}$ by the root test.

Case 3. Suppose that $R = 0$. Then $\limsup |u_n|^{1/n} = \infty$, so $\sum_{n=0}^{\infty} a_n (x - c)^n$ converges only for every $x = c$ by the root test. $\square$

Similarly, we can make use of the ratio test in order to determine the radius of convergence of the power series $\sum_{n=0}^{\infty} a_n (x - c)^n$. Let us assume that the following limit exists

$$\lim \left| \frac{u_{n+1}}{u_n} \right| = \lim \left| \frac{a_{n+1} (x - c)^{n+1}}{a_n (x - c)^n} \right| = \lim \left| \frac{a_{n+1}}{a_n} \right| |x - c|.$$

Then, by the ratio test the series $\sum_{n=0}^{\infty} a_n (x - c)^n$ converges if

$$\lim \left| \frac{a_{n+1}}{a_n} \right| |x - c| < 1,$$

that is, if

$$|x - c| < \frac{1}{\lim \left| \frac{a_{n+1}}{a_n} \right|} = \lim \left| \frac{a_n}{a_{n+1}} \right|.$$

Example 7.37

Find the domain of convergence of the series $\sum_{n=1}^{\infty} \frac{(x+2)^n}{n}$.

Solution

Applying the ratio test, we get

$$\lim \left| \frac{(x + 2)^{n+1} / (n + 1)}{(x + 2)^n / n} \right| = \lim \frac{n}{n + 1} |x + 2| = |x + 2|.$$

Thus the series converges if $|x + 2| < 1$, i.e. $-3 < x < -1$, and diverges if $|x + 2| > 1$. For $x = -3$, the series is the convergent alternating series $\sum_{n=11}^{\infty} \frac{(-1)^n}{n}$, and for $x = -1$, the series is the divergent series $\sum_{n=1}^{\infty} \frac{1}{n}$. Thus the series $\sum_{n=1}^{\infty} \frac{(x+2)^n}{n}$ converges on the interval $[-3, -1)$. $\square$

Example 7.38

Find the domain of convergence of the series $\sum_{n=1}^{\infty} \frac{x^n}{n!}$.

Solution

Applying the ratio test, we get

$$\lim \left| \frac{x^{n+1} / (n+1)!}{x^n / n!} \right| = \lim \frac{1}{n+1} |x| = 0 < 1.$$

By the ratio test, the series converges for all values of x. $\qquad\square$

Example 7.39

Find the domain of convergence of the power series $\sum_{n=1}^{\infty} \frac{(2x)^n}{n}$.

Solution

For each x, the n-th term of the given series is $u_n = \frac{(2x)^n}{n}$. Thus

$$\limsup \left| \frac{(2x)^{n+1} / (n+1)}{(2x)^n / n} \right| = \lim \left| \frac{n}{n+1} \right| |2x|.$$

By the ratio test, the series converges if $|2x| < 1$ and diverges if $|2x| > 1$; i.e. the series converges on $(-1/2, 1/2)$. For $x = -1/2$, the series is the convergent alternating series $\sum_{n=1}^{\infty} (-1)^n / n$. For $x = 1/2$, the series is the divergent harmonic series $\sum_{n=1}^{\infty} (1/n)$. We conclude that the domain of convergence of the given power series $\sum_{n=1}^{\infty} (2x)^n / n$ is the interval $[-1/2, 1/2)$. $\qquad\square$

A power series *may* or *may not* converge uniformly on its interval of convergence. However, a power series always converges uniformly on any closed interval strictly contained in its interval of convergence.

Theorem 7.40 (Abel)

Let $\sum_{n=0}^{\infty} a_n (x - c)^n$ be a power series with radius of convergence R. Then the power series converges uniformly on $[c - S, c + S]$ for any $0 < S < R$.

Proof

Consider $0 < S < R$. Since $c + S$ is in the interval $(c - R, c + R)$, the numerical series

$$\sum |a_n| S^n = \sum |a_n| \left[(c+S) - c\right]^n$$

converges. For all $x \in [c - S, c + S]$ and for each n we have

$$|a_n (x - c)^n| \le |a_n| S^n.$$

The Weierstrass M-test, Theorem 7.22, applies and shows that $\sum_{k=0}^{\infty} a_n (x - c)^n$ converges uniformly on $[c - S, c + S]$.   □

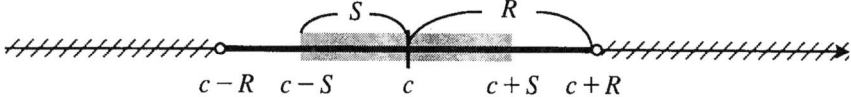

It follows from the above theorem that the sum of a power series is a continuous function on any interval lying entirely within its interval of convergence.

The following result highlights an important feature of power series.

Theorem 7.41

Let $\sum_{k=0}^{\infty} a_n (x - c)^n$ be a power series with radius of convergence R. If

$$f(x) = \sum_{n=0}^{\infty} a_n (x - c)^n \text{ for } x \in (c - R, c + R),$$

then f is differentiable and

$$f'(x) = \sum_{n=1}^{\infty} n a_n (x - c)^{n-1} \text{ for } x \in (c - R, c + R).$$

Proof

Consider $0 < S < R$. By Theorem 7.40, both

$$\sum_{n=0}^{\infty} a_n (x - c)^n$$

and

$$\sum_{n=1}^{\infty} n a_n (x - c)^{n-1}$$

converge uniformly on the closed interval $[c - S, c + S]$. Termwise differentiation is legitimate and yields the desired result.   □

Example 7.42

Suppose that both $\sum_{n=1}^{\infty} a_n (x - c)^n$ and $\sum_{n=1}^{\infty} b_n (x - c)^{n-1}$ converge on $(c - S, c + S)$, $S > 0$, to the same function f. Show that $a_n = b_n$ for all $n \in \mathbb{N}$.

Solution

It follows from Theorem 7.41 that the power series $\sum_{n=1}^{\infty} a_n (x - c)^n$ and $\sum_{n=1}^{\infty} b_n (x - c)^n$ are k times differentiable for any $k \in \mathbb{N}$ and that

$$f^{(k)}(x) = \sum_{n=k}^{\infty} \frac{n!}{(n-k)!} a_n (x - c)^{n-k} = \sum_{n=k}^{\infty} \frac{n!}{(n-k)!} b_n (x - c)^{n-k}.$$

In particular, we have $f^{(n)}(c) = n! a_n = n! b_n$. Hence $a_n = b_n$ for all n. $\square$

7.4 Taylor Series

Example 7.43

Let f be a function defined on some open interval containing a point x_0. Suppose that f is smooth enough in the sense that it possesses derivatives up to the $(n + 1)$-th order at x_0. Find a polynomial $P_n(x)$ of degree at most n such that $P_n(x_0) = f(x_0)$, $P_n'(x_0) = f'(x_0)$, $P_n''(x_0) = f''(x_0)$, ..., and $P_n^{(n)}(x_0) = f^{(n)}(x_0)$.

Solution

Let us seek $P_n(x)$ in the form

$$P_n(x) = a_0 + a_1 (x - x_0) + a_2 (x - x_0)^2 + \cdots + a_n (x - x_0)^n. \qquad (7.3)$$

Then

$$P_n'(x) = a_1 + 2a_2 (x - x_0) + 3a_3 (x - x_0)^2 + \cdots + n a_n (x - x_0)^{n-1}$$
$$P_n''(x) = 2a_2 + 3 \cdot 2a_3 (x - x_0) + \cdots + n(n-1) a_n (x - x_0)^{n-2}$$
$$\dotfill \qquad (7.4)$$
$$P_n^{(n)}(x) = n(n-1)(n-2) \cdots 2 \cdot 1 \cdot a_n.$$

It follows that

$$P_n(x_0) = a_0$$
$$P_n'(x_0) = a_1$$
$$P_n''(x_0) = 2 \cdot 1 a_2$$
$$\dots\dots\dots\dots\dots\dots\dots\dots\dots\dots\dots$$
$$P_n^{(n)}(x_0) = n(n-1)(n-2)\cdots 2 \cdot 1 \cdot a_n.$$

Whence we find that

$$a_0 = f(x_0)$$
$$a_1 = f'(x_0)$$
$$a_2 = \frac{1}{2 \cdot 1} f''(x_0)$$
$$\dots\dots\dots\dots\dots\dots\dots\dots$$
$$a_n = \frac{1}{n(n-1)(n-2)\cdots 2 \cdot 1} f^{(n)}(x_0).$$

Thus the required polynomial is

$$P_n(x) = f(x_0) + \frac{f'(x_0)}{1}(x - x_0) + \frac{f''(x_0)}{2 \cdot 1}(x - x_0)^2$$
$$+ \cdots + \frac{f^{(n)}(x_0)}{n(n-1)\cdots 2 \cdot 1}(x - x_0)^n$$
$$= \sum_{k=0}^{n} \frac{f^{(k)}(x_0)}{k!}(x - x_0)^k .$$

$\square$

The polynomial $\sum_{k=0}^{n} \frac{f^{(k)}(x_0)}{k!}(x - x_0)^k$ is called **Taylor's polynomial** of f of degree n, about x_0, and is denoted by $P_n(f, x_0)(x)$. In the event that the function f has derivatives of all orders, one can consider the infinite series

$$\sum_{n=0}^{\infty} \frac{f^{(n)}(x_0)}{n!}(x - x_0)^n = f(x_0) + \frac{f'(x_0)}{1!}(x - x_0) + \frac{f''(x_0)}{2!}(x - x_0)^2 + \dots.$$

Such a series is called the **Taylor series** of f about x_0. For the particular case when $x_0 = 0$, this series has the form

$$\sum_{n=0}^{\infty} \frac{f^{(n)}(0)}{n!}x^n = f(0) + \frac{f'(0)}{1!}x + \frac{f''(0)}{2!}x^2 + \frac{f'''(0)}{3!}x^3 + \dots,$$

and is called the **Maclaurin series** of f.

Our interest is now to discover under what conditions on the function f the Taylor polynomial of a function f is "close" to f.

If f has derivatives of all orders on (a, b), then according to Taylor's formula (Theorem 5.23, page 137) given x, x_0 in (a, b), the difference $f(x) - P_n(x)$ is given by

$$R_n(f, x_0)(x) = \frac{f^{(n+1)}(\xi)}{(n+1)!}(x - x_0)^{n+1}$$

for some ξ between x_0 and x. Three things may happen:

- $\lim_{n \to \infty} R_n(f, x_0)(x)$ does not exist as a real number. In which case, the Taylor series $\sum_{n=0}^{\infty} \frac{f^{(n)}(x_0)}{n!}(x - x_0)^n$ diverges.

- $\lim_{n \to \infty} R_n(f, x_0)(x) = 0$. In which case, $f(x) = \sum_{n=0}^{\infty} \frac{f^{(n)}(x_0)}{n!}(x - x_0)^n$ and we say that f is represented by its Taylor series for that x.

- $\lim_{n \to \infty} R_n(f, x_0)(x)$ exists but is different from 0. We say that the Taylor series of f exists but does not represent f for that x.

The second alternative turns out to be the most interesting case in many applications.

Theorem 7.44

Let f be defined on (a, b) and assume that $a < x_0 < b$. Suppose that f possesses derivatives of all orders on (a, b), and that there is a constant C such that, for each $n \in \mathbb{N}$, $\left| f^{(n)}(x) \right| \leq C$ for all $x \in (a, b)$. Then

$$f(x) = \sum_{n=0}^{\infty} \frac{f^{(n)}(x_0)}{n!}(x - x_0)^n \quad \text{for all } x \in (a, b). \tag{7.5}$$

Proof

Since $\left| f^{(n)}(x) \right| \leq C$ for all $x \in (a, b)$ and for all $n \in \mathbb{N}$, we have

$$|R_n(f, x_0)(x)| \leq \frac{C}{n!}|x - x_0|^n \leq \frac{C}{n!}(b - a)^n \quad \text{for all } x \in (a, b).$$

Hence $\lim_{n \to \infty} R_n(f, x_0)(x) = 0$ for all $x \in (a, b)$. Using (5.10) and (5.11), we see that

$$f(x) = \sum_{n=0}^{\infty} \frac{f^{(n)}(x_0)}{n!}(x - x_0)^n \quad \text{for all } x \in (a, b).$$

$\square$

It is to be emphasized that the Taylor series of a given function f does not necessarily converge to f. To indicate the association of a function f with its Taylor series, we write

$$f \rightsquigarrow \sum_{n=0}^{\infty} \frac{f^{(n)}(x_0)}{n!} (x - x_0)^n.$$

We use the notation

$$f(x) = \sum_{n=0}^{\infty} \frac{f^{(n)}(x_0)}{n!} (x - x_0)^n$$

to express that the Taylor series converges to f at the point $x \in (a, b)$.

Taylor's Expansion of sine about 0

Let $f(x) = \sin x$. Then

$$f^{(n)}(x) = \begin{cases} \cos x & \text{if } n = 1 + 4k, \ k = 0, 1, 2, \ldots; \\ -\sin x & \text{if } n = 2 + 4k, \ k = 0, 1, 2, \ldots; \\ -\cos x & \text{if } n = 3 + 4k, \ k = 0, 1, 2, \ldots; \\ \sin x & \text{if } n = 4 + 4k, \ k = 0, 1, 2, \ldots \end{cases}$$

for all n and for all x. Thus

$$f^{(n)}(0) = \begin{cases} 1 & \text{if } n = 1 + 4k, \ k = 0, 1, 2, \ldots; \\ -1 & \text{if } n = 3 + 4k, \ k = 0, 1, 2, \ldots; \\ 0 & \text{otherwise.} \end{cases}$$

Taylor's theorem applies and yields

$$\sin x - \sum_{k=0}^{\text{int}((n-1)/2)} \frac{(-1)^k}{(2k+1)!} x^{2k+1} = \frac{f^{(n)}(\xi)}{n!} x^n.$$

Since $\left| f^{(n)}(x) \right| \leq 1$, for all x and for all n, we see by Theorem 7.5 that the Taylor series about 0, or the Maclaurin series of sine $\sum_{n=0}^{\infty} \frac{(-1)^n}{(2n+1)!} x^{2n+1}$ represents $\sin x$ for all x, i.e.

$$\sin x = \sum_{n=0}^{\infty} \frac{(-1)^n}{(2n+1)!} x^{2n+1} \quad \text{for } x \in \mathbb{R}. \tag{7.6}$$

Our next figure shows some Taylor polynomials of $y = \sin x$.

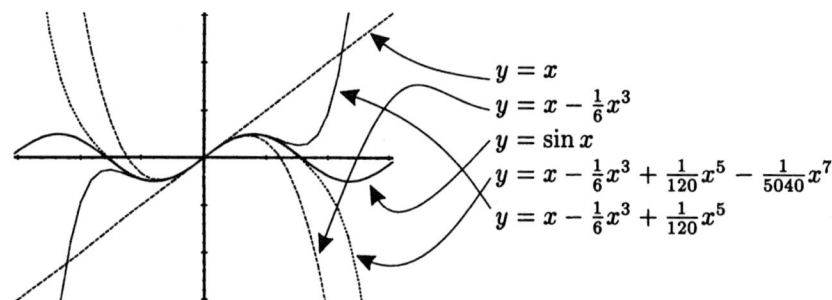

$$y = x$$
$$y = x - \tfrac{1}{6}x^3$$
$$y = \sin x$$
$$y = x - \tfrac{1}{6}x^3 + \tfrac{1}{120}x^5 - \tfrac{1}{5040}x^7$$
$$y = x - \tfrac{1}{6}x^3 + \tfrac{1}{120}x^5$$

Taylor Series of Exponential

Let $f(x) = e^x$. Then $f^{(n)}(x) = e^x$ for all n and for all x. Thus Taylor's theorem applies on any interval (a, b). Let $x_0, x \in (a, b)$. There is ξ between x and x_0 such that

$$R_n(f, x_0)(x) = \frac{e^\xi}{(n+1)!}(x - x_0)^{n+1}.$$

Since $0 < e^\xi \le e^{|b|}$, we see that $\lim_{n \to \infty} \frac{e^\xi}{(n+1)!}(x - x_0)^{n+1} = 0$. Hence the Taylor series $\sum_{k=0}^{\infty} \frac{e^{x_0}}{k!}(x - x_0)^k$ represents e^x about any point x_0 for every x, i.e.

$$e^x = \sum_{k=0}^{\infty} \frac{e^{x_0}}{k!}(x - x_0)^k \quad \text{for } x \in \mathbb{R}.$$

Our next example shows that the Taylor series of a function does not always represent the function.

Example 7.45

Let f be defined by

$$f(x) = \begin{cases} e^{\left(-x^{-2}\right)} & \text{if } x \ne 0, \\ 0 & \text{if } x = 0. \end{cases}$$

(1) Show that for all n

$$f^{(n)}(x) = \begin{cases} \frac{p_n(x)}{x^{3n}} e^{\left(-x^{-2}\right)} & \text{if } x \ne 0, \\ 0 & \text{if } x = 0. \end{cases} \tag{7.7}$$

where p_n is a polynomial of degree $< 3n$.

(2) Deduce that the Taylor series of f does not represent f.

Solution

Let $A = \{n \in \mathbb{N} : (7.7) \text{ holds}\}$. We have

$$f'(0) = \lim_{h \to 0} \frac{e^{(-1/h^2)}}{h} = \lim_{h \to 0} \frac{1/h}{e^{(1/h^2)}} = 0.$$

Hence

$$f'(x) = \begin{cases} \frac{2}{x^3} e^{(-x^{-2})} & \text{if } x \neq 0, \\ 0 & \text{if } x = 0. \end{cases}$$

Thus $1 \in A$. Suppose that $n \in A$. Let $p_n(x) = a_0 + a_1 x + \cdots + a_{3n-1} x^{3n-1}$. Then

$$f^{(n)}(x) = \left[a_0 \frac{1}{x^{3n}} + a_1 \frac{1}{x^{3n-1}} + \cdots + a_{3n-1} \frac{1}{x} \right] e^{(-x^{-2})}.$$

Hence for $x \neq 0$, we have

$$f^{(n+1)}(x) = \left[a_0 \frac{1}{x^{3n}} + a_1 \frac{1}{x^{3n-1}} + \cdots + a_{3n-1} \frac{1}{x} \right]' e^{(-x^{-2})}$$

$$+ \left[a_0 \frac{1}{x^{3n}} + a_1 \frac{1}{x^{3n-1}} + \cdots + a_{3n-1} \frac{1}{x} \right] \left[e^{(-x^{-2})} \right]'$$

$$= - \left[a_0 \frac{3n}{x^{3n+1}} + a_1 \frac{3n-1}{x^{3n}} + \cdots + a_{3n-1} \frac{1}{x^2} \right] e^{(-x^{-2})}$$

$$+ \left[a_0 \frac{1}{x^{3n}} + a_1 \frac{1}{x^{3n-1}} + \cdots + a_{3n-1} \frac{1}{x} \right] \left[\frac{2}{x^3} \right] e^{(-x^{-2})}.$$

This shows that for $x \neq 0$, $f^{(n+1)}(x) = \frac{p_{n+1}(x)}{x^{3n+1}} e^{(-x^{-2})}$ where p_{n+1} is a polynomial of degree $< 3n + 1$. On the other hand, we have

$$f^{(n+1)}(0) = \lim_{h \to 0} \frac{\frac{p_n(h)}{h^{3n}} e^{(-h^{-2})}}{h}$$

$$= \lim_{h \to 0} \frac{p_n(h)/h^{3n+1}}{e^{(1/h^2)}} = 0.$$

Hence $n + 1 \in A$. So $A = \mathbb{N}$ as desired. This completes the proof of (1).

(2) It follows that the Taylor series of $f(x) = e^{(-x^{-2})}$ about 0 is

$$\sum_{n=0}^{\infty} \frac{f^{(n)}(0)}{n!} x^n = \sum_{n=0}^{\infty} 0 \cdot x^n = 0.$$

Since $f(x) = e^{(-x^{-2})} > 0$ for $x \neq 0$, the Taylor series for f at 0 cannot represent $f(x)$ at $x \neq 0$. The proof is complete. $\square$

EXERCISES

7.1   For $x \in [0,1)$, let

$$f_n(x) = \begin{cases} nx & \text{if } x \in [0, 1/n) \\ 1 & \text{if } x \in [1/n, 1]. \end{cases}$$

   (1)   Find $f(x) = \lim f_n(x)$.

   (2)   Determine whether $f_n \to f$ uniformly on $[0,1]$.

7.2   For $x \in [0,1)$, let $f_n(x) = \frac{x^n - 1}{x - 1}$.

   (1)   Find $f(x) = \lim f_n(x)$.

   (2)   Determine whether $f_n \to f$ uniformly on $[0,1)$.

7.3   For $x \in [0, +\infty)$, let $f_n(x) = n^2 x e^{-nx}$.

   (1)   Find $f(x) = \lim f_n(x)$.

   (2)   Show that (f_n) does not converge uniformly on $[0,1]$.

7.4   Let $f_n(x) = \left(1 + \frac{x}{n}\right)^n$.

   (1)   Find $f(x) = \lim f_n(x)$.

   (2)   Determine whether $f_n \to f$ uniformly on $\mathbb{R}$.

7.5   For $x \in [0, +\infty)$, let $f_n(x) = n\left(x^{\frac{1}{n}} - 1\right)$.

   (1)   Find $f(x) = \lim f_n(x)$.

   (2)   Determine whether $f_n \to f$ uniformly on $[0, +\infty)$.

7.6   For $x \in [0,1]$, let $f_n(x) = x^n - x^{n+1}$.

   (1)   Find $f(x) = \lim f_n(x)$.

   (2)   Determine whether $f_n \to f$ uniformly on $[0,1]$.

7.7   For $x \in [0,1]$, let $f_n(x) = x^n - x^{2n}$.

   (1)   Find $f(x) = \lim f_n(x)$.

   (2)   Determine whether $f_n \to f$ uniformly on $[0,1]$.

7.8   For $x \in \mathbb{R}$, let $f_n(x) = x^n / n!$.

   (1)   Find $f(x) = \lim f_n(x)$.

   (2)   Determine whether $f_n \to f$ uniformly on $\mathbb{R}$.

7.9   Suppose that $f : \mathbb{R} \to \mathbb{R}$ is uniformly continuous. For each $n \in \mathbb{N}$, let $f_n(x) = f\left(x + \frac{1}{n}\right)$ for $x \in \mathbb{R}$. Show that f_n converges uniformly to f on $\mathbb{R}$.

7.10 Provide an example of a sequence of everywhere discontinuous functions which converges uniformly to a continuous function.

7.11 Show that if $f_n \to f$ on a closed interval $[a, b]$ and if the f_n are continuous, then f is bounded on $[a, b]$.

7.12 Show that if the f_n are continuous at c and if $f_n \to f$ uniformly on some neighborhood of c, then f is continuous at c.

7.13 Let $f(x) = \lim_{n \to \infty} f_n(x)$ for every $x \in A \subset \mathbb{R}$, where the f_n are continuous. Show that f is continuous at a point $c \in A$ if and only if for each $\varepsilon > 0$, there exists $m \in \mathbb{N}$ and $\delta > 0$ such that $x \in A$ and $|x - c| < \delta$ implies $|f_m(x) - f(x)| < \varepsilon$.

7.14 A function $f : [a, b] \to \mathbb{R}$ is said to be **Baire-1** if it is the pointwise limit of a sequence of continuous functions. Give an example of a noncontinuous Baire-1 function.

7.15 Let (f_n) be a sequence of real-valued functions on $[a, b]$. Let (a_n) and (b_n) be two sequences in $[a, b]$ such that $\lim a_n = a$ and $\lim b_n = b$. Show that if (f_n) converges uniformly to f on $[a, b]$, then

$$\lim \int_{a_n}^{b_n} f_n(x)\, dx = \int_{a}^{b} f(x)\, dx.$$

7.16 Let f be continuous on $[0, 1]$. Let $f_0 = f$ and let f_{n+1} be defined by $f_{n+1}(x) = \int_0^x f_n(t)\, dt$ for $n \in \mathbb{N}$, and $x \in [0, 1]$. Find the pointwise limit of the sequence (f_n). Is the convergence uniform? Explain.

7.17 Show that if $f_n \to f$ uniformly on a set A and $g_n \to g$ uniformly on A, then $f_n + g_n \to f + g$ on A.

7.18 Give an example of two sequences of functions (f_n) and (g_n) such that $f_n \to f$ uniformly and $g_n \to g$ uniformly but $f_n g_n \nrightarrow fg$ uniformly.

7.19 Let (f_n) and (g_n) be two uniformly bounded sequences of functions. Show that if $f_n \to f$ uniformly and $g_n \to g$ uniformly, then $f_n g_n \nrightarrow fg$ uniformly.

7.20 Show that if $f_n \to f$ uniformly on a set $[a, b]$ and g is continuous on $[a, b]$, then $f_n g \to fg$ uniformly on $[a, b]$.

7.21 Let (f_n) be a sequence of functions converging uniformly to f on (a, b). Prove that if each f_n is uniformly continuous, then f is uniformly continuous.

7.22 Let $a > 0$. Show that $\lim_{n \to \infty} \int_a^\pi \frac{\sin nx}{nx} dx = 0$. Discuss the case when $a = 0$.

7.23 Let (f_n) be a sequence of differentiable functions converging uniformly to f on $[a, b]$. Prove or disprove

(1) f is differentiable on $[a, b]$;

(2) (f_n') converges pointwise to f' $[a, b]$.

(3) Under what sufficient condition may one have uniform convergence of (f_n')?

7.24 Let (f_n) be a sequence of real-valued functions on a set A. Then (f_n) is said to be **equicontinuous** on A if

for each $\varepsilon > 0$ and $a \in A$, there exists $\delta > 0$ such that
$x \in A$ and $|x - a| < \delta$ implies $|f_n(x) - f_n(a)| < \varepsilon$ for all $n \in N$.

(1) Write out the corresponding definition for uniform equicontinuity.

(2) Show that if (f_n) is equicontinuous on $[a, b]$, then (f_n) is uniformly equicontinuous on $[a, b]$.

7.25 Let (f_n) be a sequence of real-valued functions on a set A. Suppose that

(1) (f_n) is pointwise bounded, i.e. for every $x \in A$, there exists M such that $|f_n(x)| \le M$ for all $n \in \mathbb{N}$;

(2) (f_n) is equicontinuous.

Show that (f_n) has a uniformly convergent subsequence.

7.26 Find the domain of convergence of each of the given series.

(a) $\sum_{n=0}^\infty \frac{x^n}{2^n}$ (e) $\sum_{n=1}^\infty \frac{\cos nx}{n^2}$ (i) $\sum_{n=2}^\infty \frac{x}{\sqrt{n}\ln n}$

(b) $\sum_{n=1}^\infty (-1)^{n+1} \frac{x^n}{n^2}$ (f) $\sum_{n=0}^\infty \frac{(nx)^n}{n!}$ (j) $\sum_{n=1}^\infty (x+3)^{3n-1}$

(c) $\sum_{n=0}^\infty 2^n \sin\left(\frac{x}{3^n}\right)$ (g) $\sum_{n=0}^\infty 3^{n^2} x^{n^2}$ (k) $\sum_{n=1}^\infty \frac{\ln n|x|}{n^2}$

(d) $\sum_{n=1}^\infty n(x-2)^n$ (h) $\sum_{n=1}^\infty \frac{x^{9m-8}}{9n-8}$ (l) $\sum_{n=0}^\infty \frac{x^{4n}}{4^n+1}$

7.27 Discuss the uniform convergence of the series

(a) $\sum_{n=1}^\infty \frac{x^n}{1+x^{2n}}$ (b) $\sum_{n=1}^\infty \frac{x}{n(1+nx^2)}$ (c) $\sum_{n=1}^\infty \frac{1}{1+n^2x}$

7.28 Consider the series $\sum_{n=1}^{\infty} (-1)^n \frac{x^2+n}{n^2}$.

    (1) Show that the series is uniformly convergent on $[-a, a]$ for any $a \in \mathbb{R}$.

    (2) Show that for no value of x is the series absolutely convergent.

7.29 Suppose that $\sum_{n=1}^{\infty} n |a_n|$ converges. Let $f(x) = \sum_{n=1}^{\infty} a_n \sin nx$. Show that $f'(x) = \sum_{n=1}^{\infty} n a_n \cos nx$ and that both series converge uniformly on $\mathbb{R}$.

7.30 Let $e(x) = \sum_{n=0}^{\infty} \frac{x^n}{n!}$ for $x \in \mathbb{R}$.

    (1) Show that $e' = e$.

    (2) Prove that $e(x+y) = e(x) + e(y)$ for all $x, y \in \mathbb{R}$.

7.31 Let

$$s(x) = \sum_{n=0}^{\infty} \frac{(-1)^n}{(2n+1)!} x^{(2n+1)} \text{ for } x \in \mathbb{R},$$

$$c(x) = \sum_{n=0}^{\infty} \frac{(-1)^n}{(2n)!} x^{2n} \text{ for } x \in \mathbb{R}.$$

    (1) Prove that $s' = c$ and $c' = -s$.

    (2) Prove that $\left(s^2 + c^2\right)' = 0$.

    (3) Prove that $s^2 + c^2 = 1$.

7.32 Suppose that a power series $f(x) = \sum_{n=0}^{\infty} a_n x^n$ satisfies the differential equation

$$\begin{cases} (1+x) f'(x) = \alpha f(x) \\ f(0) = 1 \end{cases}$$

where α is a given real number and $|x| < 1$.

    (1) Show that $a_0 = 1$ and $a_n = \frac{\alpha(\alpha-1)(\alpha-2)\cdots[\alpha-(n-1)]}{n!}$ for $n \geq 1$.

    (2) Prove **Newton's binomial theorem**

$$(1+x)^\alpha = 1 + \sum_{n=1}^{\infty} \frac{\alpha(\alpha-1)(\alpha-2)\cdots[\alpha-(n-1)]}{n!} x^n.$$

7.33 Use Newton's binomial theorem and integration term-by-term to find the series expansion of the function $f(x) = \arcsin x$ for $|x| < 1$.

    (1) Prove that the series converges also for $x = \pm 1$.

(2)   Deduce a formula for computing π.

7.34   Compare the interval of convergence of the series $\sum_{n=0}^{\infty} x^n/n^2$ and its derivative $\sum_{n=1}^{\infty} x^n/n$. Explain why this does not contradict the result of Theorem 7.41.

7.35   Find the derivative of the function specifying the interval of convergence.

$$\text{(a)} \sum_{n=1}^{\infty} \frac{(-1)^n x^n}{n!} \qquad \text{(b)} \sum_{n=1}^{\infty} \frac{(n!)^2 x^n}{n^n} \qquad \text{(c)} \sum_{n=1}^{\infty} \frac{(-1)^n x^{2n}}{(2n)!}$$

7.36   Find the sum of the power series specifying the interval of convergence.

$$\text{(a)} \sum_{n=1}^{\infty} n x^n \qquad \text{(b)} \sum_{n=1}^{\infty} \frac{x^n}{n+1} \qquad \text{(c)} \sum_{n=1}^{\infty} \frac{(-1)^n x^n}{n+1}$$

7.37   Find the sum of the series.

$$\text{(a)} \sum_{n=1}^{\infty} n 2^{-n} \qquad \text{(b)} \sum_{n=1}^{\infty} \frac{1}{(n+1)\, 3^n} \qquad \text{(c)} \sum_{n=1}^{\infty} \frac{(-2)^n}{n+1}$$

7.38   Show that $\sum_{n=1}^{\infty} \frac{n!}{n^n} x^n$ represents a continuous function on any interval lying entirely in $(-e, e)$.

7.39   Establish the following formulas.

$$\begin{array}{lll}
\text{(a)} & \cos x = \sum_{n=0}^{\infty} \frac{(-1)^n}{(2n)!} x^{2n} & \text{for } x \in \mathbb{R} \\
\text{(b)} & \ln(1+x) = \sum_{n=1}^{\infty} \frac{(-1)^{n+1}}{n} x^n & \text{for } -1 < x \le 1 \\
\text{(c)} & \frac{1}{1-x} = \sum_{n=0}^{\infty} x^n & \text{for } |x| < 1 \\
\text{(d)} & \arctan x = \sum_{n=1}^{\infty} \frac{(-1)^n}{2n+1} x^{2n+1} & \text{for } |x| < 1
\end{array}$$

Here $\arctan = \tan^{-1}$ is the inverse function of $\tan$.

7.40   Expand each function in powers of x specifying the interval of convergence.

$$\text{(a) } f(x) = \frac{1}{2+x} \qquad \text{(b) } f(x) = \frac{1}{1-x^2} \qquad \text{(c) } f(x) = \ln(1-x^2)$$

7.41   Expand each function in powers of $x - 1$ specifying the interval of convergence.

$$\text{(a) } f(x) = \frac{1}{1+x} \qquad \text{(b) } f(x) = \frac{1}{x^2} \qquad \text{(c) } f(x) = \ln x$$

7.42   Let ν be a nondecreasing function on $[a, b]$ and let (f_n) be a sequence in $\mathcal{R}\left([a, b], \nu\right)$. Suppose that $\sum f_n$ converges uniformly to some function f. Show that $f \in \mathcal{R}\left([a, b], \nu\right)$ and that $\int_a^b f d\nu = \sum \int_a^b f_n d\nu$.

7.43   Let the power series $\sum_{n=1}^{\infty} a_n x^n$ have a radius of convergence R. Prove that for each $x \in (-R, R)$, $\int_0^x f(t)\, dt = \sum_{n=0}^{\infty} \frac{a_n}{n+1} x^{n+1}$.

8

Local Structure on the Real Line

In previous chapters, we considered the algebraic and order structures $\mathbb{R}$ of the set of real numbers. We have seen that properties such as convergence of sequences, continuity and differentiability of functions, are all defined locally around some given point. In this chapter, we shall study $\mathbb{R}$ further by considering its features locally around each of its points. The generalization of such an approach to a more abstract setting is of great importance in analysis. In this chapter, we will discuss only subsets of the real line.

8.1 Open and Closed Sets in $\mathbb{R}$

We have seen that in the study of properties of $\mathbb{R}$ such as convergence of sequences and continuity of functions, intervals of the form

$$(a - \varepsilon, a + \varepsilon) = \{x \in \mathbb{R} : |x - a| < \varepsilon\},$$

where $a \in \mathbb{R}$ and $\varepsilon \in \mathbb{R}_+$, are often used. We have termed such an interval an ε-neighborhood of a. The term **open ball** centered at a with radius r is also used to describe such an interval. For simplicity, we often denote $(a - r, a + r)$ by $B(a, r)$. For example, we can rewrite the definition of continuity of a function f (Definition 4.16, page 104) at a point $a \in A$ by saying: for every $\varepsilon > 0$, there exists $\delta > 0$ such that

$$x \in B(a, \delta) \cap A \text{ implies } f(x) \in B(f(a), \varepsilon),$$

213

or equivalently: for every $\varepsilon > 0$, there exists $\delta > 0$ such that

$$f\left(B\left(a,\delta\right)\cap A\right)\subset B\left(f\left(a\right),\varepsilon\right).$$

Definition 8.1

A subset A of $\mathbb{R}$ is said to be **open** if for every element a of A, there exists $r > 0$ such that $B\left(a,r\right)\subset A$.

Example 8.2

Any interval of the form (a,b) is open in $\mathbb{R}$.

Solution

To see this, let $x \in (a,b)$. Consider $r = \frac{1}{2}\min\left\{|x-a|,|x-b|\right\}$. Then $x-r > a$ and $x+r < b$, i.e. $B\left(x,r\right)\subset (a,b)$. $\qquad\square$

It is clear from the definition that all open balls, $\mathbb{R}$ itself, and the empty set $\varnothing$ are open subsets of $\mathbb{R}$. One easily checks that intervals of the forms $(-\infty,a)$ and (a,∞) are also open while sets such as $\{x\},\{\frac{1}{n}:n\in\mathbb{N}\}$, $[a,b]$, $[a,b)$ are not open.

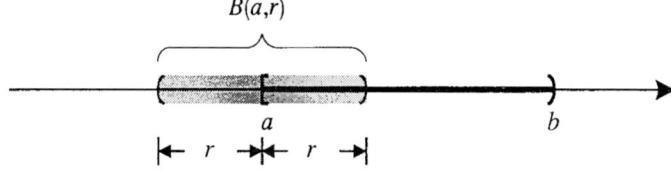

The interval $[a,b)$ is not open.

Before discussing the properties of open sets, we notice that many properties related to the notion of limit, can be expressed in terms of open sets.

Example 8.3

Show that the sequence (x_n) converges to x if and only if for every open set A containing x, there exists $N\in\mathbb{N}$ such that $n > N$ implies $x_n \in A$.

Solution

Suppose that (x_n) converges to x, and let A be an open set containing x. Then there exists $\varepsilon > 0$ such that $B\left(x,\varepsilon\right)\subset A$. For such ε, there exists $N\in\mathbb{N}$ such that $n > N$ implies $|x_n - x| < \varepsilon$. Thus

$$n > N \text{ implies } x_n \in B(x, \varepsilon) \subset A.$$

The necessity is proved.

For the converse, let $\varepsilon > 0$. By Example 8.2, the interval $(x - \varepsilon, x + \varepsilon) = B(x, \varepsilon)$ is open in $\mathbb{R}$. Thus there exists $N \in \mathbb{N}$ such that $n > N$ implies $x_n \in B(x, \varepsilon)$. Hence,

$$n > N \text{ implies } |x_n - x| < \varepsilon;$$

therefore (x_n) converges to x. $\square$

The next two results state the two fundamental properties of open sets.

Proposition 8.4

The union of any number of open sets in $\mathbb{R}$ is open.

Proof

Let $(A_i)_{i \in I}$ be a family of open sets in $\mathbb{R}$. We want to show that $A = \bigcup_{i \in I} A_i$, the union of the sets from this family is open. Fix $x \in A$. Then $x \in A_i$ for some i in I. Since A_i is open, there exists $r > 0$ such that

$$B(x, r) \subset A_i \subset A.$$

Hence A is open. $\square$

Proposition 8.5

The intersection of a finite number of open sets in $\mathbb{R}$ is open.

Proof

Let $A_1, A_2, \ldots, A_n$ be finitely many open sets in $\mathbb{R}$. We want to show that $A = \bigcap_{i=1}^{n} A_i$ is open. Fix $x \in A$. Then $x \in A_i$ for all $i = 1, 2, \ldots, n$. Since A_i is open for each i, there exists $r_i > 0$ such that $B(x, r_i) \subset A_i$. Let $r = \min \{r_1, r_2, \ldots, r_n\}$. If $y \in B(x, r)$ then $|x - y| < r \leq r_i$ for each i and thus $y \in B(x, r_i) \subset A_i$ for all i. Hence $B(x, r) \subset A$, and thus we have proved that A is open. $\square$

The following example shows that the word "finite" cannot be dropped from the hypothesis of Proposition 8.5.

Consider the family $(A_n)_{n \in \mathbb{N}}$ where $A_n = (-1/n, 1/n)$ for each $n \in \mathbb{N}$. Then the intersection $\bigcap_{n \in \mathbb{N}} A_n = \{0\}$ contains no open balls, and is therefore not open.

As a corollary to Example 8.2 and Proposition 8.4, we have the following result.

Theorem 8.6

A subset A of $\mathbb{R}$ is open if and only if it is expressible as a union of open balls.

Proof

Any open ball in $\mathbb{R}$ is an open interval, so it is an open set by Example 8.2. By Proposition 8.4, a union of open balls is open. Conversely, let U be open. Then for any $x \in U$ there exists $r_x > 0$ such that the ball $B(x, r_x)$ is contained in U. We clearly have $\bigcup_{x \in U} B(x, r_x) \subset U$. On the other hand if $y \in U$, then $y \in B(y, r_y) \subset \bigcup_{x \in U} B(x, r_x)$. This shows that $U = \bigcup_{x \in U} B(x, r_x)$ as desired. $\square$

Example 8.7

Show that every open set in $\mathbb{R}$ is the union of a countable collection of open intervals.

Solution

Let $A \subset \mathbb{R}$ be open and let $\{q_1, q_2, \ldots\}$ be an enumeration of the rationals in A. For each $n \in \mathbb{N}$, there exists m such that $B(q_n, 1/m) \subset A$. Applying the well-ordering principle (Theorem 1.34, page 18), we let

$$m_n = \min\{m \in \mathbb{N} : B(r_n, 1/m) \subset A\}.$$

Then

$$B = \bigcup_{n \in \mathbb{N}} B(q_n, 1/m_n) \subset A.$$

We also wish to show that $A \subset B$. To see this, let $x \in A$. Then there exists m such that $B(x, 1/m) \subset A$. Since $\mathbb{Q}$ is dense in $\mathbb{R}$, there exists a rational number $q \in B(x, 1/m)$ and hence $x \in B(q, 1/m)$. Since $q \in \{q_1, q_2, \ldots\}$, $q = q_n$ for some n. Now suppose that $x \notin B$. Then for all $k, x \notin B(q_k, 1/m_k)$. Thus

$$x \in B(q_n, 1/m) \setminus B(q_n, 1/m_n).$$

It follows that $1/m_n < 1/m$ which contradicts the definition of m_n. We conclude that $B = A$. $\square$

The next result is even stronger.

Theorem 8.8

Every open set in $\mathbb{R}$ is expressible as a union of a countable family of disjoint open intervals.

Proof

Let $A \subset \mathbb{R}$ be open. For each $x \in A$, there exists $t \in A$ such that $(x, t) \subset A$. Thus $b_x = \sup \{t : (x, t) \subset A\}$ exists as a real number. Similarly, we obtain the real number $a_x = \inf \{s : (s, x) \subset A\}$. Clearly, if $y \in (a_x, b_x)$, then either

- $y = x$, and thus $y \in A$; or

- $y \in (x, b_x)$, in which case there exists t_0 such that $y \in (x, t_0)$ and hence $y \in A$; or

- $y \in (a_x, x)$, in which case there exists s_0 such that $y \in (s_0, x)$ and hence $y \in A$.

Therefore, we have $x \in (a_x, b_x) \subset A$ and hence by direct inspection, we see

$$A = \bigcup_{x \in A} (a_x, b_x).$$

Moreover, let $x \in A$ and let I be any interval such that $x \in I \subset A$. Then for each $y \in I$, $(x, y) \subset A$, and thus $y \leq b_x$. Similarly, we have $a_x < y$. Hence, $I \subset (a_x, b_x)$, and therefore

$$(a_x, b_x) = \bigcup_{x \in I \subset A} I. \tag{8.1}$$

Now, suppose that $(a_x, b_x) \cap (a_y, b_y) \neq \varnothing$ for $x, y \in A$. Then $I = (a_x, b_x) \cup (a_y, b_y) \subset A$, and it follows from (8.1) that

$$(a_x, b_x) \cup (a_y, b_y) \subset (a_x, b_x) \quad \text{and} \quad (a_x, b_x) \cup (a_y, b_y) \subset (a_y, b_y).$$

Whence $(a_x, b_x) = (a_y, b_y)$. So far we have proved that A is the union of the disjoint collection $((a_x, b_x))_{x \in A}$.

By the density of $\mathbb{Q}$ in $\mathbb{R}$, we have that each interval (a_x, b_x) contains a rational number $r(x)$. By construction, the set $E = \{r(x) : x \in A\}$ is a subset of $\mathbb{Q}$. Let $\{r_1, r_2, \cdots, \}$ be an enumeration of E. It is readily seen that A is the range of the sequence (r_n). The proof is complete. $\qquad \square$

The next result follows immediately from Example 8.7. However, we present another proof.

Example 8.9 (Lindelöf)

Let $A \subset \mathbb{R}$. Suppose that $A \subset \bigcup_{i \in \mathcal{I}} E_i$ where the E_i are open. Show that there exists a countable subset $\mathcal{J}$ of $\mathcal{I}$ such that $A \subset \bigcup_{i \in \mathcal{J}} E_i$.

Solution

Let $a \in A$. Then there exists $i_a \in \mathcal{I}$ such that $a \in E_{i_a}$. Since E_{i_a} is open, there exists $\delta_{i_a} > 0$ such that $B\left(a, \delta_{i_a}\right) \subset E_{i_a}$. By the density theorem, we can find an interval I_a with rational endpoints such that $I_a \subset B\left(a, \delta_{i_a}\right) \subset E_{i_a}$. Then we have

- $A \subset \bigcup_{a \in A} I_a$ (why?), and

- there are countably many I_a (why?).

Therefore if $\{I_{i_1}, I_{i_2}, \ldots\}$ is an enumeration of the collection (I_a), then $A \subset \bigcup_{i_k \in \mathcal{J}} I_{i_k}$. For each k, we denote by E_{i_k} one of the E_i that contains I_{i_k}. Let $\mathcal{J} = \{i_1, i_2, \ldots\}$. Then $\mathcal{J}$ is countable and $\mathcal{J} \subset \mathcal{I}$. It is now easy to see that $A \subset \bigcup_{i_k \in \mathcal{J}} I_{i_k} \subset \bigcup_{i_k \in \mathcal{J}} E_{i_k}$. $\qquad \square$

Definition 8.10

A subset A of $\mathbb{R}$ is said to be **closed** if its complement $\mathbb{R} \backslash A = \{x \in \mathbb{R} : x \notin A\}$ is open.

Example 8.11

Show that the set $A = \{x \in \mathbb{R} : |x - a| \leq r\}$ is closed.

Solution

It suffices to notice that $\mathbb{R} \backslash A$ is the union of two intervals $(-\infty, a - r)$ and $(a + r, +\infty)$ which are both open. $\qquad \square$

A set of the form $[a - r, a + r] = \{x \in \mathbb{R} : |x - a| \leq r\}$, where $a \in \mathbb{R}$ and $r \in \mathbb{R}_+$, is called a **closed ball** centered at a with radius r. We shall denote such a ball by $B[a, r]$. The above example shows that a closed ball is a closed set. In fact, any closed interval of the form $[a, b]$, where a and b are real numbers, is a closed set. Indeed, $\mathbb{R} \setminus [a, b] = (-\infty, a) \cup (b, +\infty)$ is open. One verifies that intervals of the form $[a, +\infty)$, and $(-\infty, b]$, as well as $\mathbb{R}$ itself, are closed sets.

It is a good exercise to verify that

$$(a, b) = \bigcup_{n=1}^{\infty} \left[a + \frac{1}{n}, b - \frac{1}{n} \right] \text{ and } [a, b] = \bigcap_{n=1}^{\infty} \left(a - \frac{1}{n}, b + \frac{1}{n} \right).$$

Thus any open interval is the union of a sequence of closed intervals, and any closed interval is an intersection of a sequence of open intervals.

A set which can be expressed as a countable union of closed sets is called an F_σ set. A set which can be expressed as a countable intersection of open sets is called a G_δ set. Thus open intervals are F_σ and closed intervals are G_δ sets. It is clear that the complement of a G_d set (resp. F_σ set) is an F_σ (resp. G_δ set).

Example 8.12

Show that an open set in $\mathbb{R}$ is an F_σ (and a closed set is a G_δ) set.

Solution

Let A be an open set in $\mathbb{R}$. Then by Theorem 8.8, $A = \bigcup_{n=1}^{\infty} I_n$ where the I_n are disjoint intervals. Since each I_n is an F_σ set, $I_n = \bigcup_{k=1}^{\infty} F_{n,k}$ where the $F_{n,k}$ are closed sets. Then $A = \bigcup_{n=1}^{\infty} \bigcup_{k=1}^{\infty} F_{n,k} = \bigcup_{n,k \in \mathbb{N}} F_{n,k}$ shows that A is an F_σ set. $\qquad\square$

Note

We note that "open set" is not the opposite of "closed set". For example, we saw that $\mathbb{R}$ is at the same time open and closed. On the other hand, there are subsets of $\mathbb{R}$ which are neither open nor closed: such are, for example, intervals of the form $(a, b]$ and $[a, b)$ where $a < b$.

Example 8.13

Show that the intersection of any number of closed subsets of $\mathbb{R}$ is closed, and that the union of finitely many closed sets is closed.

Solution

We leave the second assertion as an exercise. Let $(F_i)_{i \in I}$ be a family of closed sets. Let $F = \bigcap_{i \in I} F_i$. Then $x \in \mathbb{R} \setminus F = \bigcup_{i \in I} (\mathbb{R} \setminus F_i)$ is open as is $\mathbb{R} \setminus F_i$ for each $i \in I$. Therefore F is a closed set. $\square$

Example 8.14

Show that if A and B are two open sets such that $A \cap B = \varnothing$ and $A \cup B = \mathbb{R}$, then either $A = \varnothing$ or $B = \varnothing$.

Solution

Suppose that A is not empty. We wish to show that $B = \varnothing$. Suppose that to the contrary that B is not empty. Let $x \in B$. Since A is open, B is closed. Then one of the sets $A \cap [x, +\infty)$ and $A \cap (-\infty, x]$ is nonempty. Suppose that $C = A \cap (-\infty, x] \neq \varnothing$. Then C is closed and bounded above. Hence $c = \max C$ exists. Since $x \notin A$, $C = A \cap (-\infty, x)$ is also open. Thus there exists $\varepsilon > 0$ such that $B(c, \varepsilon) \subset C$. This contradicts the fact that $c = \max C$. Thus B must be empty. $\square$

This last example shows a peculiar property of $\mathbb{R}$. It shows in particular that the only subsets of $\mathbb{R}$ that are both open and closed are $\varnothing$ and $\mathbb{R}$ itself. This example leads to the following definition.

Definition 8.15

A subset C of $\mathbb{R}$ is said to be **disconnected** if there are two disjoint open sets A and B such that $A \cap C \neq \varnothing$, $B \cap C \neq \varnothing$, and $C \subset A \cup B$. A subset which is not disconnected is said to be **connected**.

The empty set is of course connected. Example 8.14 asserts that $\mathbb{R}$ is connected. The set $\mathbb{N}$ is disconnected since we can take $A = (-\infty, \sqrt{2})$ and $B = (\sqrt{2}, +\infty)$ in the definition. The next figure shows that the set $(a, b] \cup [c, \infty)$ is disconnected.

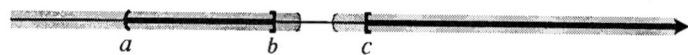

The following theorem gives a characterization of connected subsets of $\mathbb{R}$.

Theorem 8.16

Let C be a nonempty subset of $\mathbb{R}$. Then the following are equivalent.

(1) C is connected.

(2) If a, b are in C and $a < b$, then every point x such that $a < x < b$ is also in C.

Proof

Suppose that (2) does not hold. That is, suppose there exists $a < b$ in C and a point $x \in (a, b) \setminus C$. Then the intervals $(-\infty, x)$ and $(x, +\infty)$ satisfy Definition 8.15. Hence C is disconnected.

Conversely, suppose C is disconnected. Let A and B be the two open sets of Definition 8.15. Applying Theorem 8.8, we infer the existence of a countable family (A_n) of disjoint open intervals such that $A = \bigcup_n A_n$ and a countable family (B_n) of disjoint open intervals such that $B = \bigcup_n B_n$. Then there exist n and m such that $A_n \cap C \neq \varnothing$ and $B_m \cap C \neq \varnothing$. Let $x = \sup A_n$. Then $x < \inf B_m$. Now let $a \in A_n \cap C$ and $b \in B_m \cap C$. Then $a < x < b$, but $x \notin C$. The proof is finished.   □

The next corollary is now obtained without difficulty.

Corollary 8.17

A nonempty subset of $\mathbb{R}$ is connected if and only if it is an interval.

8.2 Neighborhoods and Interior Points

Definition 8.18

A subset V of $\mathbb{R}$ is said to be a **neighborhood** of a point $x \in \mathbb{R}$ if it contains an open set containing x.

Example 8.19

Show that V is a neighborhood of $x \in \mathbb{R}$ if and only if V contains an open ball with center x.

Solution

Suppose that V is a neighborhood of x. Then, by definition, there exists an open set U such that $x \in U \subset V$. Since U is open and $x \in U$, there is $r > 0$ such that $B(x, r) \subset U$. It then clearly follows that $B(x, r) \subset V$. The converse is clear since a ball with center x is open and contains x.   □

The set $A = [0, 1)$ is a neighborhood of any real number x between 0 and 1. It is not a neighborhood of any numbers $x \leq 0$ or $x \geq 1$.

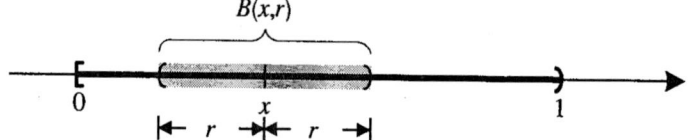

The next result describes the two fundamental properties of neighborhoods of a point.

Theorem 8.20

Let x be a real number.

(1) If V and V' are neighborhoods of x, then $V \cap V'$ is a neighborhood of x.

(2) If U is a neighborhood of x and V is a subset of $\mathbb{R}$ containing U, then V is a neighborhood of x.

Proof

(1) Suppose that V and V' are neighborhoods of $x \in \mathbb{R}$. There exist open balls $B(x, r)$ and $B(x, r')$ such that

$$B(x, r) \subset V \text{ and } B(x, r') \subset V'.$$

Let $\varepsilon = \min\{r, r'\}$. Then $B(x, \varepsilon) = B(x, r) \cap B(x, r') \subset V \cap V'$. Thus $V \cap V'$ is a neighborhood of x.

(2) If U is a neighborhood of x, then it contains a ball $B(x, r)$. Since $U \subset V$, we also have $B(x, r) \subset V$. Hence V is a neighborhood of x. □

The next result is a characterization of open sets in terms of neighborhoods.

Theorem 8.21

Let A be a subset of $\mathbb{R}$. Then A is open if and only if A is a neighborhood of each of its points.

Proof

Suppose that A is open. Let $x \in A$. Then there exists $r > 0$ such that $B(x, r) \subset A$. Thus A is a neighborhood of x.

Conversely, suppose that A is a neighborhood of each of its points. For each $x \in A$, there exists $r_x > 0$ such that $B(x, r_x) \subset A$. Let $B = \bigcup_{x \in A} B(x, r_x)$. Then B is open as a union of open sets. Clearly $B \subset A$. If $x \in A$, then $x \in B(x, r_x) \subset B$. Thus $A \subset B$ and hence $A = B$, so that A is open. $\qquad\square$

Our next example informs us that continuity of a function can be expressed in terms of neighborhoods.

Example 8.22

Show that the following assertions about a function $f : A \to \mathbb{R}$ are equivalent.

(1) f is continuous at a point $a \in A$.

(2) For every neighborhood H of $f(a)$ in $\mathbb{R}$, there exists a neighborhood G of a in $\mathbb{R}$, such that $G \cap A \subset f^{-1}(H)$.

Solution

Suppose that f is continuous at $a \in A$. Let H be a neighborhood of $f(a)$ in $\mathbb{R}$. Then there exists $\varepsilon > 0$ such that $B(f(a), \varepsilon) \subset H$. Since f is continuous, for such $\varepsilon > 0$ there exists $\delta > 0$ such that

$$x \in B(a, \delta) \cap A \quad \text{implies} \quad f(x) \in B(f(a), \varepsilon) \subset H.$$

Since the set $G = B(a, \delta)$ is clearly a neighborhood of a, we have established the necessity.

For the sufficiency, let $a \in A$ and let $\varepsilon > 0$. The set $H = B(f(a), \varepsilon)$ is a neighborhood of $f(a)$. Then, by condition (2), there exists a neighborhood G of a such that $x \in G \cap A$ implies $f(x) \in B(f(a), \varepsilon)$. Choose $\delta > 0$ such that $B(a, \delta) \subset G$. It follows that

$$x \in B(a, \delta) \cap A \quad \text{implies} \quad f(x) \in B(f(a), \varepsilon).$$

Thus f is continuous at a. The proof is complete. $\qquad\square$

Definition 8.23

A real number x is said to be an **interior point** of a set A if A is a neighborhood of x.

For example, any real number x between 0 and 1 is an interior point of $A = [0, 1)$; the numbers 0 and 1 are not interior to A.

The set of all interior points of a set A is denoted by A°. It is clear that if x is an interior point of A, then $x \in A$. But the converse is not true. For example, if $A = [0, 1)$, then $A^\circ = (0, 1)$. It could even happen that $A^\circ = \varnothing$; for example if $A = \mathbb{Q}$, then $A^\circ = \varnothing$ (show this!).

Example 8.24

Let A be a subset of $\mathbb{R}$. Show that A° is the largest open set contained in A.

Solution

Suppose that U is open and $U \subset A$. If x is in U, then x is an interior point of A, i.e. $x \in A^\circ$. Thus $U \subset A^\circ$. So every open set contained in A is contained in A°.

To see that A° is open, let $x \in A^\circ$. Then A is a neighborhood of x. There exists $r > 0$ such that $B(x, r) \subset A$. By the first part of the proof, $B(x, r) \subset A^\circ$. Hence A° is a neighborhood of x. By Theorem 8.21, A° is open. The proof is complete. $\square$

Another characterization of an open set is given by the next result.

Example 8.25

Let A be a subset of $\mathbb{R}$. Show that A is open if and only if $A = A^\circ$.

Solution

Suppose that A is open. Then the largest open set containing A is A. Therefore $A = A^\circ$.

Conversely, if $A = A^\circ$, then A is open because A° is open. The proof is finished. $\square$

Example 8.26

Let A and B be subsets of $\mathbb{R}$. Show that $(A \cap B)^\circ = A^\circ \cap B^\circ$.

Solution

The set $(A \cap B)^\circ$ is open and $(A \cap B)^\circ \subset A \cap B \subset A$. Thus $(A \cap B)^\circ \subset A^\circ$. Similarly, $(A \cap B)^\circ \subset B^\circ$. Thus $(A \cap B)^\circ \subset A^\circ \cap B^\circ$.

Conversely, $A° \subset A$ and $B° \subset B$; therefore $A° \cap B° \subset A \cap B$. Since $A° \cap B°$ is open, we have $A° \cap B° \subset (A \cap B)°$. We conclude that $(A \cap B)° = A° \cap B°$. □

Example 8.27 (Baire's Theorem)

Let (F_n) be a sequence of closed subsets of $\mathbb{R}$. Suppose that $\left(\bigcup_{n \in \mathbb{N}} F_n \right)° \neq \varnothing$. Show that there exists $n_0 \in \mathbb{N}$ such that $(F_{n_0})° \neq \varnothing$.

Solution

By way of contradiction, we suppose that for all n, $(F_n)° = \varnothing$. Let $x_0 \in \left(\bigcup_{n \in \mathbb{N}} F_n \right)°$. Then there exists $\varepsilon_0 > 0$ such that the ball $B(x_0, \varepsilon_0)$ is contained in $\bigcup_{n \in \mathbb{N}} F_n$. Since $(F_1)° = \varnothing$, then $B(x_0, \varepsilon_0) \cap F_1 = \varnothing$.

The set $(\mathbb{R} \setminus F_1) \cap B(x_0, \varepsilon_0)$ is nonempty and open; therefore we can find a point $x_1 \in (\mathbb{R} \setminus F_1) \cap B(x_0, \varepsilon_0)$, and $0 < \varepsilon_1 < 1$ such that

$$B(x_1, \varepsilon_1) \subset (\mathbb{R} \setminus F_1) \cap B(x_0, \varepsilon_0).$$

We denote by I_1 the closed bounded interval $[x_1 - \varepsilon_1/2, x_1 + \varepsilon_1/2]$. Then

$$I_1 \subset B(x_1, \varepsilon_1) \subset (\mathbb{R} \setminus F_1) \cap B(x_0, \varepsilon_0).$$

Similarly, the set $(\mathbb{R} \setminus F_2) \cap B(x_1, \varepsilon_1)$ is nonempty and open; therefore there is $0 < \varepsilon_2 < 1/2$ such that

$$I_2 \subset B(x_2, \varepsilon_2) \subset (\mathbb{R} \setminus F_2) \cap B(x_1, \varepsilon_1).$$

Again, the set $(\mathbb{R} \setminus F_3) \cap B(x_2, \varepsilon_2)$ is nonempty and open; therefore there is $0 < \varepsilon_3 < 1/3$ such that

$$I_3 \subset B(x_3, \varepsilon_3) \subset (\mathbb{R} \setminus F_3) \cap B(x_2, \varepsilon_2).$$

Continuing in this way, we construct a sequence $(I_k)_{k \in \mathbb{N}}$ of closed bounded intervals in $\mathbb{R}$, such that

$$B(x_0, \varepsilon_0) \supset I_1 \supset I_2 \supset \cdots. \tag{8.2}$$

By construction, the length of the interval I_n is less than $1/n$ for each n. According to Example 2.22, page 46, there exists an $x \in \mathbb{R}$, such that

$$\bigcap_{k \in \mathbb{N}} B \left[x_k, \frac{1}{n_k} \right] = \{x\}.$$

By the inclusions (8.2), $x \in B(x_0, \varepsilon_0) \subset \bigcup_{n \in \mathbb{N}} F_n$. On the other hand since $x \notin F_k$ for any k, then $x \notin \bigcup_{n \in \mathbb{N}} F_n$. This contradiction shows that our assumption is false. The proof is complete. □

8.3 Closure Point and Closure

Definition 8.28

A real number x is called a **closure point** of a subset A of $\mathbb{R}$ if every neighborhood of x intersects A. The set of all closure points of A is called the **closure** of A, and is denoted by A^{-}.

For example, any real number x, $0 \leq x \leq 1$, is a closure point of $A = (0,1)$. On the other hand, one easily sees that the ball $B(2, 1/2)$ does not intersect $(0,1)$; therefore the point $x = 2$ is not a closure point for $(0,1)$. In fact, later on, we will see that $A^{-} = [0,1]$.

Proposition 8.29

A real number x is a closure point of a subset A of $\mathbb{R}$ if and only if there exists a sequence (x_n) of elements of A converging to x.

Proof

Suppose that $x \in A^{-}$. Then for every $n \in \mathbb{N}$, the ball $B(x, 1/n)$ intersects A. Consider, for each n, $x_n \in B(x, 1/n) \cap A$. Then (x_n) is a sequence of elements of A, and the inequality $|x_n - x| < 1/n$ shows that $x_n \to x$.

Conversely, suppose that the sequence (x_n) of elements of A converges to some number x. Then, for every $\varepsilon > 0$, there is $N \in \mathbb{N}$ such that $n > N$ implies $|x_n - x| < \varepsilon$. Thus, for every $\varepsilon > 0$, $B(x, \varepsilon) \cap A \neq \varnothing$. Since every neighborhood V of x contains a ball with center x, $V \cap A \neq \varnothing$. Hence $x \in A^{-}$. $\qquad\square$

Theorem 8.30

Let A be a subset of $\mathbb{R}$. Then A^{-} is closed.

Proof

We show that $\mathbb{R} \backslash A^{-}$ is open. Let $x \in \mathbb{R} \backslash A^{-}$. Then x is not a closure point for A. There exists $r > 0$ such that the ball $B(x, r) \cap A = \varnothing$. Thus $B(x, r) \subset \mathbb{R} \backslash A$. Now we claim that the ball $B(x, r/3) \subset \mathbb{R} \backslash A^{-}$. Suppose that there exists $y \in B(x, r/3)$ such that $y \in A^{-}$. Then the ball $B(y, r/3)$ intersects A. Let

$z \in A \cap B(y, r/3)$. Then by the triangle inequality

$$|z - x| \le |z - y| + |y - x| < r/3 + r/3 < r.$$

Hence $z \in B(x, r) \cap A$. This contradiction proves our claim and hence the theorem. □

It is clear that any point of A is a closure point of A, hence $A \subset A^-$.

Example 8.31

Let A be a subset of $\mathbb{R}$. Show that A is closed if and only if $A = A^-$.

Solution

By the previous result if $A^- = A$, then A is closed. Conversely, suppose A is closed. Since $A \subset A^-$ always holds, we only need to show that $A^- \subset A$. Suppose that there exists $x \in A^-$ but that x is not in A. Then $x \in \mathbb{R} \setminus A$ which is an open set (since A is closed). Then, there exists $r > 0$ such that the ball $B(x, r) \subset \mathbb{R} \setminus A$. Thus the ball $B(x, r)$ does not intersect A. This means that x is not a closure point of A. Contradiction. Hence $A^- \subset A$. □

Example 8.32

Let A be a subset of $\mathbb{R}$. Then A^- is the smallest closed set containing A.

Solution

By Theorem 8.30, A^- is closed. Suppose that F is closed and $A \subset F$. Then $\mathbb{R} \setminus F$ is open. If $x \in \mathbb{R} \setminus F$, then there exists $r > 0$ such that $B(x, r) \subset \mathbb{R} \setminus F \subset \mathbb{R} \setminus A$. Thus the ball $B(x, r)$ does not intersect A. Thus $x \notin A^-$. This shows that $A^- \subset F$. □

Example 8.33

Let $A \subset \mathbb{R}$ and consider the function $f_A : \mathbb{R} \to \mathbb{R}$ defined by

$$f_A(x) = \inf \{|x - a| : a \in A\}.$$

Show that $A^- = \{x \in \mathbb{R} : f_A(x) = 0\}$. ($f_A(x)$ is usually denoted by $d(a, A)$ and is called the **distance** of x from the set A.)

Solution

For simplicity, let $B = \{x \in \mathbb{R} : f_A(x) = 0\}$. We first show that B is closed. Indeed, let (x_n) be a sequence in B converging to some number x. We know that f_A is a continuous function (see Example 4.32). Therefore, $f_A(x_n) \to f_A(x)$. But since $f_A(x_n) = 0$ for all n, we have $f_A(x) = 0$; that is $x \in B$. Hence, B is closed.

We then notice that if $a \in A$, then $f_A(a) = 0$. Therefore, $A \subset B$. On the other hand, if $x \in B$, then for every $n \in \mathbb{N}$, we can find $a_n \in A$ with the property $|x - a_n| < 1/n$. Hence, the sequence (a_n) converges to x, and so $x \in A^-$. Therefore, we have $A \subset B \subset A^-$. Since B is closed, we must have $B = A^-$. The proof is complete. $\qquad\square$

Example 8.34

Show that if A and B are two nonempty closed sets in $\mathbb{R}$ such that $A \cap B = \varnothing$, then there exists a continuous real-valued function φ defined on $\mathbb{R}$ such that $\varphi(x) = 0$, for all $x \in A$; $\varphi(x) = 1$, for all $x \in B$; and $0 \leq \varphi(x) \leq 1$, for all $x \in \mathbb{R}$.

Solution

Let f_A and f_B be both defined as in the previous example. Define $\varphi : \mathbb{R} \to \mathbb{R}$ by

$$\varphi(x) = \frac{f_A(x)}{f_A(x) + f_B(x)}.$$

Then φ is continuous. It is also clear that $0 \leq \varphi(x) \leq 1$, for all $x \in \mathbb{R}$. If $x \in A$, then $f_A(x) = 0$, and hence $\varphi(x) = 0$. If $x \in B$, then $x \notin A$, and therefore, $f_A(x) > 0$ and $f_B(x) = 0$. Hence $\varphi(x) = 1$, for all $x \in B$. $\qquad\square$

We know that in $\mathbb{R}$, a nonempty subset which is bounded above admits a supremum (completeness axiom). The next result is more precise.

Theorem 8.35

Let A be a nonempty subset of $\mathbb{R}$ that is bounded above. Then $\sup A = \max A^-$.

Proof

Let $x = \sup A$, and let V be a neighborhood of x. Then there exists $r > 0$ such that $B(x, r) \subset V$. By the definition of supremum, there exists $y \in A$ such that

$x - r < y \leq x$. Then $y \in B(x,r) \subset V$; therefore $V \cap A \neq \varnothing$. Thus $x \in A^-$.

To see that $x = \max A^-$, suppose that there exists $x' \in A^-$ such that $x < x'$. Set $r = x' - x > 0$. Then the ball $B(x',r)$ intersects A. Let $y \in B(x',r) \cap A$. Since $y > x' - r = x$, x is not an upper bound for A. We have a contradiction. Thus $x = x'$, and hence $x = \max A^-$. The proof is complete. $\square$

Definition 8.36

Let A and B be two subsets of $\mathbb{R}$. We say that A is **dense** in B if $A^- = B$.

For example, each of the intervals $(0,1)$, $[0,1)$, $(0,1]$, and $[0,1]$ is dense in $[0,1]$ while the set $\{0,1\}$ is not. $\mathbb{Q}$ is dense in $\mathbb{R}$, i.e. $\mathbb{Q}^- = \mathbb{R}$. We notice that the denseness of $\mathbb{Q}$ defined here is exactly the same as that of the one described in Theorem 1.44. Prove it!

Example 8.37

Show that a subset A is dense in $\mathbb{R}$ if and only if the only closed set containing A is $\mathbb{R}$.

Solution

Suppose that A is dense in $\mathbb{R}$. If F is a closed set containing A, then by Example 8.32, $A^- \subset F$. But since $A^- = \mathbb{R}$, we have $F = \mathbb{R}$. Conversely, we know that $A \subset A^-$ and that A^- is closed. Thus if $\mathbb{R}$ is the only closed set containing A, then necessarily $A^- = \mathbb{R}$. $\square$

Example 8.38

Show that a subset A is dense in $\mathbb{R}$ if and only if $\mathbb{R}\backslash A$ has an empty interior.

Solution

Suppose that A is dense in $\mathbb{R}$ and suppose that x is an interior point of $\mathbb{R}\backslash A$. Then there exists $r > 0$ such that the ball $B(x,r) \subset \mathbb{R}\backslash A$. Thus $B(x,r)$ does not intersect A. Therefore $x \notin A^- = \mathbb{R}$. We have a contradiction.

Conversely, suppose that $\mathbb{R}\backslash A$ has an empty interior. Let $x \in \mathbb{R}$. If $x \in A$, then $x \in A^-$. If $x \notin A$, then no ball $B(x,r)$ can be contained in $\mathbb{R}\backslash A$; therefore $x \in A^-$. Thus $A^- = \mathbb{R}$. The proof is complete. $\square$

For example, the set $A = \mathbb{R} \backslash \mathbb{Q}$ is dense in $\mathbb{R}$ because $(\mathbb{R} \backslash A)^\circ = \mathbb{Q}^\circ = \varnothing$. On the other hand, since $\mathbb{R} \backslash \mathbb{N} = (-\infty, 1) \cup \bigcup_{n=1}^{\infty} (n, n+1)$ is a nonempty open set, we see that the set $\mathbb{N}$ is not dense in $\mathbb{R}$.

8.4 Completeness and Compactness

In Chapter 1, we noticed that every Cauchy sequence of elements of $\mathbb{R}$ converges. The sequence $(1/n)$ is a Cauchy sequence in $(0, 1)$, the limit of which exists but is not in $(0, 1)$. The following notion is of great importance in analysis of the real line.

Definition 8.39

A subset A of $\mathbb{R}$ is said to be **complete** if each Cauchy sequence in A converges to a limit that belongs to A.

The following result characterizes those subsets of $\mathbb{R}$ that are complete.

Theorem 8.40

Let A be a subset of $\mathbb{R}$. Then

- if A closed, then it is complete;

- if A is complete, then it is closed.

Proof

Suppose that A is closed. Let (x_n) be a Cauchy sequence in A. Since $\mathbb{R}$ is complete, (x_n) converges to some limit x in $\mathbb{R}$. By Proposition 8.29, $x \in A^-$.

Conversely, suppose A is complete. Let $x \in A^-$. By Proposition 8.29, there exists a sequence (x_n) of elements of A converging to x. Thus the sequence (x_n) is a Cauchy sequence in A. Since A is complete (x_n) converges in A. Thus x must be an element of A. Hence $A^- = A$, i.e. A is closed. $\square$

We know that every Cauchy sequence in $\mathbb{R}$ converges to some real number. Therefore $\mathbb{R}$ is complete. It is easily checked that any finite subset of $\mathbb{R}$ is complete. However, since a nonempty proper open subset of $\mathbb{R}$ cannot be closed ($\mathbb{R}$ is connected), it is not complete. Also, since $\mathbb{Q}$ is dense in $\mathbb{R}$, it cannot be complete.

The **diameter** of a subset A of $\mathbb{R}$ is defined by the formula

$$\text{diam } A = \sup\left\{|x - y| : x, y \in A\right\}.$$

For example, $\text{diam}\,(a, b] = b - a$, $\text{diam}\,\{a\} = 0$, and $\text{diam}\,(-\infty, a) = \infty$. By convention, $\text{diam}\,\varnothing = 0$.

Our next example generalizes the result of Example 2.22 (page 46).

Example 8.41

Let (F_n) be a sequence of nonempty closed subsets of $\mathbb{R}$. Suppose that

(1) $F_1 \supset F_2 \supset \ldots \supset F_n \supset \ldots$;

(2) $\text{diam } F_n \to 0$.

Show that $F = \bigcap_n F_n$ contains exactly one point.

Solution

For each n, pick x_n in F_n. Then for each n and p, by condition (1) both x_n and x_{n+p} belong to F_n. It follows that

$$|x_{n+p} - x_n| \leq \text{diam } F_n.$$

By condition (2), the sequence (x_n) is Cauchy. Thus (x_n) converges to some real number x. Thus x is a closure point for each F_n. It follows that $x \in F$. Since $\text{diam } F \subset \text{diam } F_n$ for each n, we have $\text{diam } F = 0$. Thus F cannot contain more than one point. Hence $F = \{x\}$. $\qquad\square$

Example 8.42

Show that if (x_n) is a Cauchy sequence, then its set of values has finite diameter. (Compare with Lemma 2.38 on page 59.)

Solution

Let (x_n) be Cauchy and let $A = \{x_n : n \in \mathbb{N}\}$ be its set of values. Given $\varepsilon > 0$, we have an $N \in \mathbb{N}$ such that

$$|x_n - x_m| < \varepsilon \text{ for } n, m > N.$$

Thus $\sup\{|x_n - x_m| : n, m > N\} < \varepsilon$. Let $M = \max\{|x_n - x_m| : n, m < N\}$. It is clear that

$$\text{diam } A = \sup\{|x_n - x_m| : n, m \in \mathbb{N}\} \leq \varepsilon + M.$$

□

Not every sequence of real numbers has a convergent subsequence. For example, the sequence $(n)_{n \in \mathbb{N}}$ has no convergent subsequence. However, the Bolzano–Weierstrass Theorem 2.30 states that any bounded sequence has a convergent subsequence. This leads us to the following definition.

Definition 8.43

A subset A of $\mathbb{R}$ is said to be **sequentially compact** if every sequence in A has a subsequence that converges to an element in A.

The interval $(0, 1]$ is not sequentially compact. The sequence $\left(\frac{1}{n}\right)_{n \in \mathbb{N}}$ is a sequence of elements in $(0, 1]$. Every subsequence of $\left(\frac{1}{n}\right)_{n \in \mathbb{N}}$ converges to 0 but $0 \in (0, 1]$.

Theorem 8.44

A sequentially compact subset of $\mathbb{R}$ is complete.

Proof

Suppose that $A \subset \mathbb{R}$ is sequentially compact and let (x_n) be Cauchy in A. There is a subsequence $(x_{n_k})_{k \in \mathbb{N}}$ of (x_n) converging to some element $x \in A$. Thus for a given $\varepsilon > 0$, there is $N_1 \in \mathbb{N}$ such that

$$|x_{n_k} - x| < \varepsilon/2 \text{ whenever } k > N_1.$$

On the other hand, since (x_n) is Cauchy, there exists $N_2 \in \mathbb{N}$ such that

$$|x_n - x_m| < \varepsilon/2 \text{ whenever } n, m > N_2.$$

It follows that for $n, \, k > \max\{N_1, N_2\}$,

$$|x_n - x| \leq |x_n - x_{n_k}| + |x_{n_k} - x| < \frac{\varepsilon}{2} + \frac{\varepsilon}{2} = \varepsilon.$$

This shows that A is complete.   □

The following corollary is immediate.

Corollary 8.45

A sequentially compact subset of $\mathbb{R}$ is closed.

For our next result we recall that if $x_n \to x$, and if $y_n \to y$, then $|x_n - y_n| \to |x - y|$ (Exercise 2.27, page 63).

Theorem 8.46

A sequentially compact subset of $\mathbb{R}$ has finite diameter.

Proof

Suppose that $A \subset \mathbb{R}$ is sequentially compact. Let (x_n) and (y_n) be two sequences in A such that $|x_n - y_n| \to \operatorname{diam} A$. Since A is sequentially compact, there is a convergent subsequence $(x_{n_k})_{k \in \mathbb{N}}$ of (x_n) and a convergent subsequence (y_{n_k}) of (y_n), say $x_{n_k} \to x \in A$ and $y_{n_k} \to y \in A$. Then $|x_n - y_n| \to |x - y|$. Thus $\operatorname{diam} A = |x - y|$ is finite. $\square$

Thus a sequentially compact subset of $\mathbb{R}$ is closed and bounded. It turns out that the converse also holds.

Theorem 8.47 (Heine–Borel)

A subset of $\mathbb{R}$ is sequentially compact if and only if it is closed and bounded.

Proof

It remains to prove the sufficiency. Suppose that $A \subset \mathbb{R}$ is closed and bounded. Let (x_n) be a sequence in A. Since A is bounded, (x_n) is also bounded. By the Bolzano–Weierstrass theorem, it has a convergent subsequence, say $x_{n_k} \to x$. Since A is closed, $x \in A$. This completes the proof. $\square$

As an application, we give the following important example.

Example 8.48

Show that if A is a nonempty sequentially compact subset of $\mathbb{R}$, then $\max A$ and $\min A$ exist.

Solution

Since A is sequentially compact it is bounded, and thus $M = \sup A$ exists. For

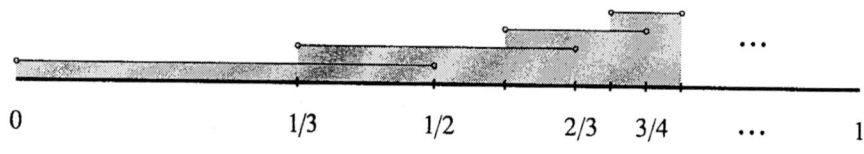

0   1/3   1/2   2/3   3/4 $\cdots$ 1

Figure 8.1   A cover of $(0,1)$

each $n \in \mathbb{N}$, there is $x_n \in A$ such that

$$M - \frac{1}{n} \leq x_n \leq M.$$

This implies that

$$\lim x_n = M.$$

Since, on the other hand A is closed, $M \in A$. Accordingly, $M = \max A$.

The proof of the fact that A has a minimum uses a similar argument and is left as an exercise. $\square$

Let A be a subset of $\mathbb{R}$. By a **cover** of A, we mean a family $(E_i)_{i \in I}$ of subsets of $\mathbb{R}$ such that $A \subset \bigcup_{i \in I} E_i$. Then we say that A is covered by $(E_i)_{i \in I}$ or $(E_i)_{i \in I}$ covers A. A cover $(E_i)_{i \in I}$ of a set A, consisting entirely of open subsets of $\mathbb{R}$ is referred to as an **open cover** of A. If J is any subset of I, the subfamily $(E_i)_{i \in J}$ is called a **subcover** of the cover $(E_i)_{i \in I}$ if we still have $A \subset \bigcup_{i \in J} E_i$. A subcover $(E_i)_{i \in J}$ is said to be finite if J is finite.

Definition 8.49

A subset A of $\mathbb{R}$ is said to be **compact** if every open cover of A contains a finite subcover.

Figure 8.1 shows how to construct a cover of $(0,1)$ by the open intervals $(0, 1/2), (1/3, 2/3), (7/12, 3/4), \ldots$. The Archimedean property can be used to show that such a family is indeed a cover of $(0,1)$. It is readily seen that such a cover admits no finite subcover. In fact, such a family ceases to be a cover of $(0,1)$ as soon as one interval is omitted. Hence the interval $(0, 1)$ is not compact.

We notice that the above argument cannot be applied to the interval $[0,1]$ since the family in question does not cover $[0,1]$; neither 0 nor 1 is in any of the intervals in consideration.

Let A be a subset of $\mathbb{R}$ and let $\varepsilon > 0$. An ε-**cover** of A is a cover consisting of ε-balls, i.e. balls of radius ε. We note that as an example, no matter how we choose $\varepsilon > 0$, the family $(B(a, \varepsilon))_{a \in A}$ is always an ε-cover of the set A.

Definition 8.50

A subset A of $\mathbb{R}$ is said to be **totally bounded** if *for every $\varepsilon > 0$, every ε-cover of A admits a finite subcover.*

We should convince ourselves that the above definition is equivalent to the following: *a subset A of $\mathbb{R}$ is totally bounded if and only if A admits a finite ε-cover for every $\varepsilon > 0$.*

It is quite easy to see that every compact set is totally bounded. However total boundedness is not enough to ensure compactness.

Theorem 8.51

A subset A of $\mathbb{R}$ is compact if and only if it is complete and totally bounded.

Proof

Suppose that A is compact and let $\varepsilon > 0$. Then $(B(a, \varepsilon))_{a \in A}$ is an open cover of A. Since A is compact, there exist finitely many elements of A, say $a_1, a_2, \ldots, a_p$ such that $A \subset \bigcup_{k=1}^{p} B(a_k, \varepsilon)$. Thus A is totally bounded.

Next, let (x_n) be Cauchy in A. Suppose that (x_n) converges to no element in A. Thus for every $a \in A$, there exists an $\varepsilon > 0$ such that

$$|x_n - a| > 2\varepsilon \tag{8.3}$$

for infinitely many n. Pick $N > 0$ such that

$$|x_n - x_m| < \varepsilon \text{ whenever } n, m \geq N.$$

We choose $n > N$ such that (8.3) holds. Then

$$\varepsilon < |x_n - a| - |x_n - x_m| \leq |x_m - a|.$$

It follows that the ball $B(a, \varepsilon)$ contains only finitely many x_n. Again, since $(B(a, \varepsilon))_{a \in A}$ is an open cover of A and since A is compact, there are finitely many elements of A, say $a_1, a_2, \ldots, a_p$ such that $A \subset \bigcup_{k=1}^{p} B(a_k, \varepsilon)$. Since each $B(a_k, \varepsilon)$ contains finitely many x_n, it follows that the sequence (x_n) has only finitely many different terms and hence it would converge in A. Contradiction. Thus A is complete.

Conversely, assume that A is complete and totally bounded. Suppose to the contrary that there exists an open covering $(E_i)_{i \in I}$ which does not contain any finite subcovering. Since A is totally bounded, we can cover A by finitely many 1/2-balls. If each of these 1/2-balls had a finite subcovering, then the union of such finite coverings would form a finite subcovering of A. Thus there exists a

1/2-ball, say $B\left(x_1, 1/2\right)$, which has no finite subcovering. Since $B\left(x_1, 1/2\right) \subset A$, $B\left(x_1, 1/2\right)$ is totally bounded. We can apply the previous argument to $B\left(x_1, 1/2\right)$ using $1/2^2$-balls. Thus there exists a $1/2^2$-ball, say $B\left(x_2, 1/2^2\right) \subset B\left(x_1, 1/2\right)$, which has no finite subcovering. By induction, we obtain a sequence (x_n) with the property that $B\left(x_n, 1/2^n\right)$ has no finite subcovering and $x_{n+1} \in B\left(x_n, 1/2^n\right)$ for each n. Thus

$$|x_{n+1} - x_n| < 1/2^n,$$

and hence,

$$|x_{n+p} - x_n| \leq |x_{n+1} - x_n| + |x_{n+2} - x_{n+1}| + \cdots + |x_{n+p} - x_{n+p-1}|$$
$$< \frac{1}{2^n} + \frac{1}{2^{n+1}} + \cdots + \frac{1}{2^{n+p-1}} < \frac{1}{2^{n-1}}.$$

It follows that (x_n) is a Cauchy sequence. Since A is complete by our hypothesis, (x_n) converges to some $y \in A$. Thus $y \in E_i$ for some i. Since E_i is open, there is $\delta > 0$ such that $B(y, \delta) \subset E_i$. Choose n large enough so that

$$|x_n - y| < \varepsilon \text{ and } \frac{1}{2^n} < \frac{\delta}{2}.$$

Then $B\left(x_n, 1/2^n\right) \subset B(y, \delta)$. Indeed if $|x_n - x| < 1/2^n$, then

$$|y - x| \leq |x_n - x| + |x_n - y| < \delta.$$

Thus $B\left(x_n, 1/2^n\right) \subset E_i$ and hence, E_i is a subcovering of $B\left(x_n, 1/2^n\right)$. Contradiction. This completes our proof. $\qquad\square$

We finish this section by summing up its main results in the following theorem.

Theorem 8.52

Let A be a subset of $\mathbb{R}$. The following assertions are equivalent

(1) A is compact;

(2) A is sequentially compact;

(3) A is complete and totally bounded;

(4) A is closed and bounded.

EXERCISES

8.1 Give an example of a sequence of open subsets of $\mathbb{R}$ whose intersection is not open.

8.2 Give an example of a sequence of closed subsets of $\mathbb{R}$ whose union is not closed.

8.3 Show that every finite subset of $\mathbb{R}$ is closed.

8.4 Prove that the interior of the intervals $[a, b]$, $(a, b]$, $[a, b)$, and (a, b) is the open interval (a, b).

8.5 Prove that the closure of the intervals $[a, b]$, $(a, b]$, $[a, b)$, and (a, b) is the closed interval $[a, b]$.

8.6 Show that if $A \subset B$, then $A^\circ \subset B^\circ$ and $A^- \subset B^-$.

8.7 Show that $A^\circ \cup B^\circ \subset (A \cup B)^\circ$. Give an example to show that equalities need not hold.

8.8 Show that

   (1) $A^- \cup B^- = (A \cup B)^-$;

   (2) $(A \cap B)^- \subset A^- \cap B^-$. Give an example to show that equalities need not hold.

8.9 Show that if B is open, then $A^- \cap B \subset (A \cap B)^-$.

8.10 Let $A \subset \mathbb{R}$. Show that

   (1) the complement of A° is the closure of $\mathbb{R} \setminus A$;

   (2) the complement of A^- is the interior of $\mathbb{R} \setminus A$.

8.11 Rewrite the definition of convergence of a sequence in terms of neighborhoods.

8.12 Rewrite the definitions of limit and continuity of functions in terms of neighborhoods.

8.13 Show that a countable set cannot be open in $\mathbb{R}$.

8.14 Show that an open set in $\mathbb{R}$ is a countable intersection of closed sets (F_σ-set).

8.15 Show that a closed set in in $\mathbb{R}$ is a countable union of open sets (G_δ-set).

8.16 The **boundary** of a set $A \subset \mathbb{R}$ is the set $\partial A = A^- \setminus A^\circ$. Describe the boundary of the sets: $[a, b]$, $(a, b]$, $[a, b)$, (a, b), $\varnothing$, $\mathbb{R}$, $\mathbb{Q}$.

8.17   Let A be a subset of $\mathbb{R}$. Show that

   (1) ∂A is closed;

   (2) $\partial A = \partial (\mathbb{R} \backslash A)$;

   (3) $\partial A^- \subset \partial A$; (Give an example to show that equality need not hold.)

   (4) $\partial A^\circ \subset \partial A$. (Give an example to show that equality need not hold.)

8.18   Let A be a subset of $\mathbb{R}$. Show that

   (1) A is open if and only if $A \cap \partial A = \varnothing$;

   (2) A is closed if and only if $\partial A \subset A$.

8.19   A point x is said to be an **accumulation point** of a set A if every neighborhood of x contains an element of A distinct from x.

   (1)   Show that every accumulation point is a closure point.

   (2)   Let A' denote the set of all accumulation points of A. A' is called the **derived set** of A. Show that $A^- = A \cup A'$ and that a set A is closed if and only if $A' \subset A$.

8.20   A point $x \in A$ is said to be **isolated** if it is not an accumulation point of A.

   (1)   Find all isolated points and accumulation points of the set $A = \left\{ \frac{1}{n} : n \in \mathbb{N} \right\} \cup \{0\}$.

   (2)   Show that if x is an isolated point of a set A, then x is an accumulation point of $\mathbb{R} \setminus A$.

8.21   Show that $(-\infty, -1) \cup (-1, 1) \cup (1, \infty)$ is dense in $\mathbb{R}$.

8.22   Write out the proof of the fact that $\mathbb{Q}$ is dense in $\mathbb{R}$.

8.23   Show that $\mathbb{R} \setminus \mathbb{Q}$ is dense in $\mathbb{R}$.

8.24   Show that A is dense in $\mathbb{R}$ if and only if every real number x is the limit of some sequence of elements of A.

8.25   Show that the closure of a connected set is connected. Compare with Example 4.28 (page 111).

8.26   Show that the union of two open connected sets is connected if and only if they have a common point.

8.27 Suppose that A is connected in $\mathbb{R}$ and let $f : A \to \mathbb{R}$.

  (1) Show that $f(A)$ is connected.

  (2) Show that if a and $b \in A$, and if $f(a) < t < f(b)$, then there exists $c \in A$ such that $f(c) = t$.

8.28 Let A and B be complete subsets of $\mathbb{R}$. Show that $A \cup B$ and $A \cap B$ are complete.

8.29 A subset A of $\mathbb{R}$ is said to be **separable** if A contains a countable dense set.

  (1) Show that $\mathbb{R}$ is separable.

  (2) Show that a compact subset of $\mathbb{R}$ is separable.

8.30 A subset A of $\mathbb{R}$ is said to be **nowhere dense** if $(A^-)^\circ = \varnothing$. If $A = \bigcup_{n=1}^{\infty} A_n$ is a countable union of nowhere dense sets A_n, then A is said to be a set of the **first category**.

  (1) Show that $\mathbb{N}$ is nowhere dense in $\mathbb{R}$.

  (2) Show that $\mathbb{N}$ and $\mathbb{Q}$ are both sets of the first category.

  (3) Show that a nonempty open set cannot be of the first category.

  (4) Show that a countable union of sets of the first category is again a set of the first category.

  (5) Show that $\mathbb{R}\backslash\mathbb{Q}$ is dense in $\mathbb{R}$ and is not of the first category.

8.31 Show that the boundary of an open or closed set is nowhere dense.

8.32 **Baire property.** Let (A_n) be a sequence of subsets of $\mathbb{R}$.

  (1) Show that if the A_n are open and dense in $\mathbb{R}$, then the set $\bigcap_{n=1}^{\infty} A_n$ is dense in $\mathbb{R}$.

  (2) Show that if $\bigcup_{n=1}^{\infty} A_n = \mathbb{R}$, then $(A_n^-)^\circ \neq \varnothing$ for some n.

8.33 Explain why the set $A = \left\{ \frac{1}{n} : n \in \mathbb{N} \right\}$ is not compact.

8.34 Show that the set $A = \left\{ \frac{1}{n} : n \in \mathbb{N} \right\} \cup \{0\}$ is compact. In general, if $x_n \to x$ in $\mathbb{R}$, show that $K = \{x_n : n \in \mathbb{N}\} \cup \{x\}$ is compact.

8.35 Show that the intersection of any family of compact sets is compact and that the union of any finite family of compact sets is compact.

8.36 Show that a closed subset of a compact set is compact.

8.37 Show that if $A \subset \mathbb{R}$ is compact and $B \subset \mathbb{R}$ is closed, then $A \cap B$ is compact.

8.38 Let A and B be subsets of $\mathbb{R}$. Show that

    (1)   if both A and B are compact, then so is $A + B = \{a + b : a \in A, b \in B\}$;

    (2)   if A is compact and B is closed, then $A + B$ is closed.

8.39 Suppose that A is compact and $x \notin A$. Show that there exist two disjoint open sets U and V such that $x \in U$ and $A \subset V$.

8.40 Suppose that A and B are compact and $A \cap B = \varnothing$. Show that there exist two disjoint open sets U and V such that $A \subset U$ and $B \subset V$.

8.41 **Finite intersection property.** Let $(A_i)_{i \in I}$ be a family of compact subsets of R. Suppose that the intersection of every finite subfamily of $(A_i)_{i \in I}$ is nonempty. Show that $\bigcap_{i \in I} A_i$ is nonempty.

9

Continuous Functions

The class of continuous functions plays an important role in analysis. In Chapter 4, we talked about the definition and some properties of continuous functions. We saw that continuity is a local property. In the present chapter, we shall view some continuity-related results in a rather more abstract way.

9.1 Global Continuity

In Chapter 4, we were mainly concerned with continuity of a function at a point. We defined a function as continuous on a subset A on $\mathbb{R}$, if it is continuous at every point of A. In this section, we shall see that the continuity of a function on a given set A can be expressed without reference to any particular element of the set A. First we show the following result.

Theorem 9.1

The following statements about a real function f are equivalent:

(1) The function f is continuous on its domain $\text{dom}\,(f)$.

(2) For each open set V in $\mathbb{R}$, there exists an open set U in $\mathbb{R}$ such that $U \cap \text{dom}\,(f) = f^{-1}\,(V)$.

Proof

Suppose that f is continuous, and let V be open in $\mathbb{R}$. Consider an element $a \in f^{-1}(V)$. Then $f(a) \in V$. Since V is open, it is a neighborhood of $f(a)$, and therefore there exists a neighborhood G of a in $\mathbb{R}$ (see Example 8.22, page 223), such that

$$x \in G \cap \text{dom}(f) \quad \text{implies} \quad f(x) \in V.$$

As a neighborhood of a, G contains an open set U_a containing a. Hence, we have

$$x \in U_a \cap \text{dom}(f) \quad \text{implies} \quad f(x) \in V.$$

Consider the set $U = \bigcup_{a \in f^{-1}(V)} U_a$. Then, U is clearly open and $U \cap \text{dom}(f) \subset f^{-1}(V)$. On the other hand, by our very definition of U, if $a \in f^{-1}(V)$, then $a \subset U$. Thus $f^{-1}(V) \subset U$. Since $f^{-1}(V) \subset \text{dom}(f)$, we have $f^{-1}(V) \subset U \cap \text{dom}(f)$, and hence, $f^{-1}(V) = U \cap \text{dom}(f)$ as desired. Thus (1) implies (2).

Conversely, suppose (2) holds. Let $a \in \text{dom}(f)$ and let $\varepsilon > 0$. Then the set $V = B(f(a), \varepsilon)$ is a open neighborhood of $f(a)$ in $\mathbb{R}$. Therefore, property (2) implies that there exists an open set U in $\mathbb{R}$ such that $U \cap \text{dom}(f) = f^{-1}(V)$. Since $f(a) \in V$, then $a \in f^{-1}(V) \subset U$. Thus the set U is a neighborhood of a. In view of Example 8.22, we can conclude that (2) implies (1).   □

We recall that given a real function f and a set $A \subset \text{dom}(f)$, the **restriction** of f to A is the function $f|_A : A \to \mathbb{R}$ defined by $f|_A(x) = f(x)$ for all $x \in A$. It is plain that $\text{dom}(f|_A) = A$. Also, we should convince ourselves that the function f is continuous on the set A if and only if the function $f|_A$ is continuous on its domain. The next corollary then easily follows, and shows that continuity on a set is in fact a "global" property.

Corollary 9.2

The following statements about a real function f are equivalent:

(1) The function f is continuous on a set $A \subset \text{dom}(f)$.

(2) For each open set V in $\mathbb{R}$, there exists an open set U in $\mathbb{R}$ such that $U \cap A = f^{-1}(V)$.

The following is a companion of Corollary 9.2.

Example 9.3

Show that $f : A \to \mathbb{R}$ is continuous if and only if for each closed set F in $\mathbb{R}$, there exists a closed set E in $\mathbb{R}$, such that $E \cap A = f^{-1}(F)$.

Solution

Suppose that f is continuous. Let F be closed in $\mathbb{R}$. Then $V = \mathbb{R} \setminus F$ is open. Thus there exists an open set U in $\mathbb{R}$ such that

$$U \cap A = f^{-1}(\mathbb{R} \setminus F) = A \setminus f^{-1}(F).$$

Consider the set $E = \mathbb{R} \setminus U$. Then E is closed and

$$f^{-1}(F) = A \setminus U \cap A = (A \setminus U) \cap A$$
$$= (\mathbb{R} \setminus U) \cap A = E \cap A.$$

This proves the necessity.

For the sufficiency, let V be open in $\mathbb{R}$. Then $F = \mathbb{R} \setminus V$ is closed. Then there exists a closed subset E of $\mathbb{R}$ such that $E \cap A = f^{-1}(F)$. Consider the open set $U = \mathbb{R} \setminus E$. We have

$$f^{-1}(V) = f^{-1}(\mathbb{R} \setminus F) = A \setminus f^{-1}(F)$$
$$= A \setminus E \cap A = U \cap A.$$

The proof is complete. $\square$

For the case where $A = \mathbb{R}$, we have

Corollary 9.4

The following assertions are equivalent for a function $f : \mathbb{R} \to \mathbb{R}$:

(1) f is continuous on $\mathbb{R}$;

(2) $f^{-1}(U)$ is open whenever U is open;

(3) $f^{-1}(F)$ is closed whenever F is closed.

The following example is known as the **Principle of extension of equalities**.

Example 9.5

Let A be dense in $\mathbb{R}$, and let f and g be continuous functions both defined on $\mathbb{R}$. Suppose that $f(x) = g(x)$ for all $x \in A$. Show that $f(x) = g(x)$ for all $x \in \mathbb{R}$.

Solution

Let $B = \{x \in \mathbb{R} : f(x) = g(x)\}$. We wish to show that $B = \mathbb{R}$. We first notice that $A \subset B \subset \mathbb{R}$. Therefore, since $A^- = \mathbb{R}$, we are done if we show that B is closed.

We notice that $B = (f - g)^{-1}(\{0\})$. Since $f - g$ is a continuous function, and the set $\{0\}$ is closed in $\mathbb{R}$, property (3) of Corollary 9.4 implies that B is closed. This concludes our proof.   □

In Chapter 4, we used the intermediate value theorem to prove that the image of an interval by a continuous function is again an interval. Therefore, the intermediate value theorem implies the preservation of connectedness. Now we shall see that the situation can be reversed, that is, the preservation of connectedness is used to prove the intermediate value theorem.

Example 9.6

Let $A \subset \mathbb{R}$ be connected. Show that if $f : A \to \mathbb{R}$ is continuous, then $f(A)$ is connected.

Solution

Suppose that $f(A)$ is disconnected. Then there exist two open sets V_1 and V_2 such that $V_1 \cap f(A)$ and $V_2 \cap f(A)$ are disjoint nonempty sets, and such that $(V_1 \cap f(A)) \cup (V_2 \cap f(A)) = f(A)$. By continuity of f, Corollary 9.2 implies the existence of two open sets U_1 and U_2 such that

$$U_1 \cap A = f^{-1}(V_1) \quad \text{and} \quad U_2 \cap A = f^{-1}(V_2).$$

Therefore, $U_1 \cap A$ and $U_2 \cap A$ are not empty,

$$
\begin{aligned}
(U_1 \cap A) \cap (U_2 \cap A) &= f^{-1}(V_1) \cap f^{-1}(V_2) \\
&= f^{-1}(V_1) \cap f^{-1}(V_2) \cap f^{-1}(f(A)) \\
&= f^{-1}[(V_1 \cap f(A)) \cap (V_2 \cap f(A))] = \varnothing
\end{aligned}
$$

and

$$
\begin{aligned}
(U_1 \cap A) \cup (U_2 \cap A) &= f^{-1}(V_1) \cup f^{-1}(V_2) \cap f^{-1}(f(A)) \\
&= f^{-1}[(V_1 \cap f(A)) \cup (V_2 \cap f(A))] \\
&= f^{-1}(f(A)) = A.
\end{aligned}
$$

Therefore, A would be disconnected. This contradiction concludes the proof.   □

Suppose that $f : A \to \mathbb{R}$ is continuous, where $A \subset \mathbb{R}$ is connected. Let y_0 be such that $\inf f < y_0 < \sup f$. We claim that $y_0 \in f(A)$. Indeed, suppose $y_0 \notin f(A)$. Consider the two open intervals $(-\infty, y_0)$ and (y_0, ∞). Then we easily check that

- $(-\infty, y_0) \cap f(A) \neq \varnothing$,
- $(y_0, \infty) \cap f(A) \neq \varnothing$,
- $[(-\infty, y_0) \cap f(A)] \cap [(y_0, \infty) \cap f(A)] = \varnothing$, and
- $[(-\infty, y_0) \cap f(A)] \cup [(y_0, \infty) \cap f(A)] = f(A)$.

That is, $f(A)$ is disconnected. This contradicts the previous example. Hence, we have proved the following form of the intermediate value theorem.

Corollary 9.7

Let $A \subset \mathbb{R}$ be connected. Suppose that $f : A \to \mathbb{R}$ is continuous. Then for every real number y_0 satisfying $\inf f < y_0 < \sup f$, there exists at least one point $x_0 \in A$ such that $f(x_0) = y_0$. (Compare with Theorem 4.26, page 110.)

9.2 Functions Continuous on a Compact Set

One of the main results of Chapter 4 is the fact that a continuous function on a closed bounded interval attains its maximum and its minimum. In this section, we shall see that this property of continuous functions extends to the more general case of compact sets. We begin with a couple of important examples.

Example 9.8

A function continuous on a compact set is bounded.

Solution

Let A be compact and let $f : A \to \mathbb{R}$ be continuous. Suppose to the contrary that f is not bounded, say above. Then for each positive integer n, there exists an $x_n \in A$ such that

$$f(x_n) > n.$$

Since A is compact, the sequence (x_n) has a subsequence (x_{n_k}) converging to an element a of A. On the one hand, we have

$$f\left(x_{n_k}\right) > n_k \geq k$$

for each k. Thus the sequence $\left(f\left(x_{n_k}\right)\right)$ is not bounded. On the other hand, since f is continuous, $f\left(x_{n_k}\right) \to f\left(a\right)$. Thus the sequence $\left(f\left(x_{n_k}\right)\right)$ is convergent, and therefore it must be bounded. This contradiction gives the desired result.   □

Example 9.9

Let $f : A \to \mathbb{R}$ be continuous where A is compact. Show that for every B closed and $B \subset A$, $f\left(B\right)$ is closed.

Solution

We first recall that $f\left(B\right) = \{y \in \mathbb{R} : y = f\left(x\right)$ for some $x \in B\}$. Let $\left(y_n\right)$ be a sequence in $f\left(B\right)$ converging to some number y. Then for each n, there exists $x_n \in B$ such that $y_n = f\left(x_n\right)$. Since A is compact, the closed subset B is compact and the sequence $\left(x_n\right)$ has a subsequence $\left(x_{n_k}\right)$ converging to some element $x \in B$. Since f is continuous at x, we have $y_{n_k} = f\left(x_{n_k}\right) \to f\left(x\right)$. Since $\left(y_{n_k}\right)$ also converges to y, we have $y = f\left(x\right)$; i.e. $y \in f\left(B\right)$. Hence, $f\left(B\right)$ is closed. The proof is complete.   □

The next result can be inferred from Example 9.8 and Example 9.9. However, we give a direct proof.

Theorem 9.10

Let $f : \mathbb{R} \to \mathbb{R}$ be continuous and let A be compact. Then $f\left(A\right)$ is compact.

Proof

Let $\left(V_i\right)_{i \in I}$ be an open cover of $f\left(A\right)$. Then since f is continuous, it follows from Corollary 9.4 that $\left(f^{-1}\left(V_i\right)\right)_{i \in I}$ is an open cover of A. Since A is compact, there exists a finite subset J of I such that $A \subset \bigcup_{i \in J} f^{-1}\left(V_i\right)$. This implies that $A \subset \bigcup_{i \in J} V_i$. Hence $f\left(A\right)$ is compact.   □

Corollary 9.11

A real function, which is continuous on a compact set, attains its maximum and its minimum.

Proof

Let A be compact and let $f : A \to \mathbb{R}$ be continuous. By Theorem 9.10, $f(A)$ is compact, hence it is sequentially compact. According to Example 8.48, $\max f(A)$ and $\min f(A)$ exist.   □

A real-valued function f defined on a set A is said to be **upper semicontinuous** at $x_0 \in A$, if for every $\varepsilon > 0$, there exists $\delta > 0$ such that

$$|x - x_0| < \delta \quad \text{implies} \quad f(x) < f(x_0) + \varepsilon.$$

Similarly f is said to be **lower semicontinuous** at $x_0 \in A$, if for every $\varepsilon > 0$, there exists $\delta > 0$ such that

$$|x - x_0| < \delta \quad \text{implies} \quad f(x_0) - \varepsilon < f(x).$$

For example, $f(x) = \text{int}(x)$ is upper semicontinuous for all x. It is clear that a function is continuous at x if and only if it is upper and lower semicontinuous at x. In fact, given a continuous function f, by increasing (resp. decreasing) the value $f(x_0)$ taken by f at x_0, one obtains a function which is upper (resp. lower) semicontinuous at x_0.

Example 9.12

Let A be compact and let $f : A \to \mathbb{R}$ be lower semicontinuous. Show that $\min f(A)$ exists.

Solution

We first show that $m = \inf f(A)$ exists as a real number. Suppose to the contrary that $\inf f(A) = -\infty$. Then there exists a sequence (x_n) in A such that $f(x_n) < -n$. Since A is compact, (x_n) has a subsequence (x_{n_k}) converging to some element $x_0 \in A$. Since f is lower semicontinuous at x_0, there exists $\delta > 0$ such that

$$x \in B(x_0, \delta) \quad \text{implies} \quad f(x_0) - 1 < f(x).$$

We can choose $N > 0$ sufficiently large enough so that $k > N$ implies

$$x_{n_k} \in B(x_0, \delta) \quad \text{and hence} \quad f(x_0) - 1 < f(x_{n_k}).$$

This is impossible since $f(x_{n_k}) < -n_k$ for all k. Hence $m = \inf f(A)$ exists.
Next we prove that $m \in A$. Let (t_n) be a sequence such that

$$f(t_n) \leq m + \frac{1}{n}.$$

Again, since A is compact, (t_n) has a subsequence (t_{n_k}) converging to some element t_0 of A. We finish the proof if we show that $f(t_0) = m$.

Suppose that $f(t_0) > m$. Since f is lower semicontinuous at t_0, for each N, there exists $\delta_N > 0$ such that

$$t \in B(t_0, \delta_N) \quad \text{implies} \quad f(t_0) - \frac{1}{N} < f(t).$$

It follows that

$$t \in B(t_0, \delta_N) \quad \text{implies} \quad m + \frac{1}{N} < f(t).$$

We can choose $N' > N$ so that $n > N'$ implies $t_n \in B(t_0, \delta_N)$, and hence $m + \frac{1}{N} < f(t_n)$. Thus for $n > N'$

$$f(t_n) \leq m + \frac{1}{n} < m + \frac{1}{N} < f(t_n).$$

This contradiction shows that $f(t_0) = m$. $\qquad\qquad\qquad\qquad\qquad\qquad$ $\square$

We leave it to the reader to state and prove the corresponding result for upper semicontinuous functions. Our next example is often referred to as the **homeomorphism theorem**.

Example 9.13

Let A be compact and let $f : A \to \mathbb{R}$ be continuous and one-to-one. Show that $f^{-1} : f(A) \to \mathbb{R}$ is continuous.

Solution

Let F be closed in $\mathbb{R}$. Then the set $F \cap A$ is compact in $\mathbb{R}$ and $\left(f^{-1}\right)^{-1}(F \cap A) = f(F \cap A)$. Since f is continuous, the set $E = f(F \cap A)$ is compact, and hence, it is closed in $\mathbb{R}$. It is easily checked that

$$E \cap f(A) = \left(f^{-1}\right)^{-1}(F).$$

The continuity of $f^{-1} : f(A) \to \mathbb{R}$ then follows from Example 9.3. $\qquad$ $\square$

Another remarkable property of continuous functions on compact sets is contained in the following

Theorem 9.14 (Heine)

A continuous function on a compact set is uniformly continuous.

Proof

Let A be compact and let $f : A \to \mathbb{R}$ be continuous. Suppose that f is not uniformly continuous. Thus there exists $\varepsilon > 0$ such that for each $n > 0$, one can find s_n and t_n in A such that

$$|s_n - t_n| < \frac{1}{n}, \text{ but } |f(s_n) - f(t_n)| \geq \varepsilon.$$

Since A is compact, the sequence (s_n) has a subsequence (s_{n_k}) converging to some element s of A. Then the inequality

$$|t_{n_k} - s| \leq |t_{n_k} - s_{n_k}| + |s_{n_k} - s|$$
$$< \frac{1}{n_k} + |s_{n_k} - s|$$

implies that (t_{n_k}) also converges to s. Since f is continuous at s, we must have

$$\lim f(s_{n_k}) = \lim f(t_{n_k}) = f(s).$$

Therefore, $\lim (f(s_{n_k}) - f(t_{n_k})) = 0$. This is in direct contradiction to the fact that $|f(s_{n_k}) - f(t_{n_k})| \geq \varepsilon$, for all k. Hence f must be uniformly continuous on A. $\qquad \square$

We emphasize that pointwise convergence of a sequence of functions does not necessarily imply its uniform convergence. A result, due to Dini, gives conditions under which pointwise convergence implies uniform convergence.

Theorem 9.15 (Dini)

Let A be compact and let (f_n) be a sequence on A such that

(1) f_n is continuous for each n;

(2) $f_n \leq f_{n+1}$ for all n;

(3) (f_n) converges pointwise on A to a continuous function f.

Then (f_n) converges uniformly on A.

Proof

Fix $\varepsilon > 0$. For each $x \in A$, there exists N_x such that

$$n \geq N_x \text{ implies } 0 \leq f(x) - f_n(x) \leq \frac{\varepsilon}{3}.$$

Since f and f_{N_x} are continuous, there exists $\delta_x > 0$ such that for $x' \in B(x, \delta_x)$,

$$|f(x') - f(x)| \le \frac{\varepsilon}{3} \quad \text{and} \quad |f_{N_x}(x') - f_{N_x}(x)| \le \frac{\varepsilon}{3}.$$

Thus

$$
\begin{aligned}
0 &\le f(x') - f_{N_x}(x') \\
&\le |f(x') - f(x)| + f(x) - f_{N_x}(x) + |f_{N_x}(x') - f_{N_x}(x)| \\
&\le \frac{\varepsilon}{3} + \frac{\varepsilon}{3} + \frac{\varepsilon}{3} = \varepsilon.
\end{aligned}
$$

Clearly $A \subset \bigcup_{x \in A} B(x, \delta_x)$. Since A is compact, there exist finitely many points $x_1, x_2, \ldots, x_m$ such that

$$A \subset B(x_1, \delta_{x_1}) \cup B(x_2, \delta_{x_2}) \cup \ldots \cup B(x_m, \delta_{x_m}).$$

Let $N = \max\{N_{x_1}, N_{x_2}, \ldots, N_{x_p}\}$. For each $x \in A$, there exists i such that $x \in B(x_i, \delta_{x_i})$. Thus

$$
\begin{aligned}
0 &\le f(x) - f_n(x) \\
&\le f(x) - f_N(x) \le f(x) - f_{N_{x_i}}(x) \le \varepsilon
\end{aligned}
$$

for all $x \in A$. Since $\varepsilon > 0$ is arbitrary, the proof is complete. $\qquad\square$

An application of Dini's theorem is the following

Example 9.16

Let (p_n) be a sequence of polynomials on $[0, 1]$ defined inductively by

$$p_1(x) = 0 \quad \text{for all } c \in [0, 1];$$

$$p_{n+1}(x) = p_n(x) + \frac{1}{2}\left[x - (p_n(x))^2\right].$$

(1) Show that $0 \le p_n(x) \le \sqrt{x}$ for all $x \in [0, 1]$.

(2) Show that (p_n) converges pointwise on $[0, 1]$.

(3) Show that (p_n) converges uniformly to $\sqrt{x}$ on $[0, 1]$.

Solution

(1) Let $A = \{n \in \mathbb{N} : 0 \le p_n(x) \le \sqrt{x} \text{ for all } x \in [0, 1]\}$. It is clear that $0 \le p_1(x) \le \sqrt{x}$. Thus $1 \in A$. Assume that $n \in A$. Then clearly $0 \le p_{n+1}(x)$ for

all $x \in [0, 1]$. On the other hand, for all $x \in [0, 1]$, we have

$$\sqrt{x} - p_{n+1}(x) = \sqrt{x} - \frac{1}{2}\left[x - (p_n(x))^2\right]$$

$$= \left[\sqrt{x} - p_n(x)\right]\left[1 - \frac{1}{2}\left(\sqrt{x} + p_n(x)\right)\right]$$

$$\geq 0.$$

Therefore $n + 1 \in A$. Hence by the principle of mathematical induction, $A = \mathbb{N}$.

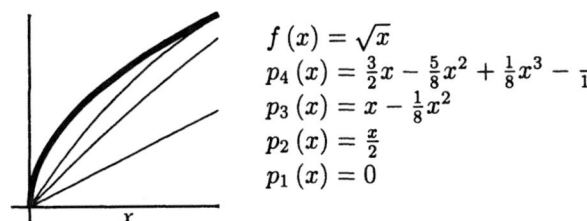

$$f(x) = \sqrt{x}$$
$$p_4(x) = \frac{3}{2}x - \frac{5}{8}x^2 + \frac{1}{8}x^3 - \frac{1}{128}x^4$$
$$p_3(x) = x - \frac{1}{8}x^2$$
$$p_2(x) = \frac{x}{2}$$
$$p_1(x) = 0$$

(2) Fix $x \in [0, 1]$. We have for every n,

$$p_{n+1}(x) - p_n(x) = \frac{1}{2}\left[x - (p_n(x))^2\right] \geq 0. \tag{9.1}$$

Thus $(p_n(x))$ is a nondecreasing sequence. Since it is bounded by $\sqrt{x}$, it is convergent to some number $f(x)$. It follows that (p_n) converges pointwise on $[0, 1]$.

(3) It easily follows from (9.1) that $(f(x))^2 = x$ for each $x \in [0, 1]$, and so since $f \geq 0$, $f(x) = \sqrt{x}$. Thus (p_n) converges pointwise to $\sqrt{x}$. Since $[0, 1]$ is compact and (p_n) is nondecreasing, Dini's theorem applies and implies that (p_n) converges uniformly to $\sqrt{x}$. $\qquad\square$

9.3 Stone–Weierstrass Theorem

In Chapter 7, we showed that power series provided one means of approximating functions by polynomials. However, this kind of approximation is limited only to those functions which have derivatives of any order. In fact, any continuous functions (not necessarily differentiable) on a closed bounded interval can also be approximated by polynomials. Such a result is known as the Stone–Weierstrass approximation theorem.

First we fix some notation. If A is a nonempty subset of $\mathbb{R}$, we denote by $\mathcal{C}(A)$ the set of all real-valued continuous functions on A. Let us recall some known properties of $\mathcal{C}(A)$. For any two f and g in $\mathcal{C}(A)$, we have $f + g$, $f \cdot g$,

$f \vee g$, and $f \wedge g \in \mathcal{C}(A)$. We also denote by $\mathcal{P}(A)$ the subset of $\mathcal{C}(A)$ consisting of polynomials on A.

Definition 9.17

A subset of $L \subset \mathcal{C}(A)$ is said to **separate the points** of A if for any two distinct points x and y of A, there exists a function f in L such that $f(x) \neq f(y)$.

Let s and t be distinct points in a given interval I. Then the polynomial $p(x) = x - s$ separates s and t as $p(s) = 0$ and $p(t) = t - s \neq 0$. Hence $\mathcal{P}(I)$ separates the points in $\mathcal{C}(I)$. The following result is more precise.

Example 9.18

Let s and t be two distinct points in an interval I. For any real numbers a and b, there exists a polynomial p defined on I such that $p(s) = a$ and $p(t) = b$.

Solution

It suffices to take the polynomial $p(x) = \frac{1}{t-s}[(b-a)(x-s) + a(t-s)]$. $\square$

This property of polynomials is shared by some subspaces of $\mathcal{C}(A)$. First we recall the following definition.

Definition 9.19

A nonempty subset $L \subset \mathcal{C}(A)$ is said to be a **linear subspace** of $\mathcal{C}(A)$ if it satisfies

(1) for every f and g in L, $f + g \in L$; and

(2) for every $f \in L$ and for each $\alpha \in \mathbb{R}$, $\alpha f \in \mathcal{C}(A)$.

For example, it is easy to see that if A is a nonempty subset of $\mathbb{R}$, then $\mathcal{P}(A)$ is a linear subspace of $\mathcal{C}(A)$. From here on, we shall denote by $\mathbf{1}$ the constant function defined by $\mathbf{1}(x) = 1$ for every $x \in A$.

Theorem 9.20

Let A be a nonempty subset of $\mathbb{R}$ and let L be a linear subspace of $\mathcal{C}(A)$. Suppose that

(1) L separates the points of A; and

(2) the constant function $\mathbf{1} \in L$.

Then for any two distinct s and t in A and real numbers a and b, there exists $f \in L$ such that $f(s) = a$ and $f(t) = b$.

Proof

Since L separates the points of A, there exists $g \in L$ such that $g(s) \neq g(t)$. Then the function

$$f(x) = \frac{1}{g(t) - g(s)} \left[(b - a) \left[g(x) - g(s) \right] + a \left[g(t) - g(s) \right] \mathbf{1}(x) \right]$$

clearly belongs to L and $f(s) = a$ and $f(t) = b$. $\square$

Definition 9.21

A nonempty subset $L \subset \mathcal{C}(A)$ is said to be a **lattice subspace** of $\mathcal{C}(A)$ if it is a linear subspace with the additional property that

for every f and g in L, $f \vee g$ and $f \wedge g$ are in L.

Theorem 9.22

Let A be a nonempty compact subset of $\mathbb{R}$ and let L be a lattice subspace of $\mathcal{C}(A)$. Suppose that

(1) L separates the points of A; and

(2) the constant function $\mathbf{1} \in L$.

Then for a given $f \in \mathcal{C}(A)$, any real number $s \in A$, and any $\varepsilon > 0$, there exists $g \in L$ such that

$$g(s) = f(s) \text{ and } f(x) - \varepsilon < g(x) \text{ for all } x \in A.$$

Proof

Let $f \in \mathcal{C}(A)$, $s \in A$, and $\varepsilon > 0$. For each $t \in A$, let f_t be the elements of L such that $f_t(s) = f(s)$ and $f_t(t) = f(t)$. Fix $\varepsilon > 0$. Since f is continuous at t, there exists $\delta_1 > 0$ such that

$$x \in B(t, \delta_1) \text{ implies } f(x) - \frac{\varepsilon}{2} < f(t). \tag{9.2}$$

Since f_t is continuous at t, there exists $\delta_2 > 0$ such that

$$x \in B(t, \delta_2) \quad \text{implies} \quad f_t(t) < f_t(x) + \frac{\varepsilon}{2}. \tag{9.3}$$

Equations (9.2) and (9.3) and the fact that $f_t(t) = f(t)$ imply that if $\delta = \min\{\delta_1, \delta_2\}$ then

$$f(x) - \varepsilon < f_t(x) \quad \text{whenever } x \in B(t, \delta).$$

Clearly, $A \subset \bigcup_{t \in A} B(t, \delta)$. Since A is compact, there exist finitely many points $t_1, t_2, \ldots, t_m$ in A such that $A \subset \bigcup_{i=1}^{m} B(t_1, \delta)$. Consider the function

$$g = f_{t_1} \vee f_{t_2} \vee \cdots \vee f_{t_m}.$$

Since L is a lattice, $g \in L$. We also have $g(s) = f(s)$. Now if $x \in A$, then $x \in B(t_p, \delta)$ for some $p \in \{t_1, t_2, \ldots, t_m\}$. Thus

$$f(x) - \varepsilon < f_{t_p}(x) \le g(x),$$

holds for all $x \in A$. This proves the theorem. $\square$

The above result is improved in the following one.

Theorem 9.23 (Stone Approximation)

Let A be a nonempty compact subset of $\mathbb{R}$ and let L be a lattice subspace of $\mathcal{C}(A)$. Suppose that

(1) L separates the points of A; and

(2) the constant function $\mathbf{1} \in L$.

Then for a given $f \in \mathcal{C}(A)$, any real number $s \in A$, and any $\varepsilon > 0$, there exists $g \in L$ such that

$$g(s) = f(s) \quad \text{and} \quad f(x) - \varepsilon < g(x) < f(x) + \varepsilon \text{ for all } x \in A.$$

Proof

Let $f \in \mathcal{C}(A)$ and fix $\varepsilon > 0$. For each $s \in A$, by Theorem 9.22 there exists $g_s \in L$ such that

$$g_s(s) = f(s) \quad \text{and} \quad f(x) - \varepsilon < g_s(x) \text{ for all } x \in A.$$

Since f is continuous at s, there exists $\delta_1 > 0$ such that

$$x \in B(s, \delta_1) \quad \text{implies} \quad f(s) - \frac{\varepsilon}{2} < f(x). \tag{9.4}$$

Since g_s is continuous at s, there exists $\delta_2 > 0$ such that

$$x \in B(s, \delta_2) \text{ implies } g_s(x) < g_s(s) + \frac{\varepsilon}{2}. \tag{9.5}$$

Equations (9.4) and (9.5) and the fact that $g_s(s) = f(s)$ imply that if $\delta = \min\{\delta_1, \delta_2\}$ then

$$g_s(x) < f(x) + \varepsilon \text{ whenever } x \in B(s, \delta).$$

Again, since A is compact, there exist finitely many points $s_1, s_2, \ldots, s_m$ in A such that $A \subset \bigcup_{i=1}^{m} B(s_i, \delta)$. Consider the function

$$g = g_{s_1} \wedge g_{s_2} \wedge \cdots \wedge g_{s_m}.$$

Since L is a lattice, $g \in L$. We also have $g(s) = f(s)$. Now if $x \in A$, then $x \in B(s_p, \delta)$ for some $p \in \{t_1, t_2, \ldots, t_m\}$. Hence,

$$f(x) - \varepsilon < g(x) < f(x) + \varepsilon \text{ for all } x \in A.$$

This completes the proof. $\square$

The Stone approximation theorem states that any continuous function defined on a nonempty compact set A, can be uniformly approximated on A by elements of any lattice subspace L, which separates points in A, and which contains the constant function $\mathbf{1}$. The Stone–Weierstrass theorem states a slightly different result: every $f \in \mathcal{C}(A)$ can be uniformly approximated on A by elements of $\mathcal{P}(A)$. As we have noticed before,

- $\mathcal{P}(A)$ is a linear subspace of $\mathcal{C}(A)$;

- $\mathcal{P}(A)$ separates the points in A;

- $\mathbf{1} \in \mathcal{P}(A)$.

Since $\mathcal{P}(A)$ is not a lattice, the Stone approximation theorem does not apply directly. Nevertheless, we shall see how the Stone approximation theorem is used to prove the Stone–Weierstrass theorem.

Definition 9.24

A nonempty subset $L \subset \mathcal{C}(A)$ is said to be an **algebra subspace** if it is a linear subspace with the additional property that $f \cdot g \in L$ whenever f and g are in L.

It is clear that if A is a compact subset of $\mathbb{R}$, then $\mathcal{P}(A)$ is an algebra subspace of $\mathcal{C}(A)$. (Prove it!)

We now formally state the Stone–Weierstrass theorem.

Theorem 9.25 (Stone–Weierstrass)

Let A be a nonempty compact subset of $\mathbb{R}$ and let L be an algebra subspace of $\mathcal{C}(A)$. Suppose that

(1) L separates the points of A;

(2) the constant function $\mathbf{1} \in L$.

Then for any $f \in \mathcal{C}(A)$ and for any $\varepsilon > 0$ there exists $g \in L$ such that

$$f(x) - \varepsilon < g(x) < f(x) + \varepsilon \quad \text{for all } x \in A. \tag{9.6}$$

Proof

Let L^- denote the set of functions f on A which can be uniformly approximated by elements of L. Since the elements of L are continuous, so are the elements of L^- (why?). Hence $L \subset L^- \subset \mathcal{C}(A)$. It follows that L^- also separates the points of A, and that $\mathbf{1} \in L^-$. Using the properties of limit, we also see that L^- is an algebra subspace of $\mathcal{C}(A)$ (prove it!).

We shall now show that L^- is a lattice subspace. To see this, let $f \in L^-$, $f \neq 0$. Set

$$a = \sup \{|f(x)| : x \in A\}.$$

(a is a real number since A is compact and f is continuous on A.) Since L^- is an algebra subspace, $f^2/a^2 \in L^-$, and more generally for any polynomial $p \in \mathcal{P}([0,1])$, the function $p(f^2/a^2) \in L^-$. Let (p_n) be the sequence of polynomials defined in Example 9.16. Then the sequence (g_n) defined by $g_n = p_n(f^2/a^2)$, belongs to L^- and converges uniformly on A to $\sqrt{f^2/a^2} = |f|/a$. It follows that $|f| \in L^-$. Now for two arbitrary elements f and g in L^-, we have

$$f \vee g = \frac{1}{2}(f + g + |f - g|) \quad \text{and} \quad f \wedge g = \frac{1}{2}(f + g - |f - g|).$$

Hence $f \vee g$ and $f \wedge g$ are both in L^-. This proves our claim.

We can then apply the Stone Approximation Theorem 9.23 to L^- to complete our proof. $\qquad \square$

Sometimes, it is helpful to write (9.6) in its equivalent form

$$\sup \{|f(x) - g(x)| : x \in A\} < \varepsilon.$$

Technically, the Stone–Weierstrass theorem says that an algebra subspace L of $\mathcal{C}(A)$, which separates points in A, and which contains the constant functions, is uniformly dense in $\mathcal{C}(A)$.

One easily checks that the polynomials satisfy the hypotheses of the Stone–Weierstrass theorem, and then one easily realizes the following celebrated Weierstrass theorem.

Corollary 9.26 (Weierstrass)[1]

A continuous real-valued function on a compact set A can be approximated uniformly on A by polynomials.

9.4 Fixed-point Theorem

Let A be a complete subset of $\mathbb{R}$. Let $f : A \to \mathbb{R}$. Then f is said to be **Lipschitz** or to satisfy the **Lipschitz condition** if there exists $0 < \alpha$ such that

$$|f(x) - f(y)| \leq \alpha |x - y|.$$

The space of all Lipschitz functions on a set A is denoted by $\mathrm{Lip}(A)$. If $0 < \alpha < 1$ in the above inequality, then f is called a **contraction**. The number α is then called the **contraction constant**.

Example 9.27

Let f be a function defined and differentiable on a closed interval I. Suppose that $\sup\{|f'(x)| : x \in I\} < 1$. Show that f is a contraction.

Solution

Let $x, y \in I$, say $x \leq y$. Applying the mean value theorem on the interval $[x, y]$, we can find $c \in [x, y]$ such that

$$f(x) - f(y) = f'(c)(x - y).$$

It easily follows that

$$|f(x) - f(y)| \leq \alpha |x - y|,$$

where $\alpha = \sup\{|f'(x)| : x \in I\}$. Hence f is a contraction. $\square$

It is easily seen that a Lipschitz function on a set A is uniformly continuous on A: indeed if $|f(x) - f(y)| \leq \alpha |x - y|$ holds for all $x, y \in A$, then on setting $\delta = \varepsilon/\alpha$, we establish the uniform continuity of f on A.

---

[1] There are several other proofs of the Weierstrass theorem. Bernstein's proof (see Appendix) provides an actual construction of the approximating polynomials.

We recall that a point a is called a **fixed point** for a function f if $f(a) = a$.

Theorem 9.28 (Fixed-point Theorem)

Let A be a complete subset of $\mathbb{R}$. A contraction $f : A \to A$ has a unique fixed point.

Proof

Suppose first that x and y are two fixed points. Then $f(x) = x$, $f(y) = y$ and we have

$$|x - y| = |f(x) - f(y)| \le \alpha |x - y|.$$

Since $0 < \alpha < 1$, the above inequality implies that $|x - y| = 0$, and so $x = y$. Hence f has at most one fixed point.

To see that f has a fixed point, choose a point $a \in A$ and define the sequence (x_n) inductively (see Figure 9.1) by

$$x_1 = a \text{ and } x_{n+1} = f(x_n) \text{ for } n = 1, 2, \ldots . \tag{9.7}$$

It follows that, for $n = 2, 3, \ldots$

$$|x_{n+1} - x_n| = |f(x_n) - f(x_{n-1})| \le \alpha |x_n - x_{n-1}|.$$

So by induction, we have

$$|x_{n+1} - x_n| \le \alpha^{n-1} |x_2 - x_1|.$$

Thus, for all n and for all p we have

$$\begin{aligned}
|x_{n+p} - x_n| &\le \sum_{i=1}^{p} |x_{n+i} - x_{n+i-1}| \tag{9.8} \\
&\le |x_2 - x_1| \sum_{i=1}^{p} \alpha^{n+i-2} \\
&\le \frac{|x_2 - x_1|}{1 - \alpha} \alpha^{n-1}.
\end{aligned}$$

Now since $0 < \alpha < 1$, $\lim \alpha^n = 0$, this inequality shows that (x_n) is a Cauchy sequence. Since A is complete, $\lim x_n = x$ exists in A. By continuity of f, we have

$$f(x) = \lim f(x_n) = \lim x_{n+1} = x,$$

so x is a fixed point. This completes the proof. $\qquad\square$

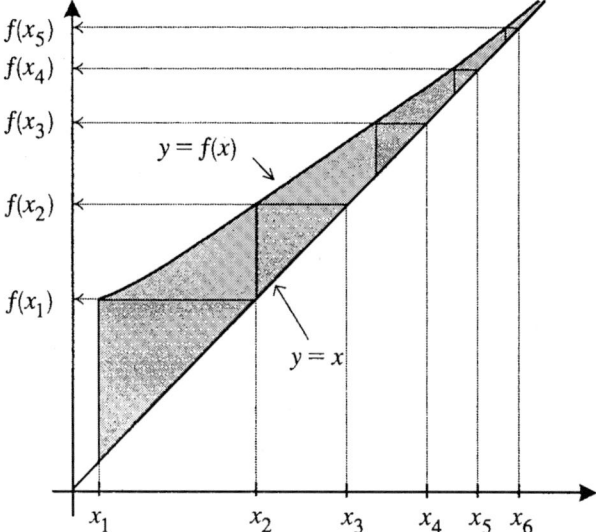

Figure 9.1   Fixed point

Notice that our proof relies heavily on the completeness of the set A. We also observe from the above proof that if f is a contraction with contraction constant α, and if (x_n) is the sequence defined as in (9.7), then (x_n) converges to a unique fixed point x with the rate of convergence estimated by (9.8). The argument used to show the existence of x such that $f(x) = x$ in the proof of the above theorem is also known as "the method of successive approximations".

9.5 Ascoli–Arzelà Theorem

In Chapter 2, the Bolzano–Weierstrass theorem asserts that every bounded sequence of real numbers has a convergent subsequence. In this section we will study the corresponding statement for sequences of functions. To do this we need a few definitions.

Definition 9.29

A family $\mathcal{F}$ of real-valued functions on a set A is said to be

(1) **pointwise bounded** if for every $x \in A$, there exists $\alpha_x > 0$ such that

$$|f(x)| < \alpha_x \text{ for each } f \in \mathcal{F};$$

(2) **uniformly bounded** if there exists $\alpha > 0$ such that

$$\sup\{|f(x)| : x \in A\} < \alpha \text{ for each } f \in \mathcal{F}.$$

Clearly, uniform boundedness implies pointwise boundedness. An example of a pointwise bounded but not uniformly bounded family is given by the singleton $\mathcal{F} = \{f\}$ where $f(x) = 1/x$ on $(0, 1)$.

Example 9.30

Show that a uniformly convergent sequence of continuous functions (f_n) on a compact set K is uniformly bounded on K.

Solution

Let (f_n) be a uniformly convergent sequence of continuous functions on K. Then (f_n) is uniformly Cauchy by Theorem 7.12. Also since K is compact, for each n

$$M_n = \sup\{|f_n(x)| : x \in K\} < \infty.$$

Consider $\varepsilon = 1$. There is N in $\mathbb{N}$ such that

$$\sup\{|f_{n+p}(x) - f_n(x)| : x \in K\} < 1 \text{ for all } n \geq N \text{ and for all } p \in \mathbb{N}.$$

In particular, $M_{N+p} < M_{N+1} + 1$, for all $p \in \mathbb{N}$. It follows that if

$$\alpha = \max\{M_1, M_2, \ldots, M_N, M_{N+1} + 1\},$$

then

$$\sup\{|f_n(x)| : x \in A\} < \alpha \text{ for each } n \in \mathbb{N}$$

as desired. $\square$

Definition 9.31

Let A be a subset of $\mathbb{R}$. A family $\mathcal{F}$ of real-valued functions on A is said to be

(1) **equicontinuous** at a point $a \in A$ if for every $\varepsilon > 0$, there exists $\delta > 0$ such that

$$x \in A \text{ and } |x - a| < \delta \text{ implies } |f(x) - f(a)| < \varepsilon, \text{ for all } f \in \mathcal{F}.$$

The family $\mathcal{F}$ is said to be equicontinuous if it is equicontinuous at every $a \in A$.

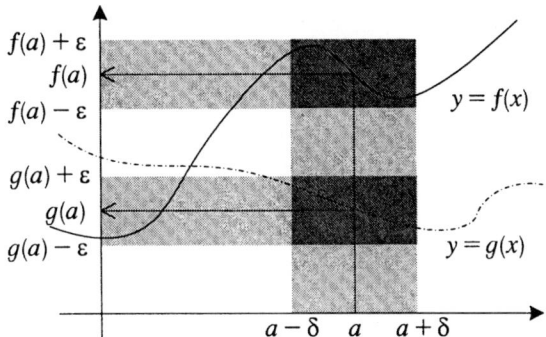

Figure 9.2   Equicontinuity at $x = a$

(2) **uniformly equicontinuous** if for every $\varepsilon > 0$, there exists $\delta > 0$ such that

$$x, y \in A \text{ and } |x - y| < \delta \text{ implies } |f(x) - f(y)| < \varepsilon, \text{ for all } f \in \mathcal{F}.$$

We notice that if the family $\mathcal{F}$ has finitely many continuous members, then it is equicontinuous (prove it). Figure 9.2 illustrates the equicontinuity of $\mathcal{F} = \{f, g\}$ at a point a. It is plain that uniform equicontinuity implies equicontinuity. It should be clear that if $\mathcal{F}$ is equicontinuous (resp. uniformly equicontinuous) on A, then each $f \in \mathcal{F}$ is continuous (resp. uniformly continuous) on A. The reader should also notice that an argument similar to that of the proof of Heine's Theorem 9.14 shows that if A is compact, then equicontinuity and uniform equicontinuity are exactly the same property. (Prove it!) We use this fact in the proof of the following

Theorem 9.32

Let K be a compact subset of $\mathbb{R}$ and let (f_n) be an equicontinuous sequence of real-valued functions. Suppose that (f_n) is pointwise bounded on K. Then (f_n) is uniformly bounded on K.

Proof

Let $\varepsilon > 0$. Since (f_n) is bounded pointwise on K, for each $x \in K$, $\varphi(x) = \sup\{|f_n(x)| : n \in \mathbb{N}\}$ exists as a real number. We claim that φ is continuous on K.

By (uniform) equicontinuity, there exists $\delta > 0$ such that

$$|f_n(x) - f_n(y)| < \varepsilon, \text{ for all } n \in \mathbb{N} \text{ whenever } x, y \in K \text{ and } |x - y| < \delta.$$

Suppose that x and y are fixed in K and such that $|x - y| < \delta$, then

$$|f_n(y)| < |f_n(x)| + \varepsilon, \text{ for all } n \in \mathbb{N}.$$

It follows that $|f_n(y)| < \varphi(x) + \varepsilon$, for all $n \in \mathbb{N}$, and hence

$$\varphi(y) < \varphi(x) + \varepsilon.$$

A similar argument (interchanging the roles of x and y) yields $\varphi(x) < \varphi(y) + \varepsilon$. We conclude that

$$|\varphi(x) - \varphi(y)| < \varepsilon \text{ whenever } x, y \in K \text{ and } |x - y| < \delta,$$

as claimed.

Since K is compact, φ is uniformly continuous on K and consequently (f_n) is uniformly bounded. This completes our proof. $\qquad\square$

Theorem 9.33

Let K be a compact subset of $\mathbb{R}$. A uniformly convergent sequence of continuous functions (f_n) on K is (uniformly) equicontinuous on K.

Proof

Let $\varepsilon > 0$. Again there is N in $\mathbb{N}$ such that

$$\sup\{|f_{n+p}(x) - f_n(x)| : x \in K\} < \varepsilon/3 \text{ for all } n \geq N \text{ and for all } p \in \mathbb{N}.$$

Also since K is compact, each f_n is uniformly compact on K. Thus for each n, there exists $\delta_n > 0$ such that

$$x, y \in K \text{ and } |x - y| < \delta_n \text{ implies } |f_n(x) - f_n(y)| < \varepsilon/3.$$

It follows that if

$$\delta = \min\{\delta_1, \delta_2, \ldots, \delta_n\},$$

then

$$x, y \in K, \text{ and } |x - y| < \delta \text{ implies } |f_n(x) - f_n(y)| < \varepsilon/3 \text{ for } n \leq N.$$

On the other hand, if $p \in \mathbb{N}$, then

$$|f_{N+p}(x) - f_{N+p}(y)| \leq |f_{N+p}(x) - f_N(x)| + |f_N(x) - f_N(y)|$$
$$+ |f_{N+p}(y) - f_N(y)|$$
$$< \frac{\varepsilon}{3} + \frac{\varepsilon}{3} + \frac{\varepsilon}{3} = \varepsilon.$$

This completes the proof. $\qquad\square$

Theorem 9.34

Let K be a compact subset of $\mathbb{R}$ and let (f_n) be an equicontinuous sequence of real-valued functions. Suppose that (f_n) converges pointwise to a function f on K. Then (f_n) converges to f uniformly on K.

Proof

Let $\varepsilon > 0$. Using equicontinuity, for each $x_0 \in K$ there exists $\delta_{x_0} > 0$ such that

$$x \in K \text{ and } x \in B(x_0, \delta_{x_0}) \text{ implies } |f_n(x) - f_n(x_0)| < \varepsilon/3, \text{ for all } n \in \mathbb{N}.$$

Since (f_n) converges pointwise to a function f on K, it follows that

$$x \in K \text{ and } x \in B(x_0, \delta_{x_0}) \text{ implies } |f(x) - f(x_0)| < \varepsilon/3, \text{ for all } n \in \mathbb{N}.$$

It is clear that $K \subset \bigcup_{x_0 \in K} B(x_0, \delta_{x_0})$. Since K is compact, there is a finite collection

$$\{B(x_1, \delta_{x_1}), B(x_2, \delta_{x_2}), \ldots, B(x_m, \delta_{x_m})\}$$

such that $K \subset \bigcup_{i=1}^{m} B(x_i, \delta_{x_i})$. For each $i \in \{1, 2, \ldots, m\}$, there exists N_i large enough that

$$n \geq N_i \text{ implies } |f_n(x_i) - f(x_i)| < \varepsilon/3.$$

It follows that

$$n \geq \max\{N_1, N_2, \ldots, N_m\} \text{ implies } |f_n(x_i) - f(x_i)| < \varepsilon/3 \text{ for all } i.$$

Therefore if $x \in K$, then $x \in B(x_i, \delta_{x_i})$ for some $i \in \{1, 2, \ldots, m\}$ and for $n \geq \max\{N_1, N_2, \ldots, N_m\}$

$$\begin{aligned}
|f_n(x) - f(x)| &\leq |f_n(x) - f_n(x_i)| \\
&\quad + |f_n(x_i) - f(x_i)| + |f(x_i) - f(x)| \\
&< \frac{\varepsilon}{3} + \frac{\varepsilon}{3} + \frac{\varepsilon}{3} = \varepsilon.
\end{aligned}$$

This completes the proof. $\square$

The conditions of the above theorem can be improved (see Exercise 9.37).

Example 9.30 and Theorem 9.33 assert that uniform convergence on compact sets implies uniform boundedness and (uniform) equicontinuity. The converse, which is established in the next theorem, can be thought of as the counterpart of the Bolzano–Weierstrass theorem for sequences of functions.

Theorem 9.35

Every uniformly bounded and equicontinuous sequence of functions on a compact set K has a uniformly convergent subsequence.

Before we prove this theorem, we notice that in light of Theorem 9.32, the uniform boundedness condition in the above result can be replaced by pointwise boundedness. The key tool for the proof is the so-called **Cantor diagonal method**.

Proof

Let (f_n) be uniformly bounded and equicontinuous on a compact set K. Then there exists $\alpha > 0$ such that

$$\sup \{|f_n(x)| : x \in A\} < \alpha \text{ for each } n \in \mathbb{N}.$$

Let $A = \{r_n : n \in \mathbb{N}\}$ be the rationals in K. Then the sequence $(f_n(r_1))_{n \in \mathbb{N}}$ is bounded and therefore admits a convergent subsequence $(f_{1n}(r_1))_{n \in \mathbb{N}}$. Then the sequence $(f_{1,n}(r_2))_{n \in \mathbb{N}}$ is also bounded and therefore admits a convergent subsequence $(f_{2,n}(r_2))_{n \in \mathbb{N}}$. Continuing in this fashion we obtain a double sequence $((f_{i,n})_{n \in \mathbb{N}})_{i \in \mathbb{N}}$ with the following properties:

(i) for each $i \in \mathbb{N}$, $(f_{i+1,n})_{n \in \mathbb{N}}$ is a subsequence of $(f_{i,n})_{n \in \mathbb{N}}$;

(ii) for each $i \in \mathbb{N}$, $(f_{i,n}(r_i))$ converges as $n \to \infty$.

$$
\begin{array}{ccccc}
f_{1,1} & f_{1,2} & \cdots & f_{1,n} & \cdots \\
f_{2,1} & f_{2,2} & \cdots & f_{2,n} & \cdots \\
\vdots & \vdots & \ddots & \vdots & \\
f_{n,2} & f_{n,2} & & f_{n,n} & \cdots \\
\vdots & \vdots & & \vdots & \ddots
\end{array}
$$

We select the "diagonal" sequence $(f_{n,n})_{n \in \mathbb{N}}$. The sequence $(f_{n,n})_{n=i}^{\infty}$ is a subsequence of $(f_{i,n})_{n \in \mathbb{N}}$, hence $(f_{n,n}(r_i))_{n \in \mathbb{N}}$ converges for each i.

Now let $\varepsilon > 0$. By (uniform) equicontinuity, there exists $\delta > 0$ such that

$$x, y \in K, \text{ and } |x - y| < \delta \text{ implies } |f_n(x) - f_n(y)| < \varepsilon/3 \text{ for } n \in \mathbb{N}.$$

Since A is dense in K, we have $K \subset \bigcup_{n=1}^{\infty} B(r_n, \delta)$. By the compactness of K, there are finitely many points $a_1, a_2, \ldots, a_k \in A$ such that $K \subset \bigcup_{j=1}^{k} B(a_j, \delta)$. Choose an integer N large enough so that for $j = 1, 2, \ldots, k$

$$|f_{n+p,n+p}(a_j) - f_{n,n}(a_j)| < \frac{\varepsilon}{3} \text{ for all } p \in \mathbb{N} \text{ and } n > N.$$

Now if $x \in K$, then $x \in B(a_j, \delta)$ for some j, and hence if $n > N$, then

$$
\begin{aligned}
|f_{n+p,n+p}(x) - f_{n,n}(x)| &\leq |f_{n+p,n+p}(x) - f_{n+p,n+p}(a_j)| \\
&\quad + |f_{n+p,n+p}(a_j) - f_{n,n}(a_j)| \\
&\quad + |f_{n,n}(a_j) - f_{n,n}(x)| \\
&< \frac{\varepsilon}{3} + \frac{\varepsilon}{3} + \frac{\varepsilon}{3} = \varepsilon
\end{aligned}
$$

for all $p \in \mathbb{N}$. This shows that the subsequence $(f_{n,n})_{n \in \mathbb{N}}$ is uniformly Cauchy, hence it is uniformly convergent as desired. $\qquad\square$

All of these results taken together imply the following theorem.

Theorem 9.36 (Ascoli–Arzelà)

Let K be a compact subset of $\mathbb{R}$ and let $\mathcal{F}$ be a family of continuous functions on K. Then the following statements are equivalent:

(1) $\mathcal{F}$ is pointwise bounded and equicontinuous on K;

(2) every sequence from $\mathcal{F}$ contains a uniformly convergent subsequence.

EXERCISES

9.1 Let $f : \mathbb{R} \to \mathbb{R}$ be continuous and let $\alpha, \beta \in \mathbb{R}$. Show that

(1) the set $A = \{x : f(x) > \alpha\}$ is open;

(2) the set $B = \{x : \alpha \leq f(x) \leq \beta\}$ is closed.

Give an example of a function $f : \mathbb{R} \to \mathbb{R}$ and an open set U such that the set $f^{-1}(U)$ is not open

9.2 Let $A \subset \mathbb{R}$ and $f : A \to \mathbb{R}$. Show that f is continuous if and only if the sets

$$\{x \in A : f(x) < \alpha\} \quad \text{and} \quad \{x \in A : f(x) > \alpha\}$$

are open for every α.

9.3 Let $f : \mathbb{R} \to \mathbb{R}$. Show that f is continuous if and only if $f(A^-) \subset [f(A)]^-$ for every $A \subset \mathbb{R}$.

9.4 Let $A \subset \mathbb{R}$ and $f : A \to \mathbb{R}$. Show that f is lower semicontinuous if and only if the set $\{x \in A : f(x) > \alpha\}$ is open for every α.

9.5 Let A be a nonempty compact subset of $\mathbb{R}$. For each $x \in \mathbb{R}$, show that there exists $a \in A$ such that

$$|x - a| = \inf \{|x - y| : y \in A\}.$$

9.6 Define the distance between two subsets A and B of $\mathbb{R}$ by $d(A, B) = \inf \{|a - b| : a \in A, \ b \in B\}$.

    (1) Give an example of two subsets A and B with $A \cap B = \varnothing$ and such that $d(A, B) = 0$.

    (2) If $A \cap B = \varnothing$, A is nonempty and compact, B is nonempty and closed, show that $d(A, B) \neq 0$.

9.7 Show that a function $f : \mathbb{R} \to \mathbb{R}$ is continuous if and only if it is continuous on every compact subset of $\mathbb{R}$.

9.8 Give an example of a function $f : \mathbb{R} \to \mathbb{R}$ which maps an open set to a closed set.

9.9 Give an example of a function $f : \mathbb{R} \to \mathbb{R}$ and a Cauchy sequence (x_n) for which $(f(x_n))$ is not Cauchy.

9.10 Let $f : \mathbb{R} \to \mathbb{R}$ be a continuous function. Show that if A is totally bounded, then f is uniformly continuous on A.

9.11 Prove that if every continuous real function attains a maximum value on A, then A is compact.

9.12 Let f and g be continuous functions both defined on $\mathbb{R}$. Show that the set $A = \{x : f(x) = g(x)\}$ is closed.

9.13 Let (f_n) be a monotone sequence of real functions on $\mathbb{R}$. Suppose that there exists a dense subset A of $\mathbb{R}$ such that $\lim f_n(x)$ exists in $\mathbb{R}$ for each $x \in A$. Show that $\lim f_n(x)$ exists in $\mathbb{R}$ for all but at most countably many x.

9.14 Let (f_n) be a sequence on compact set A and let f be continuous on A such that $x_n \to x$ in A implies $f_n(x_n) \to f(x)$. Show that (f_n) converges uniformly to f on A.

9.15 Let (f_n) be a sequence of real-valued continuous functions on $\mathbb{R}$ such that f_n converges uniformly to some function f on every compact subset of $\mathbb{R}$. Show that f is continuous.

9.16 Show that the set of points of discontinuity of a nondecreasing function on $[a, b]$ is at most countable.

9.17 Show that any continuous function on $[0, 1]$ can be uniformly approximated by functions in the algebra generated by $\{1, x^2\}$. Do we have the same results for continuous functions on $[-1, 1]$?

9.18 Let L be the linear space generated by $\{1, \sin x, \sin^2 x, \sin^3 x, \ldots\}$ defined on $[0, 1]$. That is $f \in L$ if and only if there exist finitely many numbers $a_0, a_1, \ldots, a_N$ such that $f(x) = \sum_{k=0}^{N} a_k \sin^k x$ for all $x \in [0, 1]$. Show that L satisfies the hypotheses of the Stone–Weierstrass theorem. What conclusion can be drawn?

9.19 Show that every continuous real-valued function on $[0, \pi]$ is the uniform limit of a sequence of polynomials in cosines.

9.20 Show that every continuous real-valued function on $[0, \pi]$ is the uniform limit of a sequence of functions of the form

$$a_0 + a_1 \cos x + a_2 \cos 2x + \cdots + a_n \cos nx.$$

9.21 A function of the form

$$T_n(x) = \frac{\alpha_0}{2} + \sum_{k=1}^{n} (\alpha_k \cos kx + \beta_k \sin kx)$$

is called a **trigonometric polynomial** of degree n. Show that every continuous function on $[-\pi, \pi]$ can be uniformly approximated by trigonometric polynomials.

9.22 Let $f \in C([0, 1])$. Suppose that $\int_0^1 x^n f(x) \, dx = 0$ for $n = 0, 1, 2, \ldots$. Show that $f(x) = 0$ for all $x \in [0, 1]$.

9.23 Let $f \in C([0, 1])$. Suppose that $\int_0^1 f\left(\sqrt[2n+1]{x}\right) dx = 0$ for $n = 0, 1, 2, \ldots$. Show that $f(x) = 0$ for all $x \in [0, 1]$.

9.24 Let $f \in C([0, \infty))$ be bounded. Show that

$$f = 0 \text{ if and only if } \int_0^\infty f(x) e^{-nx} dx = 0.$$

9.25 Let $f(x) = \frac{1}{2}\left(x + \frac{2}{x}\right)$.

(1) Show that f is a contraction on $A = [1, +\infty)$ with a contraction constant $\alpha = \frac{1}{2}$ and fixed point $\sqrt{2}$.

(2) Fix $a \in A$ and define the sequence (x_n) inductively by

$$x_1 = a \text{ and } x_{n+1} = f(x_n) \text{ for } n = 1, 2, \ldots.$$

Show that $\left|x_n - \sqrt{2}\right| \leq 2^{-(n-1)}$.

9.26 Let A be a subset of $\mathbb{R}$. Let $f : A \to A$ satisfy $|f(x) - f(y)| < |x - y|$ for $x \neq y$.

   (1) Show that f has at most one fixed point.

   (2) Show that if A is compact, then f has exactly one fixed point.

9.27 Show by counterexample that the fixed-point theorem for contraction fails if

   (1) A is not complete;

   (2) $\alpha \geq 1$.

9.28 Let A be a complete subset of $\mathbb{R}$ and let $f : A \to A$ be a function from A into itself. Suppose that there exists a sequence (a_n) of real numbers such that

   (1) $\lim a_n = 0$;

   (2) $|f^n(x) - f^n(y)| < \alpha_n |x - y|$ for all n and for all $x, y \in A$, where $f^n(x) = f(f(f \cdots (f(x))))$ is the n-th iterate of f at x; that is

   $$f^1(x) = f(x) \text{ and } f^{n+1}(x) = f(f^n(x)) \text{ for } n = 1, 2, \ldots .$$

   Show that f has a unique fixed point.

9.29 Show that if $f \in C([0, 1])$, then f cannot assume each value in its range exactly twice.

9.30 Let K be a compact subset of $\mathbb{R}$. Let $f : K \to K$ satisfy $|f(x) - f(y)| \geq |x - y|$ for all $x, y \in K$.

   (1) Show that f is one-to-one and that f^{-1} is continuous.

   (2) Show that f is onto. (Hint: Given $x_0 \in K$, consider the sequence (x_n) defined by $x_{n+1} = f(x_n)$ for $n = 0, 1, 2, \ldots .$)

9.31 Let (f_n) be a sequence of differentiable functions on $[a, b]$. Suppose that

   (1) there exists $x_0 \in [a, b]$ such that the sequence $(f_n(x_0))$ converges;

   (2) the sequence (f_n') is uniformly bounded on $[a, b]$.

   Show that (f_n) has a convergent subsequence.

9.32 Write down the proof of the fact that if the family of functions $\mathcal{F}$ has finitely many members, then it is equicontinuous.

9.33 Let $f \in \mathcal{C}([0, \infty))$. For each n define $f_n(x) = f(x^n)$. Show that the family $(f_n)_{n \in \mathbb{N}}$ is equicontinuous at $x = 1$ if and only if f is a constant function.

9.34 Let $\mathcal{F}$ be a family of real functions on A. Suppose that there exists $c > 0$ and $\alpha > 1$ such that $|f(x) - f(y)| \leq c |x - y|^\alpha$ for all $x, y \in A$. Show that $\mathcal{F}$ is uniformly equicontinuous.

9.35 Suppose that $\mathcal{F}$ is an equicontinuous family of real functions on A. Let $\mathcal{F}^-$ be the set of all uniform limits of sequences in $\mathcal{F}$. Show that $\mathcal{F}^-$ is equicontinuous.

9.36 Let $\mathcal{F}$ be a family of all functions $f : [0, 1] \to \mathbb{R}$ such that f' exists, is continuous and uniformly bounded. Show that $\mathcal{F}$ is uniformly equicontinuous.

9.37 Let K be a compact subset of $\mathbb{R}$ and let (f_n) be an equicontinuous sequence of real-valued functions on K. Suppose that (f_n) converges pointwise to a function f on a subset A dense in K. Show that (f_n) converges to f uniformly on K.

9.38 Let A be a subset of $\mathbb{R}$, not necessarily compact. Suppose that (f_n) is an equicontinuous sequence of real-valued functions on A. Suppose that (f_n) converges pointwise to a function f on A. Show that $f \in \mathcal{C}(A)$.

9.39 For each $f \in \mathcal{C}([a, b])$, we call the real number

$$\|f\|_\infty = \sup \{|f(x)| : x \in [a, b]\}$$

the **supremum norm** of f. A mapping $G : \mathcal{C}([a, b]) \to \mathbb{R}$ is said to be a **positive bounded linear functional** if

- $G(\alpha f + \beta g) = \alpha G(f) + \beta G(g)$ for all $\alpha, \beta \in \mathbb{R}$ and $f, g \in \mathcal{C}([a, b])$;

- there exists $M > 0$ such that $|G(f)| < M \|f\|_\infty$ for all $f \in \mathcal{C}([a, b])$;

- for each $f \in \mathcal{C}([a, b])$ with $f \geq 0$, we have $G(f) \geq 0$.

Show that the Riemann integral $\int_a^b (\bullet)\, dx$ is a positive bounded linear functional on $\mathcal{C}([a, b])$.

9.40 Show that if ν is a monotone nondecreasing function on $[a, b]$ and if G is defined for each $f \in \mathcal{C}([a, b])$ by $G(f) = \int_a^b f \, d\nu$, then G is a bounded positive linear functional on $\mathcal{C}([a, b])$.

10

Introduction to the Lebesgue Integral

Although the Riemann integral finds many useful applications, we will see in this chapter that it has some great limitations. The class of Riemann integrable functions is "incomplete" in some sense. We will build upon these limitations of the Riemann integral to introduce a more far-reaching theory of integration: the *Lebesgue integral.*[1]

10.1 Null Sets

In the previous chapter, we defined the notion of length for intervals (page 148). For an arbitrary subset A of $\mathbb{R}$, let $\mathcal{I}(A)$ be the collection of all the countable families of covers of A consisting of bounded intervals. It is plain that such a collection is not empty for any given set A. We define the **length** of A to be

$$\ell^*(A) = \inf\left\{\sum_{n=1}^{\infty} \ell(I_n) : (I_n) \in \mathcal{I}(A)\right\}. \tag{10.1}$$

It goes without saying that if I is an interval, then $\ell^*(I) = \ell(I)$. It is possible to have $\ell^*(A) = +\infty$. For example, $\ell^*(\mathbb{R}) = +\infty$. The following observations are also useful.

---

[1] Although it is not followed here, the standard approach to the development of the Lebesgue integral is by way of Measure Theory, which is appropriate for more advanced courses.

Example 10.1

Let A and B be two subsets of $\mathbb{R}$. Show that

(1) if $A \subset B$, then $\ell^*(A) \leq \ell^*(B)$;

(2) if (A_n) is a sequence of subsets of $\mathbb{R}$, then

$$\ell^*\left(\bigcup_{n=1}^{\infty} A_n\right) \leq \sum_{n=1}^{\infty} \ell^*(A_n).$$

Solution

(1) Let $(I_n) \in \mathcal{I}(B)$. Since $A \subset B$, then $(I_n) \in \mathcal{I}(A)$. Therefore $\ell^*(A) \leq \sum_{i=1}^{\infty} \ell(I_n)$. Hence $\ell^*(A) \leq \ell^*(B)$.

(2) If one of the A_n is such that $\ell^*(A_n) = +\infty$, then there is nothing to prove. Thus we assume $\ell^*(A_n) < +\infty$ for all n. Let $\varepsilon > 0$. For each n, choose $(I_{n,k})_{k \in \mathbb{N}} \in \mathcal{I}(A_n)$ such that

$$A_n \subset \bigcup_{k \in \mathbb{N}} I_{n,k} \quad \text{and} \quad \sum_{k \in \mathbb{N}} \ell(I_{n,k}) \leq \ell^*(A_n) + \frac{\varepsilon}{2^n}.$$

Let $A = \bigcup_{n=1}^{\infty} A_n$. Then clearly, the family $(I_{n,k})_{n,k \in \mathbb{N}} \in \mathcal{I}(A)$, and we have

$$\ell^*(A) \leq \sum_{n,k \in \mathbb{N}} \ell(I_{n,k}) = \sum_{n \in \mathbb{N}} \sum_{k \in \mathbb{N}} \ell(I_{n,k})$$

$$\leq \sum_{n \in \mathbb{N}} \left(\ell^*(A_n) + \frac{\varepsilon}{2^n}\right) = \sum_{n \in \mathbb{N}} \ell^*(A_n) + \varepsilon.$$

Since $\varepsilon > 0$ is arbitrary, property (2) is proved. $\qquad \square$

We now introduce a very important notion in Lebesgue integral theory.

Definition 10.2

A subset A of $\mathbb{R}$ is said to be a **null set** if for every $\varepsilon > 0$ there exists a sequence of bounded intervals $(I_n)_{n \in \mathbb{N}}$ such that

$$A \subset \bigcup_{n \in \mathbb{N}} I_n \quad \text{and} \quad \sum_{n \in \mathbb{N}} \ell(I_n) < \varepsilon.$$

Hence the relation $\ell^*(A) = 0$ is often used to indicate that the set A is a null set. It is easy to see that the empty set and any set of finite elements are null sets. Also any subset of a null set is a null set.

Example 10.3

Show that any countable subset of $\mathbb{R}$ is a null set.

Solution

Let A be a countable subset of $\mathbb{R}$. Then we can write $A = \{x_n : n \in \mathbb{N}\}$. Fix $\varepsilon > 0$. Consider for each n, the closed intervals $\left[x_n - \frac{\varepsilon}{2^{n+2}}, x_n + \frac{\varepsilon}{2^{n+2}}\right]$. Then for each n, $x_n \in \left[x_n - \frac{\varepsilon}{2^{n+2}}, x_n + \frac{\varepsilon}{2^{n+2}}\right]$ and therefore we have

$$A \subset \bigcup_{n=1}^{\infty} \left[x_n - \frac{\varepsilon}{2^{n+2}}, x_n + \frac{\varepsilon}{2^{n+2}}\right].$$

On the other hand, we also have

$$\sum_{n=1}^{\infty} \left[\left(x_n + \frac{\varepsilon}{2^{n+2}}\right) - \left(x_n - \frac{\varepsilon}{2^{n+2}}\right)\right] = \sum_{n=1}^{\infty} \frac{\varepsilon}{2^{n+1}} = \frac{\varepsilon}{2} < \varepsilon.$$

Thus A is a null set. $\square$

It is readily seen that union of two null sets is again a null set. In fact, we can state the following stronger result.

Theorem 10.4

A countable union of null sets is a null set.

Proof

Let (A_n) be countably many null sets in $[a, b]$ and let $A = \bigcup_{n=1}^{\infty} A_n$. Then for every $\varepsilon > 0$ and for each n, there exist families of closed bounded intervals $([a_{n_k}, b_{n_k}])_{k \in \mathbb{N}}$ such that

$$A_n \subset \bigcup_{k=1}^{\infty} [a_{n_k}, b_{n_k}] \quad \text{and} \quad \sum_{k=1}^{\infty} (b_{n_k} - a_{n_k}) < \frac{\varepsilon}{2^n}.$$

Now consider the family $([a_{n_k}, b_{n_k}])_{n,k \in \mathbb{N}}$. We have

$$A \subset \bigcup_{n=1}^{\infty} \left(\bigcup_{k=1}^{\infty} [a_{n_k}, b_{n_k}]\right) = \bigcup_{n,k \in \mathbb{N}} [a_{n_k}, b_{n_k}]$$

and

$$\sum_{n,k \in \mathbb{N}} (b_{n_k} - a_{n_k}) = \sum_{n=1}^{\infty} \left(\sum_{k=1}^{\infty} (b_{n_k} - a_{n_k})\right) < \sum_{n=1}^{\infty} \frac{\varepsilon}{2^n} < \varepsilon.$$

This completes our proof. $\square$

Note

An uncountable set of numbers *may* or *may not be* a null set. Any interval with endpoints a and b ($a < b$) cannot be a null set (why?). On the other hand, the so-called **Cantor ternary set**

$$\left\{ x \in \mathbb{R} : x = \sum_{n=1}^{\infty} \frac{a_n}{3^n}, \text{ where } a_n = 0 \text{ or } 2 \right\}$$

is known as an **uncountable** null set (see Appendix A.3).

Let $P(t)$ be a property depending on the variable t of a set A. We say that $P(t)$ is true **almost everywhere** or **for almost all** t in A if there exists a null set $N \subset A$ such $P(t)$ is true for all t in $A \backslash N$. For short, we write $P(t)$ is true **a.e.** if $P(t)$ is true almost everywhere. For example we say that

- two functions f and g are equal almost everywhere or f and g are **essentially equal** (denoted $f \sim g$) on $[a, b]$ if there is a null N set such that $f(t) = g(t)$ for all $t \in [a, b] \backslash N$;

- a function f is **essentially bounded** on $[a, b]$ if there exists a positive number M such that $|f(x)| \leq M$ for almost every x in $[a, b]$;

- a function f is **essentially continuous** on $[a, b]$ if f is continuous at all x in $[a, b]$ except possibly for those x's in a null set.

As examples, since the rationals form a countable set, Dirichlet's discontinuous function $\chi_{\mathbb{Q} \cap [0,1]}$ is essentially null; the function $f : \mathbb{R} \to \mathbb{R}$ defined by

$$f(t) = \begin{cases} \frac{1}{n} & \text{if } t = \frac{m}{n} \text{ is rational} \\ 0 & \text{if } t \text{ is irrational} \end{cases}$$

is continuous almost everywhere (Exercise 4.25, page 120); the function

$$f(t) = \begin{cases} t & \text{if } t \text{ is rational} \\ 1 & \text{if } t \text{ is irrational} \end{cases}$$

is essentially bounded (in fact it is essentially equal to the constant function 1).

Proposition 10.5

Let f, g and h be three functions defined on a given set. Then

(1) $f \sim f$;

(2) if $f \sim g$, then $g \sim f$;

(3) if $f \sim g$ and $g \sim h$, then $f \sim h$.

Proof

(1) and (2) are obvious.

(3) Suppose that $f \sim g$ and $g \sim h$. Then there exist two null sets N_1 and N_2 such that $f(t) = g(t)$ for all $t \notin N_1$, and $g(t) = h(t)$ for all $t \notin N_2$. Thus for $t \notin N_1 \cup N_2$, we have $f(t) = g(t) = h(t)$. Since $N_1 \cup N_2$ is a null set, we conclude that $f \sim h$. $\qquad\square$

Example 10.6

Show that if $f_1 \sim g_1$ and $f_2 \sim g_2$, then

(1) $f_1 + g_1 \sim f_2 + g_2$;

(2) $f_1 \cdot g_1 \sim f_2 \cdot g_2$.

Solution

We leave (2) as an exercise. For (1) since $f_1 \sim g_1$, there exists N_1 a null set such that $f_1(t) = g_1(t)$ for $t \notin N_1$. Similarly, since $f_2 \sim g_2$, there exists N_2 a null set such that $f_2(t) = g_2(t)$ for $t \notin N_2$. Consider the set $N = N_1 \cup N_2$. Then N is a null set (why?). We also have $f_1(t) + g_1(t) = f_2(t) + g_2(t)$ for all $t \notin N$. Thus $f_1 + g_1 \sim f_2 + g_2$. $\qquad\square$

In Chapter 7, we discussed two types of convergence of sequences of functions, namely pointwise and uniform convergence. Our next definition describes yet another type of convergence.

Definition 10.7

A sequence of functions (f_n) is said to **converge almost everywhere** to a function f on a set A if

$$\lim_{n \to \infty} f_n(t) = f(t) \text{ for almost every } t \in A.$$

It is worth noticing that unlike the usual convergence, the limit of a sequence of functions converging a.e. is not necessarily unique.

Proposition 10.8

If (f_n) converges a.e. to a function f, then (f_n) converges a.e. to any function g essentially equal to f.

Proof

Suppose that $f_n \to f$ a.e., and let $g \sim f$. Then there exist two null sets N_1 and N_2 such that $\lim_{n \to \infty} f_n(t) = f(t)$ for all $t \notin N_1$, and $g(t) = f(t)$ for all $t \notin N_2$. Thus for $t \notin N_1 \cup N_2$, we have $\lim_{n \to \infty} f_n(t) = f(t) = g(t)$. Since $N_1 \cup N_2$ is a null set, we conclude that $f_n \to g$ a.e. $\qquad \square$

In fact, as shown in the next result, the limit of a sequence of functions converging a.e. is essentially unique.

Proposition 10.9

If $f_n \to f$ a.e. and $f_n \to g$ a.e., then $f \sim g$.

Proof

Suppose that $f_n \to f$ a.e., and let $f_n \to g$ a.e. Then there exist two null sets N_1 and N_2 such that $\lim_{n \to \infty} f_n(t) = f(t)$ for all $t \notin N_1$, and

$$\lim_{n \to \infty} f_n(t) = g(t) \text{ for all } t \notin N_1.$$

Thus for $t \notin N_1 \cup N_2$, by uniqueness of limit in $\mathbb{R}$ we have $f(t) = g(t)$. Again since $N_1 \cup N_2$ is a null set, we conclude that $f \sim g$. $\qquad \square$

One should notice that the arguments used in the last two proofs are similar. Such arguments are standard for most results concerning convergence a.e. We leave as exercises the proofs of the results contained in the following proposition.

Proposition 10.10

If $f_n \to f$ a.e. and $g_n \to g$ a.e., then

(1) $f_n + g_n \to f + g$ a.e.;

(2) $f_n \cdot g_n \to f \cdot g$ a.e.;

(3) if for each n, $g_n \neq 0$ a.e. and if $g \neq 0$ a.e., then $\frac{f_n}{g_n} \to \frac{f}{g}$.

Since the empty set is a null set, it is clear that if (f_n) converges pointwise to f, then (f_n) converges almost everywhere. The converse is not true.

Example 10.11

Consider on $[0, 1]$ the function $f_n(t) = \frac{nt}{nt+1}$. Show that

(1) $f_n \nrightarrow 1$ pointwise;

(2) $f_n \to 1$ a.e.

Solution

(1) We have $f_n(0) = 0 \to 0$. So $f_n \nrightarrow 1$ pointwise.
(2) For $t > 0$, $f_n(t) = \frac{nt}{nt+1} \to 1$. Since $\{0\}$ is a null set, $f_n \to 1$ a.e.   □

Another important type of convergence of sequences of functions is described in the next definition.

Definition 10.12

A sequence of functions (f_n) in $\mathcal{R}$ is said **to converge in mean** to a function f in $\mathcal{R}$ if

$$\lim_{n \to \infty} \int |f_n - f| = 0.$$

It is always a good practice to try to compare the different types of convergence. We first prove the following technical lemma.

Lemma 10.13

Let $f \in \mathcal{S}$ such that $\int f \le K$, where $K > 0$. For each $\varepsilon > 0$, consider the set $A = \{t : f(t) > K/\varepsilon\}$. Then there exist disjoint bounded intervals $(I_k)_{k=1}^n$ such that

$$A = \bigcup_{k=1}^n I_k \quad \text{and} \quad \sum_{k=1}^n \ell(I_k) < \varepsilon.$$

Proof

We first suppose that $f = \sum_{i=1}^n \alpha_i \chi_{I_i}$ is expressed in its standard representation. Let $\varepsilon > 0$. Then $f(t) \ge K/\varepsilon$ if and only if for some i, $\alpha_i \ge K/\varepsilon$. It follows that the set $A = \{t : f(t) \geq K/\varepsilon\}$ is the disjoint union of the subfamily $(I_i)_{\alpha_i > K/\varepsilon}$ of $(I_i)_{i=1}^n$ or the empty set and

$$\sum_{\alpha_i > K/\varepsilon} \ell(I_i) \leq \sum_{\alpha_i > K/\varepsilon} \frac{\alpha_i}{K/\varepsilon} \ell(I_i) \leq \frac{1}{K/\varepsilon} \int f \leq \varepsilon$$

as desired.   □

Example 10.14

Let (f_n) be a sequence in $\mathcal{S}$ converging in mean to 0. Show that there exists a subsequence (f_{n_k}) of (f_n) converging almost everywhere to 0.

Solution

Since $f_n \to 0$ in mean, for each integer k, there exists $N_k > 0$ such that for $n > N_k$, $\int |f_n| \, dt < 2^{-2k}$. For each k, and for $n > N_k$, consider the set $A_k = \{t : |f_n(t)| > 2^{-k}\}$. Then by Lemma 10.13, $\ell^*(A_k) < 2^{-k}$. Hence the set $E_m = \bigcup_{k=m}^{\infty} A_k$ satisfies

$$\ell^*(E_m) < \sum_{k=m}^{\infty} 2^{-k} < 2^{-(m-1)}.$$

Now it is easy to see that $\{t : f_n \nrightarrow 0\} \subset E_m$ for all m, and thus

$$\ell^*(\{t : f_n \nrightarrow 0\}) \leq \ell^*(E_m) < 2^{-(m-1)}.$$

Given $\varepsilon > 0$, choose m large enough so that $2^{-(m-1)} < \varepsilon$, and hence

$$\ell^*(\{t : f_n \nrightarrow 0\}) < \varepsilon.$$

Since $\varepsilon > 0$ is arbitrarily chosen, this shows that $\{t : f_n \nrightarrow 0\}$ is a null set. Consequently, $f_n \to 0$ a.e. This completes the proof.   □

The next result states that a Riemann integrable function can be approximated a.e. by sequences of step functions.

Theorem 10.15

Let $f \in \mathcal{R}$. Then there exist a nondecreasing sequence (φ_n) and a nonincreasing sequence (ψ_n), both in $\mathcal{S}$ and both converging a.e. to f.

Proof

Let $[a, b]$ be a support interval for f. Let (P_n) be a sequence of partitions of $[a, b]$ ordered by refinement. Let ψ_n and φ_n be respectively the upper and the lower step functions associated to f on $[a, b]$. Then clearly $\varphi_n(x) \leq f(x) \leq \psi_n(x)$ for all $x \in [a, b]$, φ_n is nondecreasing, and ψ_n is nonincreasing. For each x, let $g(x) = \lim \varphi_n(x)$ and $h(x) = \lim \psi_n(x)$. Then the sequence of step functions $\psi_n(x) - \varphi_n(x) \to h - g$ for each x, and we have

$$\varphi_n(x) \leq g(x) \leq f(x) \leq h(x) \leq \psi_n(x).$$

On the other hand, since $f \in \mathcal{R}([a, b])$, we have

$$0 = \lim \left[\int_a^b \psi_n dt - \int_a^b \varphi_n dt \right] = \lim \int_a^b |\psi_n(t) - \varphi_n(t)| \, dt.$$

That is, $(\psi_n - \varphi_n)$ converges in mean to 0. By Example 10.14, a subsequence $(\psi_{n_k} - \varphi_{n_k})$ of $(\psi_n - \varphi_n)$ converges a.e. to 0. An appeal to Proposition 10.9 now shows that $h - g \sim 0$. This implies $h \sim f \sim g$. Hence $\varphi_n \to f$ a.e. and $\psi_n \to f$ a.e. The proof is finished. □

We prove another technical lemma.

Lemma 10.16

Let (f_n) be a sequence in $\mathcal{S}$ such that

(1) $0 \leq f_{n+1} \leq f_n$ a.e. for all n; and

(2) $f_n \to 0$ a.e.

Then $\lim \int f_n = 0$.

Proof

Suppose that f_1 vanishes outside the interval $[a, b]$ and that $f_1(t) \leq K$, where $K > 0$. Then the same is true for all the f_n.

Now fix $\varepsilon > 0$. Let D be the null set such that $f_n(t) \nrightarrow 0$ if $t \in D$. Since each f_n has at most finitely many points of discontinuity, all the points of discontinuity of all the f_n are countable and hence form a null set N. Thus the set $D \cup N$ is also a null set.

Let (I_n) be a sequence of open intervals of total length not greater than ε, which covers $D \cup N$. Then if $t \notin D \cup N$, there exists N_t such that $0 \leq f_{N_t}(t) < \varepsilon$, and f_{N_t} is continuous at t. Thus there exists $\delta_t > 0$ such that $|t - s| <$

δ_t implies $f_{N_t}(t) = f_{N_t}(s)$. Hence $0 \le f_{N_t}(s) \le \varepsilon$ whenever $s \in B(t, \delta_t)$. Since the sequence (f_n) is nonincreasing, we have

$$0 \le f_n(s) < \varepsilon \text{ for every } s \in B(t, \delta_t), \text{ whenever } n \ge N_t.$$

We then notice that $[a, b] \subset \bigcup_{n \in \mathbb{N}} I_n \cup \bigcup_{t \notin D \cup N} B(t, \delta_t)$. Since $[a, b]$ is compact, it can be covered with a finite subfamily, say

$$\left(E_{i_1}, E_{i_2}, \dots, E_{i_p}, B(t_1, \delta_{t_1}), B(t_2, \delta_{t_2}), \dots, B(t_n, \delta_{t_n}) \right).$$

Let $N = \max \{N_{t_1}, N_{t_2}, \dots, N_{t_n}\}$ and consider the sets

$$E = \left(E_{i_1} \cup E_{i_2} \cup \dots \cup E_{i_p} \right) \cap [a, b], \text{ and}$$
$$B = \left(B(t_1, \delta_{t_1}) \cup B(t_2, \delta_{t_2}) \cup \dots \cup B(t_n, \delta_{t_n}) \right) \cap [a, b].$$

Then $0 \le f_n(s) < \varepsilon$ whenever $s \notin B$ and $n > N$. It follows that for $n > N$,

$$\int f_n \le \int f_n \chi_E \, dt + \int f_n \chi_B \, dt$$
$$\le \varepsilon K + \varepsilon \sum_{i=1}^p \ell \left(B(t_i, \delta_{t_i}) \cap [a, b] \right)$$
$$= \varepsilon [K + (b - a)].$$

This shows that $\int f_n \to 0$. $\qquad \square$

The next example extends the result of Example 10.14 to the Riemann integrable functions.

Example 10.17

Let (f_n) be a sequence in $\mathcal{R}$ converging in mean to 0. Show that there exists a subsequence (f_{n_k}) of (f_n) converging almost everywhere to 0.

Solution

For each k, let $\left(\varphi_n^k \right)_{n \in \mathbb{N}}$ and $\left(\psi_n^k \right)_{n \in \mathbb{N}}$ be respectively a nondecreasing and a nonincreasing sequence of step functions both converging a.e. to f_k (Theorem 10.15). Consider

$$\varphi_n = \max \left\{ \varphi_n^k : k \le n \right\} \text{ and } \psi_n = \min \left\{ \psi_n^k : k \le n \right\}.$$

Then (φ_n) and (ψ_n) are sequences of step functions, the first one nondecreasing and the second one nonincreasing, and both converge a.e. to 0. In view of Lemma 10.16 (page 279), we have

$$\lim \int \varphi_n = \lim \int \psi_n = 0.$$

Therefore by Example 10.14, there exists a subsequence of (φ_n), say (φ'_n), converging a.e. to 0. By the same lemma, there exists a subsequence of (ψ'_n), say (ψ''_n), converging a.e. to 0. We notice that $\varphi''_n \leq f''_n \leq \psi''_n$ for every n and deduce that the subsequence (f''_n) of (f_n) converges a.e. to 0. $\qquad\square$

The following technical lemma is an extension of Lemma 10.13 (page 277).

Lemma 10.18

Let $f \in \mathcal{R}$ such that $\int f \leq K$, where $K > 0$. For each $\varepsilon > 0$, consider the set $A = \{t : f(t) > K/\varepsilon\}$. Then $\ell^*(A) \leq \varepsilon$.

Proof

On the basis of Theorem 10.15, let (φ_n) be a nondecreasing sequence of step functions converging a.e. to f. Let N be the null set such that $\varphi_n(x) \nrightarrow f(x)$ if $x \in N$. Fix $\varepsilon > 0$. Applying Lemma 10.13 to the step function φ_n for each n, we learn that the set

$$A_n = \left\{t : \varphi_n(t) > \frac{K}{\varepsilon(1 + 1/2^n)}\right\}$$

is such that $\ell^*(A_n) < \varepsilon(1 + 1/2^n)$. Since (φ_n) is nondecreasing the sequence (A_n) is also nondecreasing and thus by the monotone property of ℓ^*,

$$\ell^*(A_n) \leq \lim_n \ell^*(A_n) \leq \varepsilon.$$

Now if $t \notin N$, since $\varphi_n(t) \to f(t)$ there exists n such that $t \in A_n$. This shows that $A \subset N \cup \bigcup_{n \in \mathbb{N}} A_n$ and thus

$$\ell^*(A) \leq \ell^*(N) + \lim_n \ell^*(A_n) \leq \varepsilon$$

as desired. $\qquad\square$

The following result turns out to be instrumental in our development of the Lebesgue integral theory.

Theorem 10.19

Let (f_n) be a sequence in $\mathcal{R}$ such that

(1) $f_n \leq f_{n+1}$ a.e. for all n;

(2) there exists $K > 0$ such that for each n, $\int f_n \leq K$.

Then (f_n) converges a.e.

Proof

We may assume without loss of generality that $f_n \geq 0$ for otherwise we could consider the nonincreasing sequence $(f_n - f_1)$ and recall that $\int (f_n - f_1) = \int (f_n) - \int (f_1)$. We may also assume that $f_n \leq f_{n+1}$ for all n. Indeed since the set $D = \bigcup_{n=1}^{\infty} \{x : f_n(x) > f_{n+1}(x)\}$ is a null set (as a countable union of null sets), we could consider the functions $\chi_{\mathbb{R} \backslash D} f_n$ and notice that the sequence (f_n) converges a.e. if and only if so does the sequence $(\chi_{\mathbb{R} \backslash D} f_n)$.

Let $E = \{x : f_n(x) \to \infty\}$ and let $\varepsilon > 0$. Consider the set

$$E_n = \{x : f_n(x) \geq K/\varepsilon\}.$$

If $x \in E$, then for n sufficiently large, $f_n(x) \geq K/\varepsilon$. Therefore $E \subset \bigcup_{n=1}^{\infty} E_n$. According to Lemma 10.18, $\ell^*(E_n) \leq \varepsilon$ for each n. Since $f_n \leq f_{n+1}$, we have $E_n \subset E_{n+1}$ and thus $\ell^*(E_n) \leq \ell^*(E_{n+1}) \leq \varepsilon$. It follows that $\ell^*(E) \leq \varepsilon$. Since $\varepsilon > 0$ is arbitrary, E is a null set. This completes the proof. $\square$

We end this section with a remarkable result due to Lebesgue. This result gives a characterization of the Riemann integrable functions in terms of their discontinuity.

Theorem 10.20 (Lebesgue)

Let $f : [a, b] \to \mathbb{R}$ be a bounded function. Then $f \in \mathcal{R}([a, b])$ if and only if f is essentially continuous.

Proof

Assume first that $f \in \mathcal{R}([a, b])$. Let (φ_n) and (ψ_n) be the sequences of step functions given by Theorem 10.15. Then $\varphi_n(x) \uparrow f(x)$ and $\psi_n(x) \downarrow f(x)$ for all $x \in [a, b] \backslash N$ where N is some null set. In fact N is finite. Evidently for each n, as step functions, φ_n and ψ_n are essentially continuous. Let A_n and B_n be null sets such that φ_n is continuous on $[a, b] \backslash A_n$ and ψ_n is continuous on $[a, b] \backslash B_n$. Then the set $D = N \cup (\bigcup_{n=1}^{\infty} (A_n \cup B_n))$ is a null set (countable union of null sets). We claim that f is continuous on $[a, b] \backslash D$.

Let $t_0 \in [a, b] \backslash D$ and fix $\varepsilon > 0$. There exists $N > 0$ so that $n > N$ implies

$$f(t_0) - \varphi_n(t_0) < \varepsilon \quad \text{and} \quad \psi_n(t_0) - f(t_0) < \varepsilon.$$

The point t_0 must be in some subinterval $[x_{i-1}, x_i]$ of the partition P_n. Thus

$$\varphi_n(t_0) = m_i = \inf\{f(t) : t \in [x_{i-1}, x_i]\},$$
$$\psi_n(t_0) = M_i = \sup\{f(t) : t \in [x_{i-1}, x_i]\}.$$

It follows that if $|t - t_0| < \|P_n\|$, then $t \in (x_{i-1}, x_i)$ and hence

$$-\varepsilon < m_i - f(t_0) \leq f(t) - f(t_0) \leq M_i - f(s_0) < \varepsilon.$$

This proves our claim.

Conversely, suppose that f is continuous on $[a, b] \setminus N$ where N is a null set. Again, for each partition P_n of $[a, b]$, as in the proof of Theorem 10.15, let (φ_n) and (ψ_n) be the lower and the upper step functions associated to f and for each x, let $g(x) = \lim \varphi_n(x)$ and $h(x) = \lim \psi_n(x)$. If $x \notin N \cup (\bigcup_{n=1}^{\infty} P_n)$, then f is continuous at x and thus necessarily $g(x) = f(x) = h(x)$. It is then easy to see that the sequence of step functions $(\psi_n(x) - \varphi_n(x))$ is nonincreasing and converges to 0 for all $x \notin N \cup (\bigcup_{n=1}^{\infty} P_n)$. By Lemma 10.16,

$$\lim \int (\psi_n - \varphi_n) = 0.$$

This shows that $f \in \mathcal{R}([a, b])$ and completes our proof. $\square$

10.2 Lebesgue Integral

Definition 10.21

A sequence $(f_n) \subset \mathcal{R}$ is said to be **Cauchy in mean** if the f_n have a common support interval and if

for every $\varepsilon > 0$, there exists $N \in \mathbb{N}$ such that
$$n, m > N \quad \text{implies} \quad \int |f_n - f_m| < \varepsilon.$$

It is readily seen that if (f_n) is Cauchy in mean, then since

$$\left| \int (f_n - f_m) \right| \leq \int |f_n - f_m|,$$

the numerical sequence $(\int f_n)$ is Cauchy in $\mathbb{R}$, and so $\lim \int f_n$ exists. The next theorem is more precise.

Theorem 10.22

Let (f_n) in $\mathcal{R}$ be such that

(1) (f_n) is Cauchy in mean;

(2) $f_n \to 0$ a.e.

Then $\lim \int f_n = 0$.

Proof

For each k, let $\left(\varphi_n^k\right)_{n \in \mathbb{N}}$ and $\left(\psi_n^k\right)_{n \in \mathbb{N}}$ be respectively a nondecreasing and a nonincreasing sequence of step functions both converging a.e. to f_k (Theorem 10.15). Consider

$$\varphi_n = \max \left\{\varphi_n^k : k \leq n\right\} \quad \text{and} \quad \psi_n = \min \left\{\psi_n^k : k \leq n\right\}.$$

Then (φ_n) is nondecreasing while (ψ_n) is nonincreasing and both converge a.e. to 0. In view of Lemma 10.16, we have

$$\lim \int \varphi_n = \lim \int \psi_n = 0.$$

On the other hand, we notice that $\varphi_n \leq f_n \leq \psi_n$ and thus, by the monotonicity property of integrals, we have $\int \varphi_n \leq \int f_n \leq \int \psi_n$. Thus necessarily

$$\lim_{n \to \infty} \int f_n = 0$$

as asserted.   □

Corollary 10.23

If (f_n) and (g_n) are in $\mathcal{R}$, both Cauchy in mean, and both converge a.e. to f, then

$$\lim \int f_n dt = \lim \int g_n dt.$$

Proof

It suffices to apply the above theorem to the sequence $(f_n - g_n)$.   □

Let $\{r_1, r_2, \ldots\}$ be an increasing enumeration of the rationals in $[0, 1]$. For each $n \in \mathbb{N}$, let

$$f_n(t) = \begin{cases} 1 & \text{if } t \in [0, r_n] \cap \mathbb{Q} \\ 0 & \text{otherwise.} \end{cases}$$

Then for each n, $f_n(t) = 0$ except for $t = r_1, r_2, \ldots, r_n$. Then $f_n \in \mathcal{R}$ for each n. We also have $|f_{n+p}(t) - f_n(t)| = 0$ for all n, p and for each t except possibly for $t = r_{n+1}, r_{n+2}, \ldots, r_{n+p}$. Therefore $|f_{n+p} - f_n| \in \mathcal{R}([0,1])$ and $\int |f_{n+p} - f_n| = 0$. Thus (f_n) is Cauchy in mean. On the other hand, we notice that (f_n) converges (pointwise, and therefore) a.e. to the Dirichlet discontinuous function $\chi_{[0,1] \cap \mathbb{Q}}$. As noticed in the note after Definition 10.21, the quantity $\lim_{n \to \infty} \int f_n$ exists. We wish to be able to write

$$\lim_{n \to \infty} \int f_n = \int \chi_{[0,1] \cap \mathbb{Q}}.$$

Unfortunately since $\chi_{[0,1] \cap \mathbb{Q}} \notin \mathcal{R}$, the integral on the right of the above equation does not have a meaning in the theory of Riemann integrals. This example provides a motive for our definition of the Lebesgue integral.

Definition 10.24

A function $f : \mathbb{R} \to \mathbb{R}$ is said to be **(Lebesgue) integrable** if there exists a sequence (f_n) in $\mathcal{R}$ which is Cauchy in mean and converges almost everywhere to f. In such a case, the quantity

$$\lim_{n \to \infty} \int f_n$$

is called the **Lebesgue integral** of f.

The sequence (f_n) in the above definition is called a **generating sequence** for f. We notice that an integrable function possesses more than one generating sequence. For example, it is a nice exercise to show that given a generating sequence (f_n) for a function f, the sequence $(f_n + \theta_n)$, where (θ_n) is any sequence of essentially null functions, is also a generating sequence for f. Nevertheless, a glance at Corollary 10.23 reassures us that our definition of the Lebesgue integral of a function f is independent of the choice of the generating sequence used to define it.

Notation

The set of all Lebesgue integrable functions will be denoted by $\mathcal{L}(\mathbb{R})$ or simply by $\mathcal{L}$. The Lebesgue integral of a function f is then denoted by $\int f d\ell$.

It is clear that $\mathcal{R} \subset \mathcal{L}$. If A is a subset of $\mathbb{R}$, we say that a function f defined on A is Lebesgue integrable over A if the associated function defined by

$$\tilde{f}(x) = \begin{cases} f(x) & \text{if } x \in A \\ 0 & \text{otherwise} \end{cases}$$

belongs to $\mathcal{L}(\mathbb{R})$. It is common to denote by $\mathcal{L}(A)$ the set of all Lebesgue integrable functions over A. The Lebesgue integral of f over A is denoted by $\int_A f \, d\ell$. We often identify f and $\tilde{f}$ so that we may write $\mathcal{L}(A) \subset \mathcal{L}(\mathbb{R})$. We also notice that if $f \in \mathcal{L}(\mathbb{R})$, then $f \cdot \chi_A \in \mathcal{L}(A)$. It is easy to see that if $f|_A$ denotes the restriction of f to A, then

$$\int_A f|_A \, d\ell = \int f \cdot \chi_A \, d\ell.$$

For an interval $[a, b]$, we continue to use $\int_a^b$ to indicate a Riemann integral and reserve $\int_{[a,b]}$ to designate a Lebesgue integral over the interval $[a, b]$.

It is a nice exercise to prove that $f \in \mathcal{L}([a, b])$ if and only if f admits a generating sequence in $\mathcal{R}([a, b])$. The following observation is in order.

Example 10.25

Show that a Riemann integrable function f on $[a, b]$ is Lebesgue integrable on $[a, b]$, i.e. $\mathcal{R}([a, b]) \subset \mathcal{L}([a, b])$, and

$$\int_{[a,b]} f \, d\ell = \int_a^b f(t) \, dt.$$

Solution

Suppose that $f \in \mathcal{R}([a, b])$. Let $P_n = \{t_0, t_1, \ldots, t_n\}$ be a partition of $[a, b]$. For each n, let f_n be the lower step function associated to f relative to P_n. Since $f \in \mathcal{R}([a, b])$, $\lim \int f_n = \int_a^b f(t) \, dt$. Thus the sequence of real numbers $\left(\int f_n \right)$ is Cauchy. Since $f_n \uparrow f$, we have for every p,

$$\int |f_{n+p} - f_n| = \int (f_{n+p} - f_n) \to 0 \text{ as } n \to \infty.$$

Thus (f_n) is a generating sequence for f. Whence $f \in \mathcal{L}([a, b])$. Clearly, we have

$$\int_{[a,b]} f \, d\ell = \lim \int f_n = \int_a^b f(t) \, dt.$$

$\square$

The collection $\mathcal{L}$ of Lebesgue integrable functions shares many of the nice properties of $\mathcal{R}$.

Theorem 10.26 (Translation Invariance)

Let $f \in \mathcal{L}$ and for $d \in \mathbb{R}$ let $\tau_d f$ be defined by $\tau_d f(x) = f(x + d)$. Then $\tau_d f \in \mathcal{L}$ and $\int \tau_d f \, d\ell = \int f \, d\ell$.

Proof

Let (f_n) be a generating sequence for f, i.e. for each n, $f_n \in \mathcal{R}$, (f_n) is Cauchy in mean, and $f_n \to f$ a.e. Then in view of Theorem 6.11, we see that $\tau_d f_n \in \mathcal{R}$, $\tau_d f_n \to \tau_d f$ a.e., and

$$\int |\tau_d f_m - \tau_d f_n| = \int |\tau_d(f_n - f_m)| = \int |f_n - f_m|.$$

This shows that $(\tau_d f_n)$ is a generating sequence for $\tau_d f$. Clearly, we also have

$$\int \tau_d f \, d\ell = \lim_{n \to \infty} \int \tau_d f_n = \lim_{n \to \infty} \int f_n = \int f \, d\ell.$$

$\square$

The following remark is very useful.

Remark 10.27

If $f \in \mathcal{L}$ and $g \sim f$, then $g \in \mathcal{L}$ and $\int f \, d\ell = \int g \, d\ell$.

Proof

To see this, let (f_n) be a generating sequence for f; i.e. (f_n) is a sequence in $\mathcal{R}$ which is Cauchy in mean and such that $f_n \to f$ a.e. Then by virtue of Proposition 10.8, $f_n \to g$ a.e. This proves that (f_n) is also a generating sequence for g and hence proves our remark. $\square$

For example, since $\chi_{\mathbb{Q} \cap [0,1]} \sim 0$, we have $\chi_{\mathbb{Q} \cap [0,1]} \in \mathcal{L}$ and $\int \chi_{\mathbb{Q} \cap [0,1]} \, d\ell = 0$. This example also shows that $\mathcal{R}$ is a proper subspace of $\mathcal{L}$.

Once we understand its definition, many properties of the Lebesgue integral come along easily.

Example 10.28

Show that if $f \in \mathcal{L}$, then $|f| \in \mathcal{L}$ and $\left| \int f \, d\ell \right| \leq \int |f| \, d\ell$.

Solution

Suppose that $f \in \mathcal{L}$. Let (f_n) be a generating sequence for f. Then $\int ||f_{n+p}| - |f_n|| \leq \int |f_{n+p} - f_n|$ and $|f_n| \to |f|$ a.e. Thus $|f| \in \mathcal{L}$. We also have (Theorem 6.27, page 163) $|\int f_n| \leq \int |f_n|$. Thus

$$\left| \int f \, d\ell \right| = \lim \left| \int f_n \, d\ell \right| \leq \lim \int |f| \, d\ell = \int |f| \, d\ell.$$

This completes the proof. □

Using a similar argument in tandem with Theorem 6.24 (page 160), one obtains the following properties, the proofs of which are left as exercises.

Theorem 10.29

Let $f, g \in \mathcal{L}$. Then

(1) for $\alpha, \beta \in \mathbb{R}$, $\alpha f + \beta g \in \mathcal{L}$ and $\int (\alpha f + \beta g) \, d\ell = \alpha \int f \, d\ell + \beta \int g \, d\ell$;

(2) if $f \leq g$ a.e., then $\int f \, d\ell \leq \int g \, d\ell$;

(3) if $m \leq f \leq M$ a.e. where $m, M \in \mathbb{R}$, then $m \leq \int f \, d\ell \leq M$.

Theorem 10.29 may be summarized by saying that $\mathcal{L}$ is a linear space and the operator $\int (\cdot) \, d\ell$ is a positive linear operator on $\mathcal{L}$. In fact the next example shows that $\mathcal{L}$ is a lattice.

Example 10.30

Let $f, g \in \mathcal{L}$. Show that $f \vee g, f \wedge g \in \mathcal{L}$.

Solution

Let us first suppose that $g = 0$. Then $\max \{f, 0\} = f^+$. Let (φ_n) be a generating sequence for f and let $A = \{x : f(x) \geq 0\}$. Then noticing that $\varphi_n^+ \in \mathcal{R}$ for all n, and that

$$\left| \varphi_m^+ - \varphi_n^+ \right| = |\varphi_m \chi_A - \varphi_n \chi_A| = |\varphi_m - \varphi_n| \chi_A \leq |\varphi_m - \varphi_n|,$$

we have

$$\int \left| \varphi_m^+ - \varphi_n^+ \right| \leq \int |\varphi_m - \varphi_n|.$$

This implies that (φ_n^+) is Cauchy in mean. On the other hand, it is clear that (φ_n^+) converges a.e. to f^+. Thus (φ_n^+) is a generating sequence for f^+ and hence

$f^+ \in \mathcal{L}$. The general case of $f \vee g$ follows from property (1) of Theorem 10.29. It suffices to notice that $f \vee g = (f - g)^+ + g$. The case of $f \wedge g$ can be obtained by using $f \wedge g = -(-f) \vee (-g)$.   □

Before we state one of the foundations of Lebesgue theory, we first give the following

Lemma 10.31

Let (φ_n) be a sequence in $\mathcal{R}$ such that

(1) $\varphi_n \leq \varphi_{n+1}$ a.e. for all n;

(2) there exists $M > 0$ such that for each n, $\int \varphi_n \leq M$.

Then there exists $f \in \mathcal{L}$ such that $\varphi_n \to f$ a.e., and

$$\int f d\ell = \lim_{n \to \infty} \int \varphi_n.$$

Proof

On the basis of Theorem 10.19, we know that (φ_n) converges a.e. to say some function f. Since the sequence $(\int \varphi_n)$ is nondecreasing and bounded, it is convergent. Also since

$$\int |\varphi_{n+p} - \varphi_n| = \int (\varphi_{n+p} - \varphi_n),$$

we see that (φ_n) is Cauchy in mean. Hence (φ_n) is a generating sequence for f. Therefore $f \in \mathcal{L}$ and $\int f d\ell = \lim_{n \to \infty} \int \varphi_n$.   □

The following theorem generalizes the previous result to the space $\mathcal{L}$ and is attributed to B. Levi.

Theorem 10.32 (Monotone Convergence Theorem)

Let (f_n) be a sequence in $\mathcal{L}$ such that

(1) $f_n \leq f_{n+1}$ a.e. for all n;

(2) there exists $M > 0$ such that for each n, $\int f_n d\ell \leq M$.

Then there exists $f \in \mathcal{L}$ such that $f_n \to f$ a.e., and

$$\int f d\ell = \lim_{n \to \infty} \int f_n d\ell.$$

Proof

As in the proof of Theorem 10.19 we may and do assume without loss of generality that $f_n(x) \le f_{n+1}(x)$ for all x. Let $f(x) = \lim f_n(x)$.

For each k, choose a generating sequence $\left(\varphi_n^{(k)}\right)_{n \in \mathbb{N}}$ for f_k in $\mathcal{R}$. Fix $\varepsilon > 0$. For $k = 1$, pick N_1 such that

$$\int \varphi_n^{(1)} \le \int f_1 d\ell + \varepsilon \quad \text{whenever} \quad n > N_1.$$

Choose $n_1 > N_1$ such that $\left|\varphi_{n_1}^{(1)}(x) - f_1(x)\right| < \frac{1}{2^1}$ holds for all x not in a null set E_1. For $k = 2$, we can choose $N_2 > N_1$ such that

$$\int \varphi_n^{(2)} dx \le \int f_2 d\ell + \varepsilon \quad \text{whenever} \quad n > N_2.$$

Since $f_k \le f_{k+1}$, we can choose $n_2 > N_2$ such that

$$\varphi_{n_1}^{(1)}(x) \le \varphi_{n_2}^{(2)}(x) \quad \text{and} \quad \left|\varphi_{n_2}^{(2)}(x) - f_2(x)\right| < \frac{1}{2^2}$$

holds for all x not in a null set E_2. Continuing inductively in this fashion, we construct a nondecreasing sequence $\left(\varphi_{n_k}^{(k)}\right)_{k \in \mathbb{N}}$ in $\mathcal{R}$ such that

$$\int \varphi_{n_k}^{(k)} dx \le \int f_k d\ell + \varepsilon \tag{10.2}$$

and such that

$$\varphi_{n_k}^{(k)}(x) \le \varphi_{n_{k+1}}^{(k+1)}(x) \quad \text{and} \quad \left|\varphi_{n_k}^{(k)}(x) - f_k(x)\right| < \frac{1}{2^k} \tag{10.3}$$

for all $x \notin E = \bigcup_{k=1}^{\infty} E_k$. Since E is a null set, (10.3) implies that $\varphi_{n_k}^k \to f$ a.e. as $k \to \infty$. Also since inequality (10.2) holds for all k, $\int \varphi_{n_k}^{(k)} < M + \varepsilon$. Thus the sequence $\left(\varphi_{n_k}^{(k)}\right)_{k \in \mathbb{N}}$ satisfies the conditions of Lemma 10.31. Hence

$$\int f d\ell = \lim_{k \to \infty} \int \varphi_{n_k}^k \le \lim_{k \to \infty} \int f_k d\ell + \varepsilon.$$

Since $\varepsilon > 0$ is arbitrary, we conclude

$$\int f d\ell = \lim_{k \to \infty} \int f_k d\ell.$$

$\square$

It is a good exercise to show that the monotone convergence theorem holds for a nonincreasing sequence of Lebesgue integrable functions (f_n) satisfying $\int f_n d\ell \ge M$ for all n.

Example 10.33

Show that if $(f_n) \subset \mathcal{L}$ and if there exists $F \in \mathcal{L}$ such that $|f_n(t)| \leq F(t)$ a.e., then $\sup f_n \in \mathcal{L}$ and $\inf f_n \in \mathcal{L}$.

Solution

Let $g_n(t) = \max\{f_1(t), f_2(t), \ldots, f_n(t)\}$. Then (g_n) converges a.e. to $\sup f_n$. On the other hand since $\mathcal{L}$ is a lattice (Example 10.30), $g_n \in \mathcal{L}$ for all n. It is clear that (g_n) is nondecreasing, and $\int g_n d\ell \leq \int F d\ell$. By the monotone convergence theorem, (g_n) converges a.e. to some function $g \in \mathcal{L}$. Proposition 10.9 now implies that $\sup f_n = g$ a.e.; hence $\sup f_n \in \mathcal{L}$. We also deduce that $\inf f_n = -\sup(-f_n) \in \mathcal{L}$.   □

Another important result of Lebesgue theory is known as Fatou's lemma.

Theorem 10.34 (Fatou's Lemma)

Let (f_n) be a sequence in $\mathcal{L}$ such that

(1) $f_n \geq 0$ a.e. for all n;

(2) $\liminf \int f_n d\ell < \infty$.

Then $\liminf f_n \in \mathcal{L}$ and

$$\int \liminf f_n d\ell \leq \liminf \int f_n d\ell.$$

Proof

Let $g_m = \inf\{f_{m+1}, f_{m+2}, \ldots\}$. We have by Example 10.33, $g_m \in \mathcal{L}$. Then since $0 \leq g_m \leq f_n$ whenever $n > m$,

$$\int g_m d\ell \leq \int f_n d\ell \text{ for } m < n.$$

Thus

$$\int g_m d\ell \leq \liminf \int f_n d\ell.$$

Since (g_m) is nondecreasing and $\lim_m g_m = \liminf f_n$, the monotone convergence theorem implies that

$$\int \liminf f_n d\ell \leq \lim \int g_m d\ell \leq \liminf \int f_n d\ell.$$

□

Example 10.35

Let $f \in \mathcal{L}$ and let (f_n) be a sequence in $\mathcal{L}$ such that

(1) $f_n \to f$ a.e.;

(2) $\int |f| \, d\ell = \lim_{n \to \infty} \int |f_n| \, d\ell$.

Show that $\lim_{n \to \infty} \int |f_n - f| \, d\ell = 0$.

Solution

We notice that $0 \leq |f_n| + |f| - |f_n - f|$ a.e. Then by Fatou's lemma and condition (2) of the example, we have

$$2 \int |f| \, d\ell = \int \lim \left(|f_n| + |f| - |f_n - f| \right) d\ell$$

$$\leq \liminf \int \left(|f_n| + |f| - |f_n - f| \right) d\ell$$

$$= \lim \int |f_n| \, d\ell + \int |f| \, d\ell + \liminf \int \left(-|f_n - f| \right) d\ell$$

$$= 2 \int |f| \, d\ell - \limsup \int |f_n - f| \, d\ell.$$

Since $\int |f| \, d\ell < \infty$, we have $0 \leq -\limsup \int |f_n - f| \, d\ell$. Then necessarily

$$\limsup \int |f_n - f| \, d\ell = \liminf \int |f_n - f| \, d\ell = \lim \int |f_n - f| \, d\ell = 0,$$

as desired. $\square$

Example 10.36

Let $f \in \mathcal{L}$, $f \geq 0$ and $\int f d\ell = 0$. Show that $f \sim 0$.

Solution

Let (φ_n) be a generating sequence for f. Then $(|\varphi_n|)$ is a generating sequence for $|f| = f$ and

$$0 = \int f d\ell = \int |f| \, d\ell = \lim \int |\varphi_n|.$$

According to Example 10.17, a subsequence (φ_{n_k}) of (φ_n) converges a.e. to 0. Proposition 10.9 now ensures the desired conclusion. $\square$

We saw (Theorem 7.13 on page 184, and Exercise 6.21 on page 173) that under certain conditions, it is allowed to interchange the limit operator and the integral operator for a sequence of continuous or Riemann integrable functions on a closed bounded interval $[a, b]$. Our next result also gives sufficient conditions for the interchangeability of the two mentioned operators for Lebesgue integrable functions. This result is one the most significant results of Lebesgue integration theory.

Theorem 10.37 (Lebesgue's Dominated Convergence Theorem)

Let (f_n) be a sequence in $\mathcal{L}$ such that

(1) $|f_n| \leq g$ a.e. for all n and for some fixed $g \in \mathcal{L}$;

(2) $f_n \to f$ a.e.

Then $f \in \mathcal{L}$ and $\int f d\ell = \lim_{n \to \infty} \int f_n d\ell$.

Proof

Clearly $|f| \leq g$ a.e. We leave it as an exercise to show that $f \in \mathcal{L}$ (Exercise 10.16, page 309). We can apply Fatou's lemma and Theorem 10.29 to the sequence $(g + f_n)$ to obtain

$$\int g d\ell + \int f d\ell = \int (g + f) \, d\ell = \int \liminf (g + f_n) \, d\ell$$
$$\leq \liminf \int (g + f_n) \, d\ell$$
$$= \liminf \left(\int g d\ell + \int f_n d\ell \right)$$
$$= \int g d\ell + \liminf \int f_n d\ell.$$

Therefore $\int f d\ell \leq \liminf \int f_n d\ell$.

Similarly, Fatou's lemma applies to the sequence $(g - f_n)$ and gives

$$\int g d\ell - \int f d\ell = \int (g - f) \, d\ell = \int \liminf (g - f_n) \, d\ell$$
$$\leq \liminf \int (g - f_n) \, d\ell$$
$$= \liminf \left(\int g d\ell - \int f_n d\ell \right)$$
$$= \int g d\ell - \limsup \int f_n d\ell.$$

Hence $\limsup \int f_n d\ell \leq \int f d\ell$. Therefore, $\lim \int f_n d\ell$ exists and

$$\int f d\ell = \lim \int f_n d\ell$$

holds.   □

We end this section with some interesting applications of the Lebesgue dominated convergence theorem.

Theorem 10.38

Let I be an interval and $[a, b] \subset \mathbb{R}$. Let $f : [a, b] \times I \to \mathbb{R}$ be a function such that

(1) for each $t \in I$, $f(\cdot, t) \in \mathcal{L}([a, b])$;

(2) there exists $g \in \mathcal{L}([a, b])$ such that $|f(x, t)| \leq g(x)$ for almost every $x \in [a, b]$ and for all $t \in I$ and for some fixed $g \in \mathcal{L}([a, b])$;

(3) there exists a function h such that $\lim_{t \to t_0} f(x, t) = h(x)$ a.e. for some $t_0 \in I^-$ (possibly $\pm\infty$).

Then $h \in \mathcal{L}([a, b])$ and

$$\int_{[a,b]} h d\ell = \lim_{t \to t_0} \int_{[a,b]} f(x, t) \, d\ell(x).$$

Proof

Suppose that f satisfies the hypotheses of the theorem. Let (t_n) be a sequence in I converging to t_0. Set $h_n(x) = f(x, t_n)$. Then

- $h_n \in \mathcal{L}([a, b])$;

- $|h_n| \leq g$ a.e. for each n;

- $h_n \to h$ a.e.

Thus (h_n) satisfies the hypotheses of the Lebesgue dominated convergence theorem. We conclude that $h \in \mathcal{L}([a, b])$ and

$$\int_{[a,b]} h d\ell = \lim_{n \to \infty} \int_{[a,b]} f(x, t_n) \, d\ell(x).$$

                                                                                              □

Next we recall that if $f : [a, b] \times (\alpha, \beta) \to \mathbb{R}$ is a function and $t_0 \in (\alpha, \beta)$, then the partial derivative of f at t_0, if it exists, is given by

$$\frac{\partial f}{\partial t}(x, t_0) = \lim_{t \to t_0} \frac{f(x, t) - f(x, t_0)}{t - t_0}.$$

Theorem 10.39

Let $f : [a, b] \times (\alpha, \beta) \to \mathbb{R}$ be a function such that

(1) for each $t \in (\alpha, \beta)$, $f(\cdot, t) \in \mathcal{L}([a, b])$;

(2) for some t_0, $\frac{\partial f}{\partial t}(x, t_0)$ exists for almost all $x \in [a, b]$;

(3) there exists $g \in \mathcal{L}([a, b])$ and a $\delta > 0$ such that

$$t \in B(t_0, \varepsilon) \quad \text{implies} \quad \left| \frac{f(x, t) - f(x, t_0)}{t - t_0} \right| \leq g(x)$$

for almost every $x \in [a, b]$.

Then $\frac{\partial f}{\partial t}(\cdot, t_0) \in \mathcal{L}([a, b])$ and the function $F : [a, b] \times (\alpha, \beta) \to \mathbb{R}$ defined by $F(t) = \int_{[a,b]} f(x, t) \, d\ell(x)$ is differentiable at t_0 and

$$F'(t) = \int_{[a,b]} \frac{\partial f}{\partial t}(x, t_0) \, d\ell(x).$$

Proof

Set $\frac{\partial f}{\partial t}(x, t_0) = 0$ at each point x where $\frac{\partial f}{\partial t}(x, t_0)$ does not exist. Then

$$\lim_{t \to t_0} \frac{f(x, t) - f(x, t_0)}{t - t_0} = \frac{\partial f}{\partial t}(x, t_0)$$

exists for almost all $x \in [a, b]$. Thus Theorem 10.38 applies and implies that $\frac{\partial f}{\partial t}(\cdot, t_0) \in \mathcal{L}([a, b])$ and

$$F'(t_0) = \lim_{t \to t_0} \frac{F(t) - F(t_0)}{t - t_0} = \int \frac{\partial f}{\partial t}(x, t_0) \, d\ell(x).$$

$\square$

10.3 Improper Integral

We saw in the previous section that the Lebesgue integral is a generalization of the Riemann integral. In this section we will discuss integrals of functions defined on a nonclosed or unbounded interval.

Definition 10.40

A real-valued function f is said to be **improperly Riemann integrable** on $[a, b)$ if $f \in \mathcal{R}\left([a, c]\right)$ for every $c < b$ and

$$\lim_{c \to b^-} \int_a^c f(x)\, dx$$

exists. Similarly, f is said to be improperly Riemann integrable on $(a, b]$ if $f \in \mathcal{R}\left([c, b]\right)$ for every $c > a$ and

$$\lim_{c \to a^+} \int_c^b f(x)\, dx$$

exists.

In both cases, we will respectively adhere to the following notations

$$f \in \mathcal{R}\left([a, b)\right) \quad \text{and} \quad \int_a^{b^-} f(x)\, dx = \lim_{c \to b^-} \int_a^c f(x)\, dx,$$
$$f \in \mathcal{R}\left((a, b]\right) \quad \text{and} \quad \int_{a^+}^b f(x)\, dx = \lim_{c \to a^+} \int_c^b f(x)\, dx.$$

Let us notice that the first part of the definition makes sense when $b = \infty$, and the second part when $a = -\infty$. In which cases, we denote

$$
\begin{aligned}
f \in \mathcal{R}\left([a, \infty)\right) \quad &\text{and} \quad \int_a^\infty f(x)\, dx = \lim_{c \to \infty} \int_a^c f(x)\, dx, \\
f \in \mathcal{R}\left((-\infty, b]\right) \quad &\text{and} \quad \int_{-\infty}^b f(x)\, dx = \lim_{c \to -\infty} \int_c^b f(x)\, dx.
\end{aligned}
\tag{10.4}
$$

Example 10.41

Show that $\int_0^{1^-} \frac{dx}{\sqrt{1-x^2}} = \frac{\pi}{2}$ and $\int_1^\infty \frac{dx}{x^2} = 1$.

Solution

We have for $0 < c < 1$,

$$\int_0^c \frac{dx}{\sqrt{1-x^2}} = \arcsin c.$$

Since $\lim_{c \to 1^-} \arcsin c = \pi/2$, we obtain

$$\int_0^{1^-} \frac{dx}{\sqrt{1-x^2}} = \frac{\pi}{2}.$$

Likewise for $c < \infty$, we have

$$\int_1^c \frac{dx}{x^2} = 1 - \frac{1}{c}.$$

Thus

$$\int_1^{\infty} \frac{dx}{x^2} = 1.$$

$\square$

Let us discuss the Riemann integrability of the function $f(x) = 1/\sqrt{|x|}$. Clearly, $\int_{-1}^{1} \frac{dx}{\sqrt{|x|}}$ does not exist as a Riemann integral. So $f \notin \mathcal{R}([-1,1])$. However, if $-1 < c < 0$, then

$$\int_{-1}^{c} \frac{dx}{\sqrt{|x|}} = -\int_1^{-c} \frac{dt}{\sqrt{t}} = -\left[2\sqrt{t}\right]_1^{-c} = 2 - 2\sqrt{-c}.$$

It follows that $\int_{-1}^{0^-} \frac{dx}{\sqrt{|x|}} = 2$. A similar argument shows that $\int_{0^+}^{1} \frac{dx}{\sqrt{|x|}} = 2$. Thus the improper integral $\left(\int_{-1}^{0^-} + \int_{0^+}^{1}\right) \frac{dx}{\sqrt{|x|}}$ exists since $f \in \mathcal{R}([-1,0))$ and $f \in \mathcal{R}((0,1])$. We say that f is improperly Riemann integrable on $[-1,0)\cup(0,1]$ and we write $f \in \mathcal{R}([-1,0) \cup (0,1])$ and $\left(\int_{-1}^{0^-} + \int_{0^+}^{1}\right) \frac{dx}{\sqrt{|x|}} = 4$.

Note

One should clearly understand the difference between the notations $\int_a^b$ and $\left(\int_a^{c^-} + \int_{c^+}^{b}\right)$.

For the Lebesgue integral, the above formulas for (improper) integrals do not hold. For example it is *not true* that $\int_{[a,\infty)} f \, d\ell = \lim_{n\to\infty} \int_{[a,n)} f \, d\ell$. We have already noticed that $\mathcal{R}([a,b]) \subset \mathcal{L}([a,b])$. The fact that $[a,b]$ is closed and bounded is crucial for this inclusion to hold. First we prove the following result.

Theorem 10.42

Let $f : [a,\infty) \to \mathbb{R}$ be Riemann integrable on every closed bounded subinterval of $[a,\infty)$. Then $f \in \mathcal{L}([a,\infty))$ if and only if $|f| \in \mathcal{R}([a,\infty))$. Moreover in such a case

$$\int_{[a,\infty)} f\,d\ell = \int_a^\infty f\,(x)\,dx \quad \text{and} \quad \int_{[a,\infty)} |f|\,d\ell = \int_a^\infty |f\,(x)|\,dx.$$

Proof

Suppose that $f \in \mathcal{L}\,([a,\infty))$. Then $f^+ \in \mathcal{L}\,([a,\infty))$. Let (r_n) be a sequence in $[a,\infty)$ diverging to ∞. For each n, consider

$$f_n\,(x) = \begin{cases} f^+\,(x) & \text{if } x \in [a,r_n], \\ 0 & \text{if } x > r_n. \end{cases}$$

Then for each n, since $f_n \in \mathcal{R}\,([a,r_n])$, $f_n \in \mathcal{L}\,([a,r_n]) \subset \mathcal{L}\,([a,\infty))$ and $\int f_n\,d\ell = \int_a^{r_n} f^+\,dx$. Also $f_n \to f^+$ pointwise and $0 \le f_n \le f^+$. Thus by the Lebesgue dominated convergence theorem

$$\int_{[a,\infty)} f^+\,d\ell = \lim \int_{[a,\infty)} f_n\,d\ell = \lim \int_a^{r_n} f^+\,(x)\,dx.$$

In a similar way, we prove that $f^- \in \mathcal{L}\,([a,\infty))$ and

$$\int_{[a,\infty)} f^-\,d\ell = \lim \int_a^{r_n} f^-\,(x)\,dx.$$

Since $f = f^+ - f^-$ and $|f| = f^+ + f^-$, both $\int_a^\infty f\,(x)\,dx$ and $\int_a^\infty |f\,(x)|\,dx$ exist and

$$\int_{[a,\infty)} f\,d\ell = \int_a^\infty f\,(x)\,dx \quad \text{and} \quad \int_{[a,\infty)} |f|\,d\ell = \int_a^\infty |f\,(x)|\,dx.$$

Conversely, assume that $\int_a^\infty |f\,(x)|\,dx$ exists. Consider for each n the function $f_n\,(x) = |f\,(x)|\,\chi_{[a,a+n]}$. Then (f_n) is a nonnegative and nondecreasing sequence of functions converging to $|f|$. For each n, $f_n \in \mathcal{R}\,([a,a+n]) \subset \mathcal{L}\,([a,\infty))$ and

$$\int_{[a,\infty)} f_n\,d\ell = \int_a^{a+n} |f\,(x)|\,dx \le \int_a^\infty |f\,(x)|\,dx.$$

By Fatou's lemma, $|f| \in \mathcal{L}\,([a,\infty))$ and hence $f \in \mathcal{L}\,([a,\infty))$. This completes the proof. $\qquad\square$

The next example shows that the function $f\,(x) = \frac{\sin x}{x} \in \mathcal{R}\,((0,\infty))$. However, $f \notin \mathcal{L}\,((0,\infty))$, since $|f| \notin \mathcal{R}\,((0,\infty))$.

Example 10.43

Show that $\int_{0+}^{\infty} \frac{\sin x}{x} dx$ exists but $\int_{0+}^{\infty} \left| \frac{\sin x}{x} \right| dx$ does not exist.

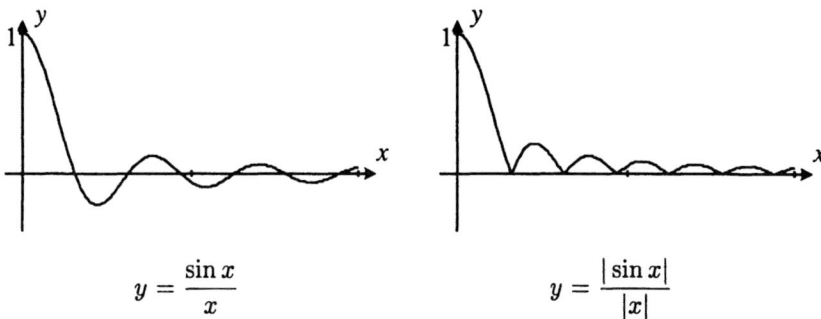

$$y = \frac{\sin x}{x} \qquad\qquad y = \frac{|\sin x|}{|x|}$$

Solution

Since $\frac{\sin x}{x}$ is continuous on $(0, 1]$ and $\lim_{x \to 0} \frac{\sin x}{x} = 1$, the integral $\int_{0+}^{1} \frac{\sin x}{x} dx$ exists. To see that $\int_{1}^{\infty} \frac{\sin x}{x} dx$ exists, let $c > 0$. Then integration by parts yields

$$\int_{1}^{c} \frac{\sin x}{x} dx = \cos 1 - \frac{\cos c}{c} + \int_{1}^{c} \frac{\cos x}{x^2} dx.$$

Now since $\left| \frac{\cos x}{x^2} \right| \leq \frac{1}{x^2}$ for $x \geq 1$, we have

$$\int_{1}^{c} \frac{\cos x}{x^2} dx \leq \int_{1}^{c} \left| \frac{\cos x}{x^2} \right| dx \leq \int_{1}^{c} \frac{1}{x^2} dx = 1 - \frac{1}{c}.$$

It follows that $\int_{1}^{\infty} \frac{\cos x}{x^2} dx$, and hence $\int_{1}^{\infty} \frac{\sin x}{x} dx$ exists, i.e. $\frac{\sin x}{x} \in \mathcal{R}((0, \infty))$. To see that $\int_{0+}^{\infty} \left| \frac{\sin x}{x} \right| dx$ diverges, we fix N and we write

$$\int_{0+}^{N\pi} \left| \frac{\sin x}{x} \right| dx = \sum_{n=0}^{N-1} \int_{n\pi}^{(n+1)\pi} \left| \frac{\sin x}{x} \right| dx.$$

Letting $x = u + n\pi$, we obtain

$$\sum_{n=0}^{N-1} \int_{n\pi}^{(n+1)\pi} \left| \frac{\sin x}{x} \right| dx = \sum_{n=0}^{N-1} \int_{0}^{\pi} \frac{\sin u}{u + n\pi} du.$$

For $u \in [0, \pi]$, $1/(u + n\pi) \geq 1/(n+1)\pi$. Thus

$$\int_{0}^{\pi} \frac{\sin u}{u + n\pi} du \geq \int_{0}^{\pi} \frac{\sin u}{(1+n)\pi} du = \frac{2}{(n+1)\pi}.$$

Since $\sum_{n=0}^{N-1} \frac{2}{(n+1)\pi} \to \infty$ as $N \to \infty$, so does $\sum_{n=0}^{N\pi} \int_{n\pi}^{(n+1)\pi} \left| \frac{\sin x}{x} \right| dx$. This implies the divergence of the integral $\int_{0+}^{\infty} \left| \frac{\sin x}{x} \right| dx$. $\square$

Example 10.44

Show that $\int_{0+}^{\infty} \frac{e^{-x} - e^{-xt}}{x} dx = \ln t$ for $t > 0$.

Solution

Let

$$f(x, t) = \begin{cases} \frac{e^{-x} - e^{-xt}}{x} & \text{if } x > 0 \text{ and } t > 0, \\ t - 1 & \text{if } x = 0 \text{ and } t > 0. \end{cases}$$

Then f is continuous on $[0, \infty) \times (0, \infty)$. We note that $\int_{0+}^{1} |f(x, t)| dx$ exists for $t > 0$ and $|f(x, t)| \leq e^{-x} + e^{-tx}$ for $x > 1$ and for $t > 0$. It follows that for $t > 0$, the function $|f(\cdot, t)| \in \mathcal{R}([0, \infty))$. Hence $f(\cdot, t) \in \mathcal{L}([a, \infty))$ and

$$F(t) = \int_{[a, \infty)} f(x, t) dx = \int_{0+}^{\infty} \frac{e^{-x} - e^{-xt}}{x} dx.$$

Next, we note that $\frac{\partial f}{\partial t}(x, t) = e^{-tx}$ for $x > 0$ and $t > 0$. Also for every $a > 0$, we have

$$0 \leq e^{-tx} \leq e^{-ax} \text{ for } t > a \text{ and for } x \geq 0.$$

Using Theorem 10.39 (replacing $[a, b]$ with $[0, \infty)$) we have

$$F'(t) = \int_{0+}^{\infty} \frac{\partial f}{\partial t}(x, t) dx = \int_{0+}^{\infty} e^{-tx} dx = \frac{1}{t}$$

for all $t > 0$. Thus $F(t) = \ln t + C$. Since $F(1) = 0$, it follows that $C = 0$ and so

$$\int_{0+}^{\infty} \frac{e^{-x} - e^{-xt}}{x} dx = \ln t.$$

$\square$

10.4 Important Inequalities

In this section, we will discuss some very important inequalities concerning Lebesgue integrable functions. We begin by establishing the following facts.

Lemma 10.45

Let $0 < \alpha < 1$. If a and b are nonnegative real numbers, then

$$a^{\alpha} b^{1-\alpha} \leq \alpha a + (1 - \alpha) b \tag{10.5}$$

with equality only if $a = b$.

Proof

The inequality is trivial if $b = 0$. So we suppose that $b \neq 0$. Consider the function f defined on $[0, \infty)$ by

$$f(t) = \alpha t - t^\alpha.$$

Then

$$f'(t) = \alpha \left(1 - t^{\alpha-1}\right).$$

Since $\alpha - 1 < 0$, we have

$$f'(t) < 0 \quad \text{for } 0 \leq t < 1,$$
$$f'(t) > 0 \quad \text{for } t > 1.$$

Thus for $t \neq 1$, we have $f(t) \geq f(1) = \alpha - 1$. It follows that

$$t^\alpha \leq 1 - \alpha + \alpha t.$$

Now if we let $t = a/b$ and multiply by b, we obtain the desired inequality. $\square$

We now use the result of the above lemma to establish the so-called **Hölder's inequality for sums**.

Theorem 10.46

Let $p > 1$, $q > 1$ be such that $1/p + 1/q = 1$.[2] Suppose that $(a_i)_{i=1}^n$ and $(b_i)_{i=1}^n$ are real numbers. Then

$$\sum_{i=1}^n |a_i b_i| \leq \left(\sum_{i=1}^n |a_i|^p\right)^{1/p} \left(\sum_{i=1}^n |b_i|^q\right)^{1/q}. \tag{10.6}$$

Proof

For the sake of simplicity of notation, we let

$$A_n = \left(\sum_{i=1}^n |a_i|^p\right)^{1/p} \quad \text{and} \quad B_n = \left(\sum_{i=1}^n |b_i|^q\right)^{1/q}$$

for each $n \in \mathbb{N}$. We suppose that $A_n > 0$ and $B_n > 0$; otherwise (10.6) is trivial. In inequality (10.5), if we substitute $1/p$ for α, we have

$$a^{1/p} b^{1/q} \leq \frac{1}{p} a + \frac{1}{q} b.$$

---

[2] Two numbers satisfying this relation are said to be **conjugate exponents**.

Letting $A = a^{1/p}$ and $B = b^{1/q}$ yields

$$AB \leq \frac{1}{p}A^p + \frac{1}{q}B^q. \qquad (10.7)$$

Now set $A = \frac{a_i}{A_n}$ and $B = \frac{b_i}{B_n}$. It follows that

$$\frac{|a_i b_i|}{A_n B_n} \leq \frac{1}{p}\left(\frac{a_i}{A_n}\right)^p + \frac{1}{q}\left(\frac{b_i}{B_n}\right)^q.$$

Summing over $i = 1, 2, \ldots, n$, we obtain

$$\frac{\sum_{i=1}^n |a_i b_i|}{A_n B_n} \leq \frac{1}{p}\frac{\sum_{i=1}^n |a_i|^p}{A_n^p} + \frac{1}{q}\frac{\sum_{i=1}^n |b_i|^q}{B_n^q} = 1.$$

Hence $\sum_{i=1}^n |a_i b_i| \leq A_n B_n$. The inequality is established. $\qquad \square$

The extension of the above inequality to infinite sums requires an easy argument. We then obtain Hölder's inequality for series.

Corollary 10.47

Let $p > 1$, $q > 1$ be such that $1/p + 1/q = 1$. Suppose that (a_n) and (b_n) are two sequences of real numbers such that both the series $\sum |a_n|^p$ and $\sum |b_n|^q$ are convergent. Then the series $\sum a_n b_n$ is absolutely convergent and

$$\sum |a_n b_n| \leq \left(\sum |a_n|^p\right)^{1/p} \left(\sum |b_n|^q\right)^{1/q}.$$

Proof

From Hölder's inequality for sums (10.6), it follows that for each $m < n$

$$\sum_{i=1}^m |a_i b_i| \leq \left(\sum_{i=1}^n |a_i|^p\right)^{1/p} \left(\sum_{i=1}^n |b_i|^q\right)^{1/q}.$$

Now let $n \to \infty$. We obtain for any m

$$\sum_{i=1}^m |a_i b_i| \leq \left(\sum_{i=1}^\infty |a_i|^p\right)^{1/p} \left(\sum_{i=1}^\infty |b_i|^q\right)^{1/q}.$$

Since both the series $\sum |a_n|^p$ and $\sum |b_n|^q$ are convergent, the series $\sum |a_n b_n|$ is bounded and hence the series $\sum a_n b_n$ is absolutely convergent. The desired inequality easily follows. $\qquad \square$

Hölder's inequality (10.6) is also used to prove **Minkowski's inequality for sums**.

Theorem 10.48

Let $p > 1$. Suppose that $(a_i)_{i=1}^n$ and $(b_i)_{i=1}^n$ are real numbers. Then

$$\left(\sum_{i=1}^n |a_i + b_i|^p \right)^{1/p} \leq \left(\sum_{i=1}^n |a_i|^p \right)^{1/p} + \left(\sum_{i=1}^n |b_i|^p \right)^{1/p}.$$

Proof

Again for the sake of simplicity, we let

$$A_n = \left(\sum_{i=1}^n |a_i|^p \right)^{1/p} \quad \text{and} \quad B_n = \left(\sum_{i=1}^n |b_i|^p \right)^{1/p}.$$

We also assume that $\sum_{i=1}^n |a_i + b_i|^p > 0$ because otherwise the desired inequality is trivial.

Let $q > 1$ such that $1/p + 1/q = 1$. Then $1 + p/q = p$ and hence for each i, we have

$$\sum_{i=1}^n |a_i + b_i|^p = \sum_{i=1}^n |a_i + b_i| \, |a_i + b_i|^{p/q}$$
$$\leq \sum_{i=1}^n |a_i| \, |a_i + b_i|^{p/q} + \sum_{i=1}^n |b_i| \, |a_i + b_i|^{p/q}.$$

Applying Hölder's inequality to each sum on the right-hand side in the above inequality, we obtain

$$\sum_{i=1}^n |a_i| \, |a_i + b_i|^{p/q} \leq A_n \left(\sum_{i=1}^n |a_i + b_i|^p \right)^{1/q}$$

and

$$\sum_{i=1}^n |b_i| \, |a_i + b_i|^{p/q} \leq B_n \left(\sum_{i=1}^n |a_i + b_i|^p \right)^{1/q}.$$

Now we add the last two inequalities. We obtain

$$\sum_{i=1}^n |a_i + b_i|^p \leq (A_n + B_n) \left(\sum_{i=1}^n |a_i + b_i|^p \right)^{1/q}.$$

As $1 - 1/q = 1/p$, the desired inequality is obtained by dividing the last inequality by $\left(\sum_{i=1}^n |a_i + b_i|^p \right)^{1/q}$. $\qquad\square$

Minkowski's inequality can also be extended to the sums of series. The proof uses a similar argument to that of the proof of Corollary 10.47. We leave the details to the reader.

Corollary 10.49

Let $p > 1$. Suppose that (a_n) and (b_n) are two sequences of real numbers such that the series $\sum |a_n|^p$ and $\sum |b_n|^p$ are both convergent. Then the series $\sum |a_n + b_n|^p$ is convergent and

$$\left(\sum_{i=1}^{\infty} |a_i + b_i|^p \right)^{1/p} \leq \left(\sum_{i=1}^{\infty} |a_i|^p \right)^{1/p} + \left(\sum_{i=1}^{\infty} |b_i|^p \right)^{1/p}.$$

Notation

Recall (Exercise 3.23, page 92) that a sequence (a_n) is p-**summable** $(1 \leq p < \infty)$ if $\sum |a_n|^p < \infty$. We denote by ℓ^p the set of all p-summable sequences of real numbers. If $(a_n) \in \ell^p$, we define

$$\|(a_n)\|_p = \left(\sum_{n=1}^{\infty} |a_n|^p \right)^{1/p}.$$

Let (a_n) and (b_n) be two p-summable sequences. The following facts are easily verified:

- $\|(a_n)\|_p \geq 0$;
- $\|(a_n)\|_p = 0$ if and only if $a_n = 0$ for each n;
- $\|\alpha (a_n)\|_p = |\alpha| \|a_n\|_p$ for $\alpha \in \mathbb{R}$;
- $\|(a_n + b_n)\|_p \leq \|(a_n)\|_p + \|(a_n)\|_p$ (Minkowski's inequality).

For example, an easy p-series test shows that the sequence $(1/(n^\alpha)) \in \ell^p$ if and only if $\alpha > 1/p$.

Before we establish Hölder's and Minkowski's inequalities for integrals, we give the following definition.

Definition 10.50

Let $0 < p < \infty$. A function f is said to be p-**integrable** if $|f|^p \in \mathcal{L}$. The collection of all p-integrable functions will be denoted by $\mathcal{L}^p$.

Theorem 10.51 (Hölder's Inequality)[3]

Let $p > 1$, $q > 1$ be such that $1/p + 1/q = 1$. Suppose that $f \in \mathcal{L}^p$ and $g \in \mathcal{L}^q$. Then $fg \in \mathcal{L}$ and

$$\int |fg| \, d\ell \leq \left(\int |f|^p \, d\ell \right)^{1/p} \left(\int |g|^q \, d\ell \right)^{1/q} .$$

Proof

For the sake of simplicity of notation, we let

$$F = \left(\int |f|^p \, d\ell \right)^{1/p} \quad \text{and} \quad G = \left(\int |g|^q \, d\ell \right)^{1/q} .$$

We suppose $F \neq 0$ and $G \neq 0$; otherwise the inequality is trivial. Letting $A = |f(x)| / F$ and $B = |g(x)| / G$ in (10.7) we have

$$\frac{|f(x)| \, |g(x)|}{FG} \leq \frac{1}{p} \frac{|f(x)|^p}{F^p} + \frac{1}{q} \frac{|g(x)|^q}{G^q} .$$

Since both of the terms on the right-hand side of the above inequality are integrable, it follows by Theorem 10.29 that $fg \in \mathcal{L}$. Integrating both sides we obtain

$$\frac{1}{FG} \int |fg| \, d\ell \leq \frac{1}{p} + \frac{1}{q} = 1.$$

The desired inequality is established.   □

Theorem 10.52 (Minkowski's Inequality)

Let $p > 1$. Suppose that $f, g \in \mathcal{L}^p$. Then $f + g \in \mathcal{L}^p$ and

$$\left(\int |f + g|^p \, d\ell \right)^{1/p} \leq \left(\int |f|^p \, d\ell \right)^{1/p} + \left(\int |g|^p \, d\ell \right)^{1/p} .$$

Proof

We let $F = \left(\int |f|^p \, d\ell \right)^{1/p}$ and $G = \left(\int |g|^p \, d\ell \right)^{1/p}$. Since

$$|f + g|^p \leq [2 \sup \{|f|, |g|\}]^p \leq 2^p \{|f|^p + |g|^p\}$$

---

[3] The particular case $p = q = 2$ is known as the **Cauchy–Bunyakovskiĭ–Schwarz inequality**.

it follows from Theorem 10.29 that $|f + g|^p \in \mathcal{L}$. Also

$$|f + g|^p = |f + g|^{p-1} |f + g|$$
$$\leq |f + g|^{p-1} |f| + |f + g|^{p-1} |g| .$$

We notice that if q is the conjugate exponent of p, then $p = (p - 1) q$ and hence $|f + g|^{(p-1)/q} \in \mathcal{L}$. Hence Hölder's inequality applies and implies

$$\int |f + g|^{p-1} |f| \, d\ell \leq F \left(\int |f + g|^{(p-1)/q} \, d\ell \right)^{1/q} ,$$

$$\int |f + g|^{p-1} |g| \, d\ell \leq G \left(\int |f + g|^{(p-1)/q} \, d\ell \right)^{1/q} .$$

We then obtain

$$\left(\int |f + g|^p \, d\ell \right) \leq (F + G) \left(\int |f + g|^p \, d\ell \right)^{1/q} .$$

If $\int |f + g|^p \, d\ell = 0$, the desired inequality is trivial. If $\int |f + g|^p \, d\ell \neq 0$, we can divide both sides of the above inequality by $\left(\int |f + g|^p \, d\ell \right)^{1/q}$ and subsequently obtain Minkowski's inequality. $\square$

Notation

If $f \in \mathcal{L}^p$, the p-integral of f is commonly denoted by

$$\|f\|_p = \left(\int |f|^p \, d\ell \right)^{1/p} .$$

With the above notation Minkowski's inequality is precisely the triangle inequality for $\|\cdot\|_p$

$$\|f + g\|_p \leq \|f\|_p + \|g\|_p .$$

We end this section with an important result. We first notice that the notions of convergence stated in Definition 10.12 and Definition 10.21 can be extended to Lebesgue integrable functions. Namely for $1 \leq p < \infty$, we say that a sequence (f_n) in $\mathcal{L}^p$ **converges** (resp. **is Cauchy**) in p-**mean** if $\int |f_n - f|^p \, d\ell \to 0$ (resp. $\int |f_n - f_m|^p \, d\ell \to 0$) as $n, m \to \infty$.

Theorem 10.53 (Riesz–Fischer)

Let $1 \leq p < \infty$. Every sequence of functions (f_n) in $\mathcal{L}^p$ which is Cauchy in p-mean is convergent in p-mean.

Proof

Let (f_n) be Cauchy in p-mean. There exists a subsequence (g_n) of (f_n) such that

$$\int |g_{n+1} - g_n|^p \, d\ell < \left(\frac{1}{2^n}\right)^p$$

for each n.

We consider the sequence (h_n) defined by

$$h_1 = 0;$$
$$h_n = |g_1| + |g_2 - g_1| + \cdots + |g_n - g_{n-1}|$$

for $n > 1$. Then (h_n) is a nondecreasing sequence of nonnegative functions. Using Minkowski's inequality, we have for each n

$$\left(\int (h_n)^p \, d\ell\right)^{1/p} \leq \left(\int |g_1|^p \, d\ell\right)^{1/p} + \sum_{i=2}^{n} \left(\int |g_i - g_{i-1}|^p \, d\ell\right)^{1/p}$$

$$\leq \left(\int |g_1|^p \, d\ell\right)^{1/p} + \sum_{i=2}^{\infty} \left(\int |g_i - g_{i-1}|^p \, d\ell\right)^{1/p}$$

$$\leq \left(\int |g_1|^p \, d\ell\right)^{1/p} + \sum_{i=2}^{n} \left(\frac{1}{2^i}\right) \leq \left(\int |g_1|^p \, d\ell\right)^{1/p} + 1.$$

The monotone convergence theorem applies and implies that there exists a function h such that $h^p \in \mathcal{L}$ and that $h_n \to h$ a.e.

Now we notice that

$$|g_{n+k} - g_n| = \left|\sum_{i=1}^{k} g_{n+i} - g_{n+i-1}\right|$$

$$\leq \sum_{i=1}^{k} |g_{n+i} - g_{n+i-1}| = h_{n+k} - h_n.$$

It follows that (g_n) converges a.e. to some function g. Since

$$|g_n| = \left|g_1 + \sum_{i=2}^{n} (g_i - g_{i-1})\right|$$

$$\leq h_n \leq h$$

a.e., then $|g| < h$ a.e. Hence $|g|^p \in \mathcal{L}$. It follows that $|g - g_n| \leq 2h$ a.e. and $\lim |g - g_n|^p = 0$. An appeal to the Lebesgue dominated convergence theorem now implies that $\lim \int |g - g_n|^p \, d\ell = 0$. The proof is complete. $\qquad\square$

EXERCISES

10.1   Show that the family of bounded intervals in (10.1) may be replaced
       (without changing ℓ^*) by a family of

    (1)   closed bounded intervals;

    (2)   open bounded intervals;

    (3)   half-open bounded intervals;

    (4)   closed bounded intervals with length less than δ, for a given
       $\delta > 0$.

10.2   Prove or disprove: if $A \subset \mathbb{R}$, then $\ell^*(A) = \ell^*(A^-)$.

10.3   Show that $\ell^*(\mathbb{R}) \neq 0$ and deduce that $\mathbb{R}$ is uncountable.

10.4   Let $A \subset \mathbb{R}$ and let $t \in \mathbb{R}$. Show that $\ell^*(A + t) = \ell^*(A)$, where
       $A + t = \{a + t : a \in A\}$.

10.5   Let $\varphi \in \mathcal{S}$ and let $k > 0$.

    (1)   Show that the set $A = \{x \in \mathbb{R} : \varphi(x) \geq k\}$ is a finite union
       of disjoint bounded intervals.

    (2)   If $\varphi \geq 0$, show that $\ell^*(A) \leq \frac{1}{k} \int \varphi$.

10.6   Show that if $f \in \mathcal{C}([a, b])$ and $f \sim 0$, then $f \equiv 0$.

10.7   Show that if (φ_n) is a sequence in $\mathcal{S}$ and converges increasingly to
       $\varphi \in \mathcal{S}$, then $\lim \int \varphi_n = \int \varphi$.

10.8   Prove or disprove: if $f \in \mathcal{L}$, then $\int f d\ell = 0$ if and only if $f \sim 0$.

10.9   Let $f \in \mathcal{L}([0, \infty))$. Suppose that $\int_{[0,t)} f d\ell = 0$ for each $t \geq 0$. Show
       that $f \sim 0$.

10.10 Let $f \in \mathcal{L}$ such that $f(x) > 0$ a.e. Suppose that $A \subset \mathbb{R}$ such that
       $\chi_A \in \mathcal{L}$. Show that $\int f\chi_A d\ell = 0$ implies $\ell^*(A) = 0$.

10.11 Let $f \in \mathcal{L}$. Show that for each $\varepsilon > 0$, there exists $\delta > 0$ such that
       $A \subset \mathbb{R}$, $\chi_A \in \mathcal{L}$ and $\ell^*(A) < \delta$ implies $\left| \int f\chi_A d\ell \right| < \varepsilon$.

10.12 Show that if $f \in \mathcal{L}$, then for every interval I, $f\chi_I \in \mathcal{L}$. Deduce that
       if $\varphi \in \mathcal{R}$, then $f\varphi \in \mathcal{L}$. Give an example of functions $f, g \in \mathcal{L}$ where
       $fg \notin \mathcal{L}$.

10.13 Show that if f is essentially bounded and $g \in \mathcal{L}$, then $fg \in \mathcal{L}$.
       Moreover if M is any essential bound of g, then

$$\int |fg| \, d\ell \leq M \int |f| \, d\ell.$$

10.14 Let $f \in \mathcal{L}([a,b])$. Define $F(t) = \int_{[a,t]} f d\ell$ for $t \in [a,b]$. Show that $F' \sim f$.

10.15 Let $(f_n) \subset \mathcal{L}$ be such that $0 \le f_{n+1} \le f_n$ a.e. for each n. Show that $f_n \to 0$ if and only if $\int f_n d\ell \to 0$.

10.16 Let (f_n) be a sequence in $\mathcal{L}$ such that $|f_n| \le g$ a.e. for all n and for some fixed $g \in \mathcal{L}$ and that $f_n \to f$ a.e. Show that $f \in \mathcal{L}$.

10.17 Let (f_n) be a sequence of nonnegative functions in $\mathcal{L}$. Suppose that

$$\sum_{n=1}^{\infty} \int_I f_n d\ell < \infty,$$

where I is an interval of $\mathbb{R}$. Show that $\sum f_n$ converges almost everywhere on I, and

$$\sum_{n=1}^{\infty} \int_I f_n d\ell = \int_I \sum_{n=1}^{\infty} f_n d\ell.$$

10.18 **Riemann–Lebesgue lemma.** Show that if $f \in \mathcal{L}$, then

$$\lim_{n \to \infty} \int f(x) \cos(nx) \, d\ell(x) = \lim_{n \to \infty} \int f(x) \sin(nx) \, d\ell(x) = 0.$$

(Hint: First consider the case where $f = \chi_{[a,b]}$.)

10.19 Consider the function $\varphi : [0,1] \to \mathbb{R}$ defined by

$$\varphi(x) = \begin{cases} 2^{-k} & \text{if } 2^{-k-1} < x \le 2^{-k} \quad (k = 0,1,2,\ldots) \\ 0 & \text{otherwise.} \end{cases}$$

(1) Construct an increasing sequence of set function (φ_n) converging a.e. to φ and such that $(I(\varphi_n))_n$ is bounded.

(2) Deduce that $\varphi \in \mathcal{L}([0,1])$ and evaluate $\int_{[0,1]} \varphi d\ell$.

(3) Let $F(x) = \int_{[0,1]} \varphi \chi_{[0,x]} d\ell$. Evaluate $F(x) = xF(x) - \frac{1}{3}[A(x)]^2$, where $A(x) = 2^{-(\ln x^{-1}/\ln 2)}$.

10.20 **Rademacher functions.** Consider for each $n \in \mathbb{N}$, the step function $r_n : [0,1] \to \mathbb{R}$ defined by

$$r_n(x) = \begin{cases} (-1)^{k-1} & \text{if } \frac{k-1}{2^n} \le x < \frac{k}{2^n} \quad (k = 0,1,2,\ldots,2^n) \\ -1 & \text{if } x = 1. \end{cases}$$

(1) Verify that $\int_0^1 r_n(x) r_m(x) \, dx = 0$ if $n \neq m$ and $\int_0^1 r_n^2(x) \, dx = 1$.

(2) Show that if $f \in \mathcal{L}([0,1])$, then

$$\lim_{n \to \infty} \int_{[0,1]} f(x) r_n(x) d\ell(x) = 0.$$

10.21 If F is differentiable everywhere on $[a,b]$, then show that $F(t) - F(a) = \int_{[a,t]} F' d\ell$.

10.22 Show that if $f(x) = \frac{\cos nx}{x^2+1}$, then $f \in \mathcal{L}([a,\infty))$.

10.23 Show that if $f(x) = \frac{\ln x}{x^2}$, then $f \in \mathcal{L}([1,\infty))$ and $\int_{[1,\infty)} f d\ell = 1$.

10.24 Show that $\lim_{n \to \infty} \int_0^n \left(1 + \frac{x}{n}\right)^n e^{-2x} dx = 1$.

10.25 Show that $\int_0^\infty x^{2n} e^{-x^2} dx = \frac{(2n)!}{2^{2n} n!} \frac{\sqrt{\pi}}{2}$. (Hint: $\int_0^\infty e^{-x^2} dx = \frac{\sqrt{\pi}}{2}$.)

10.26 Show that $\int_0^\infty e^{-tx^2} dx = \frac{1}{2}\sqrt{\frac{\pi}{t}}$ for all $t > 0$.

10.27 **Fresnel's integral.** Let $f(x) = \cos\left(x^2\right)$ and $g(x) = \sin\left(x^2\right)$.

(1) Show that both f and g belong to $\mathcal{R}([a,\infty))$.

(2) Show that neither f nor g is in $\mathcal{L}([a,\infty))$.

10.28 Let $f \in \mathcal{L}^p$ and let $\varepsilon > 0$. Show that $\ell^*\{t : |f(t)| \geq \varepsilon\} \leq \frac{1}{\varepsilon^p} \int |f|^p d\ell$.

10.29 Let $f \in \mathcal{L}$. Show that the function $\varphi : [0,\infty) \to \mathbb{R}$ defined by

$$\varphi(x) = \sup \left\{ \int |f(t+s) - f(t)| d\ell(t) : |s| < x \right\}$$

is continuous at $t = 0$.

10.30 Let $f \in \mathcal{L}$. Show that the function $\varphi : [0,\infty) \to [0,\infty]$ defined by

$$\varphi(x) = \ell^*\{t : |f(t)| \geq x\}$$

is Lebesgue integrable and that $\int_{[0,\infty)} \varphi d\ell = \|f\|_1$.

10.31 Let $f \in \mathcal{L}$. Show that

$$\lim_{t \to 0} \int |f(x) - f(x+t)| d\ell(x) = 0.$$

10.32 Let $f \in \mathcal{L}([a,b])$ and let $\varepsilon > 0$. Show that

(1) there exists $g \in \mathcal{R}([a,b])$ such that $\|f - g\|_1 < \varepsilon$;

(2) there exists $g \in \mathcal{C}([a,b])$ such that $\|f - g\|_1 < \varepsilon$;

(3) there exists a step function φ such that $\|f - \varphi\|_1 < \varepsilon$.

10.33   Give an example of a sequence of continuous functions on $[0,1]$ converging in mean but not uniformly.

10.34   Let (f_n) be a sequence defined by $f_n(x) = n^2 x e^{-nx}$. Show that $f_n(x) \to 0$ for all $x \geq 0$ but (f_n) does not converge in 2-mean.

10.35   Let (f_n) be a sequence defined by $f_n(x) = n^{1/4}$. Show that $f_n(x) \to 0$ uniformly but (f_n) does not converge in 2-mean.

10.36   Let $f : [0,1] \to \mathbb{R}$ be defined by $f(t) = t^\alpha$. Determine for which value of p the function f is p-integrable and evaluate its p-integral.

10.37   Show that if $1 \leq p \leq q < \infty$, then every q-integrable function on $[a,b]$ is p-integrable, i.e. $\mathcal{L}^q([a,b]) \subset \mathcal{L}^p([a,b])$.

10.38   Let $f \in \mathcal{L}([0,1])$. Suppose that there exists $M > 0$ such that $f(x) > M$ a.e. Show that $\ln(f) \in \mathcal{L}([0,1])$ and that $\int \ln f \, d\ell \leq \ln \int f \, d\ell$.

10.39   Let $f \in \mathcal{L}^1 \cap \mathcal{L}^2$. Show that

(1) $f \in \mathcal{L}^p$ for each $1 \leq p \leq 2$; and

(2) $\lim_{p \to 1+} \int |f|^p \, d\ell = \int |f| \, d\ell$.

10.40   Give an example of a sequence (f_n) in $\mathcal{L}([0,1])$ converging in mean but not uniformly.

10.41   Show that if $f, g \in \mathcal{L}^2$, then $fg \in \mathcal{L}^1$.

10.42   Show that if $f, g \in \mathcal{L}^2$, then so are the functions $f + g$, αf for $\alpha \in \mathbb{R}$.

10.43   Let $f, g \in \mathcal{L}^2([a,b])$. The **inner product** of f and g is defined to be the number

$$\langle f, g \rangle = \frac{1}{2} \int_a^b fg \, d\ell.$$

Verify that for $f, g, h \in \mathcal{L}^2([a,b])$

(1) $\langle f, f \rangle = \|f\|_2^2$;

(2) $\langle f, f \rangle = 0$ if and only if $f = 0$ a.e.;

(3) $\langle f, g \rangle = \langle f, g \rangle$;

(4) $\langle f + h, g \rangle = \langle f, g \rangle + \langle h, g \rangle$;

(5) $\langle \alpha f, g \rangle = \alpha \langle f, g \rangle$ for each $\alpha \in \mathbb{R}$.

10.44   Let (f_n) be in $\mathcal{L}^2([0,1])$. Suppose that $\sup\{\|f_n\| : n \in \mathbb{N}\} < \infty$. Show that $f_n/n \to 0$ a.e.

11
Elements of Fourier Analysis

The Stone–Weierstrass theorem is concerned about approximation of continuous functions by polynomials. In this chapter,[1] we discuss yet another type of approximation which applies to functions that are not necessarily continuous. The applications of the kind of approximation we are going to study here are of considerable importance especially in physics and engineering.

11.1 Fourier Series

In this section, we will consider the trigonometric functions

$$1, \ \cos x, \ \sin x, \ \cos 2x, \ \sin 2x, \ \ldots, \ \cos nx, \ \sin nx, \ \ldots$$

and will be mainly concerned with series of functions of the form

$$\frac{a_0}{2} + (a_1 \cos x + b_1 \sin x) + (a_2 \cos 2x + b_2 \sin 2x) + \ldots \qquad (11.1)$$

where the a_n and b_n are real numbers. Such series are called **trigonometric series**.

First, we lay down the following facts about the trigonometric functions.

---

[1] Although the results in this chapter may be seen as purely computational, we chose to include them because they serve as examples of application of series and integrals.

Theorem 11.1

The trigonometric functions

$$1, \ \cos x, \ \sin x, \ \cos 2x, \ \sin 2x, \ \ldots, \ \cos nx, \ \sin nx, \ \ldots$$

satisfy the following relations

$$\int_{-\pi}^{\pi} \cos kx \cos nx \, dx = \begin{cases} 0 & \text{if } k \neq n, \\ 2\pi & \text{if } k = n = 0, \\ \pi & \text{if } k = n \neq 0; \end{cases}$$

$$\int_{-\pi}^{\pi} \sin kx \sin nx \, dx = \begin{cases} 0 & \text{if } k \neq n, \\ 0 & \text{if } k = n = 0, \\ \pi & \text{if } k = n \neq 0; \end{cases}$$

$$\int_{-\pi}^{\pi} \sin kx \cos nx \, dx = 0.$$

The above properties are referred to by saying that the family of trigonometric functions $\{1, \cos x, \sin x, \cos 2x, \sin 2x, \ldots\}$ forms an **orthogonal system**. The proof of these properties follows at once from the following trigonometric formulas (verify!)

$$\cos kx \cos nx = \frac{1}{2} \left[\sin (k + n) x - \sin (k - n) x \right],$$

$$\sin kx \sin nx = \frac{1}{2} \left[\cos (k + n) x - \cos (k - n) x \right],$$

$$\cos kx \sin nx = \frac{1}{2} \left[\sin (k + n) x + \sin (k - n) x \right].$$

It is readily seen that for each n and x we have

$$\cos (nx + 2\pi) = \cos nx \quad \text{and} \quad \sin (nx + 2\pi) = \sin nx.$$

Therefore, if the series (11.1) converges, then its sum s must satisfy

$$s (x + 2\pi) = s (x).$$

Functions with such a property are termed as **periodic** with period 2π. For this reason, it is enough to study the convergence of a trigonometric series on intervals of length 2π.

Lemma 11.2

If the trigonometric series $\frac{a_0}{2} + \sum_{n=1}^{\infty} (a_n \cos nx + b_n \sin nx)$ converges uniformly on $[-\pi, \pi]$ to a function f, then for any $k \in \mathbb{N}$, both of the series

$$\frac{a_0 \cos kx}{2} + \sum_{n=1}^{\infty} (a_n \cos kx \cos nx + b_n \cos kx \sin nx),$$

$$\frac{a_0 \sin kx}{2} + \sum_{n=1}^{\infty} (a_n \sin kx \cos nx + b_n \sin kx \sin nx)$$

converge uniformly on $[-\pi, \pi]$, respectively to $f(x) \cos kx$ and $f(x) \sin kx$.

Proof

Let $\varepsilon > 0$. There exists $N \in \mathbb{N}$ such that $n > N$ implies

$$\left| f(x) - \frac{a_0}{2} - \sum_{i=1}^{n} (a_i \cos ix + b_i \sin ix) \right| < \varepsilon$$

for all $x \in [-\pi, \pi]$. Hence for every k, we have

$$\left| f(x) \cos kx - \frac{a_0}{2} - \sum_{i=1}^{n} (a_i \cos kx \cos ix + b_i \cos kx \sin ix) \right|$$

$$\leq |\cos kx| \left| f(x) - \frac{a_0}{2} - \sum_{i=1}^{n} (a_i \cos ix + b_i \sin ix) \right| < \varepsilon$$

since $|\cos kx| \leq 1$, for all $x \in [-\pi, \pi]$. The convergence of the second series is obtained in a similar fashion. $\square$

Theorem 11.3

Suppose that the trigonometric series $\frac{a_0}{2} + \sum_{n=1}^{\infty} (a_n \cos nx + b_n \sin nx)$ converges uniformly on $(-\pi, \pi)$ to an integrable function f. Then for $n = 0, 1, 2, \ldots$,

$$a_n = \frac{1}{\pi} \int_{[-\pi, \pi]} f(x) \cos nx d\ell(x) \quad \text{and} \quad b_n = \frac{1}{\pi} \int_{[-\pi, \pi]} f(x) \sin nx d\ell(x).$$

$$(11.2)$$

Proof

Since

$$f(x) = \frac{a_0}{2} + \sum_{n=1}^{\infty} (a_n \cos nx + b_n \sin nx) \tag{11.3}$$

and since the convergence is uniform on $(-\pi, \pi)$, integration term by term is legitimate on $(-\pi, \pi)$. Thus we have

$$\int_{[-\pi, \pi]} f d\ell = \int_{-\pi}^{\pi} \frac{a_0}{2} dx + \sum_{n=1}^{\infty} \left(\int_{-\pi}^{\pi} a_n \cos nx dx + \int_{-\pi}^{\pi} b_n \sin nx dx \right).$$

In view of Theorem 11.1, we obtain

$$a_0 = \frac{1}{\pi} \int_{[-\pi,\pi]} f \, d\ell.$$

To find the coefficient a_k for $k \neq 0$, we multiply both sides of (11.3) by $\cos kx$:

$$f(x) \cos kx = \frac{a_0 \cos kx}{2} + \sum_{n=1}^{\infty} (a_n \cos nx \cos kx + b_n \sin nx \cos kx).$$

By Lemma 11.2, the above convergent series is uniform on $(-\pi, \pi)$. Once again termwise integration is permitted and we have

$$\int_{[-\pi,\pi]} f(x) \cos nx \, d\ell(x) = \int_{-\pi}^{\pi} \frac{a_0 \cos kx}{2} dx + \sum_{n=1}^{\infty} \int_{-\pi}^{\pi} a_n \cos nx \cos kx \, dx$$

$$+ \sum_{n=1}^{\infty} \int_{-\pi}^{\pi} b_n \sin nx \cos kx \, dx.$$

Theorem 11.1 implies that all the integrals on the right are equal to zero, with the exception of

$$\int_{-\pi}^{\pi} a_k \cos^2 kx \, dx = a_k \pi.$$

Whence

$$a_k = \frac{1}{\pi} \int_{[-\pi,\pi]} f(x) \cos nx \, d\ell(x).$$

The coefficients b_k are obtained in a similar fashion by multiplying both sides of (11.3) by $\sin kx$:

$$b_k = \frac{1}{\pi} \int_{[-\pi,\pi]} f(x) \sin nx \, d\ell(x),$$

as desired.   □

The coefficients determined by formulas (11.2) are called the **Fourier co-efficients** of the functions f. The stage is now set for introducing the following definition.

Definition 11.4

Let f be a 2π-periodic integrable function. The **Fourier series** of f is the trigonometric series, the coefficients of which are given by the Fourier coefficients of f.

Notation

We denote the Fourier series of f by $\mathcal{F}(f)$, i.e.

$$\mathcal{F}(f) = \frac{a_0}{2} + \sum_{n=1}^{\infty} (a_n \cos nx + b_n \sin nx)$$

where

$$a_n = \frac{1}{\pi} \int_{[-\pi,\pi]} f(x) \cos nx\, d\ell(x) \text{ and } b_n = \frac{1}{\pi} \int_{[-\pi,\pi]} f(x) \sin nx\, d\ell(x)$$

for $n = 0, 1, 2, \ldots$. To indicate the association of a 2π-periodic function f with its Fourier series, we write $f \rightsquigarrow \mathcal{F}(f)$ or

$$f \rightsquigarrow \frac{a_0}{2} + \sum_{n=1}^{\infty} (a_n \cos nx + b_n \sin nx).$$

We continue to use the notation $f(x) = \mathcal{F}(f)(x)$ or

$$f(x) = \frac{a_0}{2} + \sum_{n=1}^{\infty} (a_n \cos nx + b_n \sin nx)$$

to express that the Fourier series converges at the point x and its sum is $f(x)$.

Example 11.5

Determine the Fourier series of the 2π-periodic function defined by $f(x) = x$, $-\pi < x \leq \pi$.

Solution

Direct computations yield

$$a_k = \frac{1}{\pi} \int_{-\pi}^{\pi} x \cos kx\, dx$$

$$= \frac{1}{\pi} \left[x \frac{\sin kx}{k} \Big|_{-\pi}^{\pi} - \frac{1}{k} \int_{-\pi}^{\pi} \sin kx\, dx \right] = 0,$$

$k = 0, 1, 2, \ldots$, whereas

$$b_k = \frac{1}{\pi} \int_{-\pi}^{\pi} x \sin kx\, dx$$

$$= \frac{1}{\pi} \left[x \frac{\cos kx}{k} \Big|_{-\pi}^{\pi} + \frac{1}{k} \int_{-\pi}^{\pi} \cos kx\, dx \right] = (-1)^{k+1} \frac{2}{k},$$

$n = 1, 2, \ldots$. Thus we get

$$x \rightsquigarrow 2 \left[\sin x - \frac{1}{2} \sin 2x + \frac{1}{3} \sin 3x - \cdots \right].$$

□

The following is an example of a function f whose Fourier series converges to the function f itself. Consider the function f defined by

$$f(x) = \begin{cases} -1 & \text{for } -\pi < x < 0 \\ 1 & \text{for } 0 \leq x \leq \pi. \end{cases}$$

Its Fourier series is easily obtained by direct computation

$$f \rightsquigarrow \frac{4}{\pi} \left(\sin x + \frac{\sin 3x}{3} + \frac{\sin 5x}{5} + \cdots \right).$$

The following figures represent respectively the graphs of $y = \frac{4}{\pi} \sin x$, $y = \frac{4}{\pi} \left(\sin x + \frac{\sin 3x}{3} \right)$, $y = \frac{4}{\pi} \left(\sin x + \frac{\sin 3x}{3} + \frac{\sin 5x}{5} \right)$, and $y = \frac{4}{\pi} \left(\sin x + \frac{\sin 3x}{3} + \frac{\sin 5x}{5} + \frac{\sin 7x}{7} \right)$ and show the pointwise convergence of the Fourier series of f to the function f itself.

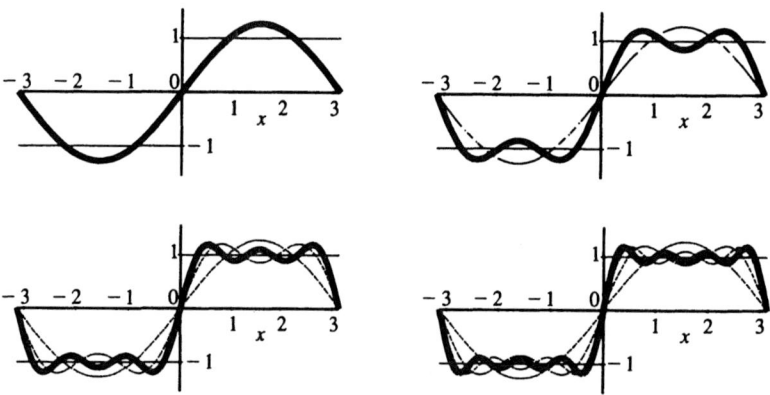

It is clear that such convergence could not be uniform (why?).

Note

It is to be emphasized that the Fourier series of a given integrable function f does not necessarily converge to f. It is of interest to know:

- Under what conditions for a given function f is it possible to find a trigonometric series convergent to f?

- Is every convergent trigonometric series the Fourier series of some function?

- Under what conditions for a given function f is it possible that $\mathcal{F}(f) = f$?

Theorem 11.3 states that if a trigonometric series converges uniformly on the interval $(-\pi, \pi)$ to a function f, then it must be the Fourier series of the function f. In looking for a trigonometric series which has a given function f as its sum, it is natural to first consider the Fourier series. We finish this section by stating without proof the following more general result.

Theorem 11.6

Let f be a 2π-periodic integrable function. Suppose that

$$f(x) = \frac{\alpha_0}{2} + \sum_{n=1}^{\infty} (\alpha_n \cos nx + \beta_n \sin nx)$$

for every x, except possibly at a finite number of points within one period. Then

$$\frac{\alpha_0}{2} + \sum_{n=1}^{\infty} (\alpha_n \cos nx + \beta_n \sin nx) = \mathcal{F}(f).$$

11.2 Convergent Trigonometric Series

It follows immediately from the Weierstrass M-test that if the series

$$\sum (|a_n| + |b_n|)$$

converges, then the corresponding trigonometric series converges absolutely and uniformly. In this section, we exhibit other sufficient (but not necessary) conditions for the convergence of a trigonometric series. First we recall useful trigonometric identities, the proofs of which are left as exercises:

$$\begin{aligned}
\tfrac{1}{2} + \cos x + \cos 2x + \cdots + \cos nx &= \tfrac{\sin\left(n+\frac{1}{2}\right)x}{2\sin\frac{1}{2}x}, \text{ for } x \neq 2k\pi; \\
\sin x + \sin 2x + \cdots + \sin nx &= \tfrac{\cos\frac{x}{2}-\cos\left(n+\frac{1}{2}\right)x}{2\sin\frac{1}{2}x}, \text{ for } x \neq 2k\pi.
\end{aligned} \tag{11.4}$$

The following result is a consequence of Dirichlet's test (Theorem 7.24, page 189).

Theorem 11.7

Let (a_n) and (b_n) be both nonnegative, both nonincreasing, and both converging to zero. Then the trigonometric series $\frac{a_0}{2} + \sum_{n=1}^{\infty} (a_n \cos nx + b_n \sin nx)$ converges uniformly on any interval $[a, b]$ not containing points of the form $x = 2k\pi$ $(k = 0, \pm1, \pm2, \ldots)$.

Proof

By periodicity, it is enough to show the result for an interval $[a, b]$ strictly contained in $[0, 2\pi]$. Fix $[a, b] \subsetneq [0, 2\pi]$. Let $x \in [a, b]$. It readily follows from the first identity of (11.4) that

$$\left| \frac{1}{2} + \cos x + \cos 2x + \cdots + \cos nx \right| \leq \frac{1}{2 \sin (x/2)}$$

for each n. Since $0 < a \leq x \leq b \leq 2\pi$, we have

$$\sin (x/2) \geq \min (\sin (a/2), \sin (b/2)) > 0.$$

Setting $M = 2/\min (\sin (a/2), \sin (b/2))$, we see that Dirichlet's test is applicable to the series $\frac{a_0}{2} + \sum_{n=1}^{\infty} a_n \cos nx$ and implies its uniform convergence.

Considering the second relation of (11.4), arguments similar to the above show that the series $\sum_{n=1}^{\infty} b_n \sin nx$ also converges uniformly on $[a, b]$. The uniform convergence of the series $\frac{a_0}{2} + \sum_{n=1}^{\infty} (a_n \cos nx + b_n \sin nx)$ on the interval $[a, b]$ clearly follows. $\qquad \square$

For example, series such as $\sum_{n=1}^{\infty} (\sin nx)/\sqrt{n}$, $\sum_{n=1}^{\infty} np^n \cos nx$ where $|p| < 1$, and $\sum_{n=1}^{\infty} (\sin nx + \cos nx)/n^2$ converge uniformly on intervals $[a, b]$ not containing $2k\pi$, $k = 0, \pm1, \pm2, \ldots$. It is worth noticing that in these examples, only the third series is absolutely convergent.

The result of the previous theorem has an immediate consequence.

Corollary 11.8

If (a_n) and (b_n) are both nonnegative, both nonincreasing, and both converging to zero, then the 2π-periodic function

$$f(x) = \frac{a_0}{2} + \sum_{n=1}^{\infty} (a_n \cos nx + b_n \sin nx)$$

is continuous for all x, except possibly for the values $x = 2k\pi$, $k = 0, \pm1, \pm2, \ldots$.

Proof

Let $x_0 \neq 2k\pi$, $k = 0, \pm 1, \pm 2, \dots$. We include x_0 in an interval $[a, b]$ not containing points of the form $x = 2k\pi$, $k = 0, \pm 1, \pm 2, \dots$. Since by Theorem 11.7, the series $\frac{a_0}{2} + \sum_{n=1}^{\infty} (a_n \cos nx + b_n \sin nx)$ converges uniformly on $[a, b]$, its limit f must be continuous on $[a, b]$ (by Theorem 7.10, page 181) and in particular it is continuous at x_0. This completes the proof. $\qquad\square$

Example 11.9

Let (a_n) be positive, nonincreasing, and converging to zero. Show that the trigonometric series $\frac{a_0}{2} + \sum_{n=1}^{\infty} (-1)^n a_n \cos nx$ converges uniformly on any interval $[a, b]$ not containing points of the form $x = (2k + 1)\pi$, $k = 0, \pm 1, \pm 2, \dots$.

Solution

For each n, we have

$$\cos n (t - \pi) = \cos n\pi \cos nt + \sin n\pi \sin nt = (-1)^n \cos nt.$$

It follows that

$$\frac{a_0}{2} + \sum_{n=1}^{\infty} a_n \cos n (t - \pi) = \frac{a_0}{2} + \sum_{n=1}^{\infty} (-1)^n a_n \cos nt.$$

Applying Theorem 11.7 to the series $\frac{a_0}{2} + \sum_{n=1}^{\infty} a_n \cos n (t - \pi)$, we obtain that such a series converges uniformly on any interval $[a, b]$ such that $t - \pi = 2k\pi \notin [a, b]$, $k = 0, \pm 1, \pm 2, \dots$. It follows that the series $\frac{a_0}{2} + \sum_{n=1}^{\infty} (-1)^n a_n \cos nt$ converges uniformly on any interval $[a, b]$ such that $t = (2k + 1)\pi \notin [a, b]$, $k = 0, \pm 1, \pm 2, \dots$. $\qquad\square$

Example 11.10

Show that the function $f(x) = \sum_{n=1}^{\infty} (-1)^n \frac{\sin nx}{n}$ is continuous on any interval $[a, b]$ not containing points of the form $x = (2k + 1)\pi$, $k = 0, \pm 1, \pm 2, \dots$.

Solution

For each n, we notice that

$$\sin n (t - \pi) = \sin nt \cos n\pi - \cos nt \sin n\pi = (-1)^n \sin nt.$$

It follows that

$$\sum_{n=1}^{\infty} \frac{1}{n} \sin n \, (t - \pi) = \sum_{n=1}^{\infty} (-1)^n \frac{\sin nt}{n}.$$

Applying Corollary 11.8, we realize that the function $g\,(t) = \sum_{n=1}^{\infty} \frac{1}{n} \sin n\,(t - \pi)$ is continuous on any interval $[a,b]$ such that $t - \pi = 2k\pi \notin [a,b]$, $k = 0, \pm 1, \pm 2, \ldots$. It follows that the function $f\,(t) = \sum_{n=1}^{\infty} (-1)^n \frac{\sin nt}{n}$ is continuous on any interval $[a,b]$ such that $t = (2k+1)\,\pi \notin [a,b]$, $k = 0, \pm 1, \pm 2, \ldots$. The proof is complete.   □

In fact, Example 11.5 implies that f is nothing else but the 2π-periodic function defined by $f\,(x) = -x/2$, $-\pi < x \leq \pi$.

11.3 Convergence in 2-mean

We emphasize that uniform convergence to a given function f is a sufficient but not necessary condition for a trigonometric series to be exactly the Fourier series of the function f. A uniform convergence condition on the trigonometric series requires necessarily that the limit be continuous. In the present section, we will discuss convergence in 2-mean of Fourier series to functions not necessarily continuous.

We first fix some notation. We denote by $\mathcal{L}^2\,(2\pi)$ the set of all 2π-periodic functions in $\mathcal{L}^2$. A function of the form

$$T_n\,(x) = \frac{\alpha_0}{2} + \sum_{k=1}^{n} (\alpha_k \cos kx + \beta_k \sin kx)$$

is called a **trigonometric polynomial** of degree n (see Exercise 9.21, page 267). Given $f \in \mathcal{L}^2\,(2\pi)$, we denote by $S_n\,(f)\,(x)$ the trigonometric polynomial whose coefficients a_k, b_k are the Fourier coefficients of f. In other words, $S_n\,(f)\,(x)$ is the n-th partial sum of the Fourier series $\mathcal{F}\,(f)$.

Lemma 11.11

Let $f \in \mathcal{L}^2\,(2\pi)$ and let

$$T_n\,(x) = \frac{\alpha_0}{2} + \sum_{k=1}^{n} (\alpha_k \cos kx + \beta_k \sin kx)$$

be any trigonometric polynomial of degree n. Then

$$\|f - T_n\|_2^2 = \|f\|_2^2 - \pi \left[\frac{a_0^2}{2} + \sum_{k=1}^{n} \left(a_k^2 + b_k^2 \right) \right]$$

$$+ \pi \left[\frac{(\alpha_0 - a_0)^2}{2} + \sum_{k=1}^{n} \left((\alpha_k - a_k)^2 + (\beta_k - b_k)^2 \right) \right].$$

Before we proceed to the proof of this lemma, let us notice its geometric interpretation.

From among all the trigonometric polynomials of degree n, the one which minimizes the expression $\|f - T_n\|_2^2$ is obtained by choosing the α_k and β_k to be the Fourier coefficients a_k and b_k.

We then have the so-called Bessel's identity:

$$\|f - S_n(f)\|_2^2 = \|f\|_2^2 - \pi \left[\frac{a_0^2}{2} + \sum_{k=1}^{n} \left(a_k^2 + b_k^2 \right) \right]. \qquad (11.5)$$

Noticing that the left side is nonnegative, we also obtain Béssel's inequality:

$$\|f\|_2^2 \geq \pi \left[\frac{a_0^2}{2} + \sum_{k=1}^{n} \left(a_k^2 + b_k^2 \right) \right], \quad \text{for all } n.$$

Proof (of Lemma 11.11)

By direct computation, we have

$$\|f - T_n\|_2^2 = \int_{[-\pi,\pi]} (f - T_n)^2 \, d\ell$$

$$= \int_{[-\pi,\pi]} f^2 d\ell - 2 \int_{[-\pi,\pi]} f T_n d\ell + \int_{[-\pi,\pi]} (T_n)^2 \, d\ell.$$

It is easy to see that

$$\int_{[-\pi,\pi]} f T_n d\ell = \frac{a_0}{2} \int_{[-\pi,\pi]} f d\ell + \sum_{k=1}^{n} \alpha_k \int_{[-\pi,\pi]} f(x) \cos kx d\ell$$

$$+ \sum_{k=1}^{n} \beta_k \int_{[-\pi,\pi]} f(x) \sin kx d\ell.$$

Moreover, using the identities of Theorem 11.1, we have

$$\int_{[-\pi,\pi]} (T_n)^2 \, d\ell = \pi \left[\frac{\alpha_0^2}{2} + \sum_{k=1}^{n} \left(\alpha_k^2 + \beta_k^2 \right) \right].$$

The desired equality is established by inserting the last two relations into the first one and by adding and subtracting the quantity

$$\pi \left[\frac{a_0^2}{2} + \sum_{k=1}^{n} \left(a_k^2 + b_k^2 \right) \right].$$

□

Next we prove the following fundamental uniqueness lemma.

Lemma 11.12

Let $g \in \mathcal{L}^2 (2\pi)$. Suppose that

(1) $\int_{[-\pi,\pi]} g\left(x \right) d\ell = 0$;

(2) $\int_{[-\pi,\pi]} g\left(x \right) \cos kx d\ell = 0, \ k \in \mathbb{N}$;

(3) $\int_{[-\pi,\pi]} g\left(x \right) \sin kx d\ell = 0, \ k \in \mathbb{N}$.

Then $g \sim 0$.

Proof

We first notice that $\mathcal{L}^2 \left([-\pi, \pi] \right) \subset \mathcal{L}^1 \left([-\pi, \pi] \right)$. Hence $g \in \mathcal{L}^2 (2\pi)$ implies $g \in \mathcal{L}^1 (2\pi)$. Let (f_n) be a generating sequence for g. Then $f_n \to g$ a.e. Thus for given $\varepsilon > 0$ and $\delta > 0$, there exists n such that

$$\left| g\left(x \right) - f_n \left(x \right) \right| < \frac{\varepsilon}{4}$$

except for a set N_1 with $\ell^* \left(N_1 \right) < \delta/3$. For this function f_n, there exists a step function φ_n such that

$$\left| \varphi_n \left(x \right) - f_n \left(x \right) \right| < \frac{\varepsilon}{4}$$

except for a set N_2 with $\ell^* \left(N_2 \right) < \delta/3$. For this step function φ_n, there exists a step function f such that

$$\left| \varphi_n \left(x \right) - f \left(x \right) \right| < \frac{\varepsilon}{4}$$

except for a set N_3 with $\ell^* \left(N_3 \right) < \delta/3$. But every continuous function on $[-\pi, \pi]$ can be uniformly approximated by trigonometric polynomials (see Exercise 9.21, page 267). Thus there exists a trigonometric polynomial T_n such that

$$\sup \left\{ \left| f \left(x \right) - T_n \left(x \right) \right| : x \in [-\pi, \pi] \right\} < \frac{\varepsilon}{4}.$$

We notice that $gT_n \in \mathcal{L}^1\left([-\pi,\pi]\right)$. Let $A = [-\pi,\pi] \setminus N_1 \cup N_2 \cup N_3$. Then

$$\left|\int_A (g^2 - gT_n)\, d\ell\right| \leq \int_A |g| \cdot |T_n - f|\, d\ell + \int_A |g| \cdot |\varphi_n - f|\, d\ell$$
$$+ \int_A |g| \cdot |f_n - \varphi_n|\, d\ell + \int_A |g| \cdot |f_n - g|\, d\ell$$
$$< \int_A |g|\, d\ell \left(\frac{\varepsilon}{4} + \frac{\varepsilon}{4} + \frac{\varepsilon}{4} + \frac{\varepsilon}{4}\right) = \varepsilon \int_A |g|\, d\ell.$$

Our hypothesis clearly implies that $\int_{[-\pi,\pi]} g(x) T_n(x)\, d\ell = 0$. Therefore

$$\left|\int_{[-\pi,\pi]} g^2\, d\ell\right| = \left|\int_{[-\pi,\pi]} (g^2 - gT_n)\, d\ell\right|$$
$$\leq \left|\int_{[-\pi,\pi]\setminus A} (g^2 - gT_n)\, d\ell\right| + \left|\int_A (g^2 - gT_n)\, d\ell\right|$$
$$< \left|\int_{[-\pi,\pi]\setminus A} (g^2 - gT_n)\, d\ell\right| + \varepsilon \int_{[-\pi,\pi]} |g|\, d\ell.$$

Since both $\varepsilon > 0$ and $\delta > 0$ are arbitrary, we get $\int_{[-\pi,\pi]} g^2\, d\ell = 0$. This implies, according to Example 10.36 (page 292), that g^2, and hence g, is almost everywhere equal to zero as asserted. $\square$

We now establish the following

Theorem 11.13

Let $f \in \mathcal{L}^2(2\pi)$ and let $(S_n(f))$ be the sequence of partial sums of the Fourier series for f. Then
$$\lim_{n\to\infty} \|f - S_n(f)\|_2 = 0.$$

Proof

It follows from Bessel's identity that for arbitrary $m < n$,

$$\|S_n(f) - S_m(f)\|_2^2 = \sum_{k=m+1}^{n} \left(a_k^2 + b_k^2\right).$$

On the other hand, Bessel's inequality implies that the series $\sum_{k=1}^{\infty} \left(a_k^2 + b_k^2\right)$ converges. It follows that the sequence $(S_n(f))$ is Cauchy in 2-mean. The Riesz–Fisher Theorem 10.53 now implies that $(S_n(f))$ is convergent in 2-mean to some function $\tilde{f} \in \mathcal{L}^2(2\pi)$. We are done if we show that $f \sim \tilde{f}$.

In view of the uniqueness Lemma 11.12, all we have to show is that the Fourier coefficients of $\tilde{f}$ coincide with those of f. Indeed, an appeal to Hölder's inequality yields

$$\left| \frac{1}{\pi} \int_{[-\pi,\pi]} \tilde{f}(x) \cos kx d\ell - a_k \right| = \left| \frac{1}{\pi} \int_{[-\pi,\pi]} \left[\tilde{f}(x) - S_n(f)(x) \right] \cos kx d\ell \right|$$

$$\leq \left\| \tilde{f} - S_n(f) \right\|_2 \left\| \cos kx \right\|_2,$$

where $n \geq k$. Since $\lim_{n \to \infty} \left\| \tilde{f} - S_n(f) \right\|_2 = 0$, we have

$$\frac{1}{\pi} \int_{[-\pi,\pi]} \tilde{f}(x) \cos kx d\ell = a_k.$$

A similar calculation shows that $\frac{1}{\pi} \int_{[-\pi,\pi]} \tilde{f}(x) \sin kx d\ell = b_k$. This ends the proof. $\qquad \square$

In particular it follows that if $f \in \mathcal{L}^2(2\pi)$, then Bessel's identity (11.5) becomes

$$\|f\|_2^2 = \pi \left[\frac{a_0^2}{2} + \sum_{k=1}^{\infty} \left(a_k^2 + b_k^2 \right) \right].$$

In technical terms, Theorem 11.13 states that the orthogonal family of trigonometric functions $\{1, \cos x, \sin x, \cos 2x, \sin 2x, \ldots\}$ is **complete** in $\mathcal{L}^2(2\pi)$. Also Theorem 11.13 implies that:

Any function $f \in \mathcal{L}^2(2\pi)$ is completely defined (except for its values on a null set) by its Fourier series, regardless of whether or not the series converges in the ordinary sense.

Our next example shows that not every convergent trigonometric series is the Fourier series of an $\mathcal{L}^2(2\pi)$ function.

Example 11.14

Show that the series $\sum_{n=1}^{\infty} (\sin nx)/\sqrt{n}$ converges for all $x \in [-\pi, \pi]$ but it cannot be the Fourier series of any $f \in \mathcal{L}^2(2\pi)$.

Solution

That the series converges at all $x \in [-\pi, \pi]$ follows from Dirichlet's test. Suppose that there exists $f \in \mathcal{L}^2(2\pi)$ with Fourier coefficients $1/\sqrt{k}$, $k = 1, 2, \ldots$. Then according to Bessel's inequality, we have for each n

$$\sum_{k=1}^{n} \frac{1}{\left(\sqrt{k} \right)^2} \leq \frac{1}{\pi} \|f\|_2^2.$$

It is readily seen that the expression on the left side is the partial sum of the divergent harmonic series $\sum 1/n$. Therefore our assumption must be false. □

The next example is another consequence of Theorem 11.13.

Example 11.15

Let $f, g \in \mathcal{L}^2(2\pi)$ and suppose that $\mathcal{F}(f) = \frac{a_0}{2} + \sum_{n=1}^{\infty}(a_n \cos nx + b_n \sin nx)$ and $\mathcal{F}(g) = \frac{\alpha_0}{2} + \sum_{n=1}^{\infty}(\alpha_n \cos nx + \beta_n \sin nx)$. Show that

$$\int_{[-\pi,\pi]} fg d\ell = a_0\alpha_0 + \sum_{k=1}^{\infty}(a_n\alpha_n + b_n\beta_n).$$

Solution

Since $f, g \in \mathcal{L}^2(2\pi)$, $f + g$ and $f - g$ belong to $\mathcal{L}^2(2\pi)$. Moreover, we have

$$\mathcal{F}(f + g) = \frac{a_0 + \alpha_0}{2} + \sum_{n=1}^{\infty}[(a_n + \alpha_n)\cos nx + (b_n + \beta_n)\sin nx],$$

$$\mathcal{F}(f - g) = \frac{a_0 - \alpha_0}{2} + \sum_{n=1}^{\infty}[(a_n - \alpha_n)\cos nx + (b_n - \beta_n)\sin nx].$$

It follows from Theorem 11.13 that

$$\frac{\|f + g\|_2^2}{\pi} = \frac{(a_0 + \alpha_0)^2}{2} + \sum_{n=1}^{\infty}\left[(a_n + \alpha_n)^2 + (b_n + \beta_n)^2\right],$$

$$\frac{\|f - g\|_2^2}{\pi} = \frac{(a_0 - \alpha_0)^2}{2} + \sum_{n=1}^{\infty}\left[(a_n - \alpha_n)^2 + (b_n - \beta_n)^2\right].$$

Subtracting term by term and using the identity $(a + b)^2 - (a - b)^2 = 4ab$, we obtain

$$4\int_{[-\pi,\pi]} fg d\ell = 4a_0\alpha_0 + \sum_{k=1}^{\infty}4(a_n\alpha_n + b_n\beta_n).$$

□

We finish this section with another important consequence of Theorem 11.13.

Example 11.16

If $f \in \mathcal{L}^2(2\pi)$, then show that the Fourier series of f can be integrated term by term, whether or not it converges.

Solution

Suppose that $a < b$ in $[-\pi, \pi]$. Let $S_n(f)(x) = \frac{a_0}{2} + \sum_{k=1}^{n} (a_k \cos kx + b_k \sin kx)$ be the sequence of the partial sums of the Fourier series for f. We wish to show that

$$\lim_{n \to \infty} \left(\int_{[a,b]} f \, d\ell - \int_{[a,b]} \frac{a_0}{2} d\ell + \sum_{k=1}^{n} \int_{[a,b]} (a_k \cos kx + b_k \sin kx) \, d\ell \right) = 0$$

$$(11.6)$$

Using the Cauchy–Schwartz inequality, we have

$$\left| \int_{[a,b]} [f - S_n(f)] \, d\ell \right| \le \int_{[a,b]} |f - S_n(f)| \, d\ell \qquad (11.7)$$

$$\le \int_{[-\pi,\pi]} |f - S_n(f)| \, d\ell$$

$$\le \|f - S_n(f)\|_2 \|1\|_2 \,.$$

Equation (11.6) is obtained by noticing that according to Theorem 11.13, the last term in (11.7) approaches 0 as $n \to \infty$. $\qquad\square$

11.4 Pointwise Convergence

We know that convergence in p-mean does not imply convergence at every point. The study of pointwise convergence of Fourier series deserves particular attention.

We begin by establishing an important result known as the Riemann–Lebesgue lemma.

Theorem 11.17 (Riemann–Lebesgue Lemma)

Let $f \in \mathcal{L}$. Then

$$\lim_n \int f(t) \cos nt \, d\ell(t) = \lim_n \int f(t) \sin nt \, d\ell(t) = 0.$$

Proof

It is left as an exercise to show that for every $\varepsilon > 0$, there exists a step function $\varphi = \sum_{i=1}^{m} \alpha_i \chi_{I_i}$ in its standard representation such that

$$\int |f - \varphi| \, d\ell < \frac{\varepsilon}{2}.$$

Now let $(a_i, b_i) = I_i$. Then

$$\left| \int \chi_{I_i}(t) \cos ntd\ell(t) \right| = \left| \int_{a_i}^{b_i} \cos ntdt \right|$$

$$= \left| \frac{\sin nb_i - \sin na_i}{n} \right| \leq \frac{2}{n}.$$

It follows that

$$\lim_n \int \varphi(t) \cos ntd\ell(t) = 0.$$

Hence there exists an $N > 0$ such that $n > N$ implies

$$\left| \int \varphi(t) \cos ntd\ell(t) \right| < \frac{\varepsilon}{2}.$$

Therefore for $n > N$

$$\left| \int f(t) \cos ntd\ell(t) \right| = \left| \left[\int f(t) \cos ntd\ell(t) - \int \varphi(t) \cos ntd\ell(t) \right] \right.$$

$$\left. + \int \varphi(t) \cos ntd\ell(t) \right|$$

$$\leq \int |f - \varphi| \, d\ell + \left| \int \varphi(t) \cos ntd\ell(t) \right|$$

$$< \frac{\varepsilon}{2} + \frac{\varepsilon}{2} = \varepsilon.$$

Since $\varepsilon > 0$ is arbitrary, we obtain $\lim_n \int f(t) \cos ntd\ell(t) = 0$. In a similar fashion, we can show that $\lim_n \int f(t) \sin ntd\ell(t) = 0$. $\qquad \square$

We can rephrase the above theorem by saying that

The Fourier coefficients of any integrable function approach zero as $n \to \infty$.

Next we derive a formula which will be needed later on.

Lemma 11.18

The n-th partial sum of a Fourier series of a 2π-periodic integrable function f on $[-\pi, \pi]$ is given by

$$S_n(f)(x) = \frac{1}{\pi} \int_{[-\pi,\pi]} f(x+t) D_n(t) \, d\ell(t)$$

where D_n is the n-th **Dirichlet kernel**, defined by

$$D_n(t) = \frac{1}{2} + \sum_{k=1}^{n} \cos kt = \begin{cases} \frac{\sin\left(n+\frac{1}{2}\right)t}{2\sin\frac{1}{2}t} & \text{if} \quad 0 < |t| \leq \pi \\ n + \frac{1}{2} & \text{if} \quad t = 0. \end{cases}$$

Proof

By definition,

$$S_n\left(f\right)\left(x\right) = \frac{a_0}{2} + \sum_{k=1}^{n}\left(a_k\cos kx + b_k\sin kx\right)$$

where for $k = 0, 1, 2, \ldots$,

$$a_k = \frac{1}{\pi}\int_{[-\pi,\pi]} f\left(t\right)\cos kt d\ell\left(t\right) \text{ and } b_k = \frac{1}{\pi}\int_{[-\pi,\pi]} f\left(t\right)\sin kt d\ell\left(t\right).$$

Therefore

$$S_n\left(f\right)\left(x\right) = \frac{1}{2\pi}\int_{[-\pi,\pi]} f d\ell + \frac{1}{\pi}\sum_{k=1}^{n}\left[\int_{[-\pi,\pi]} f\left(t\right)\cos kx \cos kt d\ell\left(t\right)\right.$$

$$\left. + \int_{[-\pi,\pi]} f\left(t\right)\sin kx \sin kt d\ell\left(t\right)\right]$$

$$= \frac{1}{\pi}\int_{[-\pi,\pi]} f\left(t\right)\left[\frac{1}{2} + \sum_{k=1}^{n}\cos k\left(x - t\right)\right] d\ell\left(t\right).$$

The transformation $t = x + s$ yields

$$S_n\left(f\right)\left(x\right) = \frac{1}{\pi}\int_{[-\pi,\pi]} f\left(x + s\right)\left[\frac{1}{2} + \sum_{k=1}^{n}\cos ks\right] d\ell\left(s\right).$$

The desired formula now follows from

$$\cos s + \cos 2s + \cdots + \cos ns = \frac{\sin\left(n + \frac{1}{2}\right)s - \sin\frac{1}{2}s}{2\sin\frac{1}{2}s}.$$

$\square$

We are now set to introduce the following test for pointwise convergence of a Fourier series.

Theorem 11.19 (Dini's Test)

Let f be a 2π-periodic integrable function and consider

$$\varphi_x\left(t\right) = f\left(x + t\right) - f\left(x - t\right) - 2f\left(t\right).$$

Suppose that $\varphi_x\left(t\right)/t$, where $x \in [-\pi, \pi]$ is integrable as a function of t in a neighborhood of 0. Then

$$\lim_n S_n\left(f\right)\left(x\right) = f\left(x\right).$$

Proof

Noticing that the Dirichlet kernel is an even function, we have

$$S_n\left(f\right)\left(x\right) = \frac{1}{\pi} \int_{[-\pi,\pi]} f\left(x+t\right) D_n\left(t\right) d\ell\left(t\right)$$

$$= \frac{1}{\pi} \int_{[0,\pi]} \left[f\left(x+t\right) + f\left(x-t\right)\right] D_n\left(t\right) d\ell\left(t\right).$$

Realizing that

$$\frac{1}{\pi} \int_{[-\pi,\pi]} D_n\left(t\right) d\ell\left(t\right) = 1$$

we obtain

$$S_n\left(f\right)\left(x\right) - f\left(x\right) = \frac{1}{\pi} \int_{[0,\pi]} \left[f\left(x+t\right) + f\left(x-t\right) - 2f\left(t\right)\right] D_n\left(t\right) d\ell\left(t\right)$$

$$= \frac{1}{\pi} \int_{[0,\pi]} \varphi_x\left(t\right) \frac{\sin\left(n+\frac{1}{2}\right)t}{2\sin\frac{1}{2}t} d\ell\left(t\right)$$

$$= \frac{1}{\pi} \int_{[0,\pi]} \frac{\varphi_x\left(t\right)}{t} \frac{t}{2\sin\frac{1}{2}t} \sin\left(n+\frac{1}{2}\right) t d\ell\left(t\right).$$

Since the function $\frac{t}{2\sin\frac{1}{2}t}$ is continuous on $[0,\pi]$, it is left as an exercise to show that the function $g\left(t\right) = \frac{\varphi_x\left(t\right)}{t} \frac{t}{2\sin\frac{1}{2}t}$ is integrable on $[0,\pi]$. Hence we are done if we show that

$$\lim_n \int_{[0,\pi]} g\left(t\right) \sin\left(n+\frac{1}{2}\right) t d\ell\left(t\right) = 0.$$

This easily follows from the Riemann–Lebesgue lemma if one notices that

$$\sin\left(n+\frac{1}{2}\right) t = \sin nt \cos\frac{1}{2}t + \sin\frac{t}{2}\cos nt$$

and therefore

$$\int_{[0,\pi]} g\left(t\right) \sin\left(n+\frac{1}{2}\right) t d\ell\left(t\right) = \int_{[0,\pi]} \left[g\left(t\right)\cos\frac{1}{2}t\right] \sin nt d\ell\left(t\right)$$

$$+ \int_{[0,\pi]} \left[g\left(t\right)\sin\frac{t}{2}\right] \cos nt d\ell\left(t\right).$$

This completes the proof.   □

Corollary 11.20

Let f be a 2π-periodic integrable function. Suppose that

(1) $f\left(x^+\right)$ and $f\left(x^-\right)$ exist;

(2) $\lim_{h \to 0^-} \frac{f(x+h)-f(x)}{h}$ and $\lim_{h \to 0^+} \frac{f(x+h)-f(x)}{h}$ exist.

Then

$$\lim_n S_n(f)(x) = \frac{f(x^+) + f(x^-)}{2}.$$

Proof

It suffices to notice that if the conditions are satisfied, then the function

$$\frac{\varphi_x(t)}{t} = \frac{f(x+t) - f(x)}{t} + \frac{f(x-t) - f(x)}{t}$$

is integrable. $\square$

The condition of Dini's test requires that

$$\frac{f(x+t) - f(x-t) - 2f(t)}{t}$$

where $x \in [-\pi, \pi]$ is integrable as a function of t in a neighborhood of 0. Thus the conclusion of Corollary 11.20 is dependent on the behavior of the function f in an arbitrarily small neighborhood of the point x. An important proposition known as **Riemann's localization principle** thus follows.

Theorem 11.21

Let f and g be two 2π-periodic integrable functions on $[-\pi, \pi]$. Suppose that $f = g$ on a neighborhood of some point x. Then their Fourier series simultaneously either converge or diverge at x.

EXERCISES

11.1   Write out the proofs of the identities (11.4).

11.2   Write out the full proof of the orthogonal property in Theorem 11.1.

11.3   Determine the Fourier series of the 2π-periodic function defined by

(a) $f(x) = \begin{cases} 0 & \text{if } -\pi < x < 0 \\ 1 & \text{if } 0 \le x \le \pi \end{cases}$ (b) $f(x) = x^2, \, 0 < x < 2\pi$

(c) $f(x) = \begin{cases} \cos x & \text{if } 0 < x < \pi \\ 0 & \text{if } \pi \le x \le 2\pi \end{cases}$ (d) $f(x) = \frac{\pi - x}{2}, \, 0 < x < 2\pi$

11.4 Determine for which values of x each of the following series converges.

(a) $\sum_{n=1}^{\infty} \frac{(-1)^n \cos nx}{\sqrt{n}+n}$ (b) $\sum_{n=1}^{\infty} \frac{\sin 3nx}{n}$ (c) $\sum_{n=1}^{\infty} \frac{\cos(2n+3)x}{n+2}$

11.5 Determine for which values of x each of the sums of the series in Exercise 11.4 is continuous.

11.6 Show that if an integrable $2l$-periodic function f is such that $f(x) = \frac{a_0}{2} + \sum_{n=1}^{\infty} \left(a_n \cos \frac{n\pi}{l} x + b_n \sin \frac{n\pi}{l} x \right)$, then

$$a_n = \frac{1}{l} \int_{[-l,l]} f(x) \cos \frac{n\pi}{l} x \, d\ell(x)$$

and

$$b_n = \frac{1}{l} \int_{[-l,l]} f(x) \sin \frac{n\pi}{l} x \, d\ell(x).$$

11.7 Show that if f is a 2π-periodic integrable function, then for any $\lambda \in \mathbb{R}$ and for $n = 0, 1, 2, \ldots$,

$$a_n = \frac{1}{\pi} \int_{\lambda}^{\lambda+2\pi} f(x) \cos nx \, dx \text{ and } b_n = \frac{1}{\pi} \int_{\lambda}^{\lambda+2\pi} f(x) \sin nx \, dx.$$

11.8 Expand in Fourier series the $2l$-periodic function f defined by

(1) $f(x) = |x|$ for $-l \le x \le l$.

(2) $f(x) = e^x$ for $-l \le x \le l$.

11.9 Show that the Fourier series of an odd function contains only sines and the Fourier series of an even function contains only cosines.

11.10 Let $f : [0, \pi] \to \mathbb{R}$ be integrable. The **even extension** of f is the 2π-periodic function f_e defined by

$$f_e(x) = \begin{cases} f(x) & \text{for} \quad x \in [0, \pi] \\ f(-x) & \text{for} \quad x \in [-\pi, 0). \end{cases}$$

Show that

(1) f_e is an even function. The Fourier series of f_e is called the Fourier **cosine series** of f.

(2) $f_e \rightsquigarrow \frac{a_0}{2} + \sum_{n=1}^{\infty} a_n \cos nx$, where $a_n = \frac{2}{\pi} \int_{[0,\pi]} f(t) \cos nt \, d\ell(t)$.

11.11 Expand each of the functions in cosine series.

(a) $f(x) = x$, for $x \in [0, 2\pi]$ \qquad (b) $f(x) = \sin x$ for $x \in [0, \pi]$

(c) $f(x) = \begin{cases} 1 & \text{for } 0 \le x < \frac{\pi}{2} \\ 0 & \text{for } \frac{\pi}{2} \le x \le \pi \end{cases}$ \qquad (d) $f(x) = x(\pi - x)$ for $x \in [0, \pi]$

11.12 Let $f : [0, \pi] \to \mathbb{R}$ be integrable. The **odd extension** of f is the 2π-periodic function f_o defined by

$$f_o(x) = \begin{cases} f(x) & \text{for } x \in [0, \pi] \\ -f(-x) & \text{for } x \in [-\pi, 0). \end{cases}$$

Show that

   (1) f_o is an odd function. The Fourier series of f_o is called the Fourier **sine series** of f.

   (2) $f_o \rightsquigarrow \frac{a_0}{2} + \sum\limits_{n=1}^{\infty} b_n \sin nx$, where $b_n = \dfrac{2}{\pi} \displaystyle\int_{[0,\pi]} f(t) \sin nt d\ell(t)$.

11.13 Expand each of the functions in sine series.

(a) $f(x) = x^2$, for $x \in [0, \pi]$ \qquad (b) $f(x) = \cos x$ for $x \in [0, \pi]$

(c) $f(x) = \begin{cases} 1 & \text{for } 0 \le x < \frac{\pi}{2} \\ 0 & \text{for } \frac{\pi}{2} \le x \le \pi \end{cases}$ \qquad (d) $f(x) = \pi - x$ for $x \in [0, \pi]$

11.14 Suppose that the series $\sum_{n=1}^{\infty} (|a_n| + |b_n|)$ converges.

   (1)   Show that the trigonometric series

$$\frac{a_0}{2} + \sum_{n=1}^{\infty} (a_n \cos nx + b_n \sin nx)$$

converges absolutely and uniformly.

   (2)   Deduce that its sum f is continuous.

   (3)   Show that the series $\frac{a_0}{2} + \sum_{n=1}^{\infty} (a_n \cos nx + b_n \sin nx) = \mathcal{F}(f)$.

11.15 Let f be a 2π-periodic function and have Fourier coefficients (a_n) and (b_n). Show that

$$\frac{1}{4\pi} \int_0^{2\pi} [f(x+h) - f(x-h)]^2 \, dx = \sum_{n=1}^{\infty} (a_n^2 + b_n^2) \sin^2(nh).$$

11.16   Let f be a 2π-periodic integrable function. Suppose that there exist numbers $c > 0$ and $\alpha > 0$ such that $|f(x) - f(x_0)| \leq c|x - x_0|^{\alpha}$ holds for all x in some neighborhood of x_0. Show that the $\mathcal{F}(f)(x_0) = f(x_0)$.

11.17   Suppose that f is a 2π-periodic integrable function and $f(x) \rightsquigarrow \frac{a_0}{2} + \sum_{n=1}^{\infty} (a_n \cos nx + b_n \sin nx)$. Let $\sigma_n(f)$ be the arithmetic means of the partial sums, i.e. for each n,

$$\sigma_n(f)(x) = \frac{1}{n}(S_0(f)(x) + S_1(f)(x) + \cdots + S_{n-1}(f)(x)).$$

Show that

   (1) $\sigma_n(f)(x) = \frac{a_0}{2} + \sum_{k=1}^{n-1} \frac{n-k}{n}(a_k \cos kx + b_k \sin kx)$;

   (2) $\sigma_n(f)(x) = \frac{1}{n\pi} \int_{[-\pi,\pi]} f(x+t) \frac{\sin^2(nt/2)}{2\sin^2(t/2)} d\ell(t)$.

11.18   Show that the partial sums of the Fourier series of a 2π-periodic integrable function f is Césaro summable at every point of continuity and at every point of jump discontinuity. (Hint: show that $\lim_n \sigma_n(f)(x) = \frac{f(x^+)+f(x^-)}{2}$ whenever $f(x^+)$ and $f(x^-)$ exist and where $\sigma_n(f)$ is as in Exercise 11.17.)

11.19   Suppose that $f \in \mathcal{L}^2(2\pi)$ and

$$f(x) \rightsquigarrow \frac{a_0}{2} + \sum_{n=1}^{\infty} (a_n \cos nx + b_n \sin nx).$$

Show that

$$\|f - \sigma_n(f)\|_2^2 = \pi \left[\frac{1}{n^2} \sum_{k=1}^{n-1} k^2 (a_k^2 + b_k^2) + \sum_{k=n}^{\infty} (a_k^2 + b_k^2) \right],$$

where $\sigma_n(f)$ is as in Exercise 11.17.

11.20   Let (a_n) and (b_n) be both positive, both nonincreasing, and both convergent to zero. Show that

   (1)   the series $\frac{a_0}{2} + \sum_{n=1}^{\infty} (-1)^n [a_n \cos nx + b_n \sin nx]$ converges uniformly on any interval $[a, b]$ which does not contain points of the form $x = 2k\pi$ ($k = 0, \pm 1, \pm 2, \ldots$);

   (2)   the sum $f(x) = \frac{a_0}{2} + \sum_{n=1}^{\infty} (-1)^n [a_n \cos nx + b_n \sin nx]$ is continuous for all x, except possibly for the values $x = 2k\pi$ ($k = 0, \pm 1, \pm 2, \ldots$).

11.21   Suppose that (a_n) and (b_n) are both positive, both nonincreasing, and both convergent to zero. Let $m, p \in \mathbb{N}$. Show that the series

$$\sum_{n=1}^{\infty} [a_n \cos (p + (n - 1) m) x + b_n \sin (p + (n - 1) m) x]$$

converges uniformly on any interval $[a, b]$ not containing points of the form $x = 2k\pi/m$ $(k = 0, \pm 1, \pm 2, \ldots)$.

11.22   Suppose that (a_n) and (b_n) are both positive, both nonincreasing, and both convergent to zero. Let $m, p \in \mathbb{N}$. Show that the series

$$\sum_{n=1}^{\infty} (-1)^n [a_n \cos (p + (n - 1) m) x + b_n \sin (p + (n - 1) m) x]$$

converges uniformly on any interval $[a, b]$ not containing points of the form $x = 2k\pi/m$ $(k = 0, \pm 1, \pm 2, \ldots)$.

11.23   Show that for $x \in [0, \pi]$

(1) $x (\pi - x) = \frac{\pi^2}{6} - \left(\frac{\cos 2x}{1^2} + \frac{\cos 4x}{2^2} + \frac{\cos 6x}{3^2} + \cdots \right)$;

(2) $x (\pi - x) = \frac{\pi}{8} \left(\frac{\sin x}{1^3} + \frac{\sin 3x}{3^3} + \frac{\sin 5x}{5^3} + \cdots \right)$.

11.24   Use the sine series of $f (x) = 1$, for $x \in [0, \pi]$ to show that $1 - \frac{1}{3} + \frac{1}{5} - \frac{1}{7} + \cdots = \frac{\pi}{4}$.

11.25   Expand the function

$$f (x) = \begin{cases} 2 - x & \text{for } 0 \leq x < 4 \\ 2 - x & \text{for } 4 \leq x \leq 8 \end{cases}$$

in Fourier series of period 8.

11.26   Expand the function

$$f (x) = \begin{cases} x & \text{for } 0 \leq x < 1 \\ 2 - x & \text{for } 1 \leq x \leq 2 \end{cases}$$

(1)   in sine series;

(2)   in cosine series.

11.27   Suppose that the Fourier series associated to an integrable 2π-periodic function f converges uniformly on $(-\pi, \pi)$.[2] Show directly that

$$\int_{[-\pi, \pi]} |f (x)|^2 \, dx = \frac{a_0^2}{2} + \sum_{n=1}^{\infty} (a_n^2 + b_n^2) .$$

---

[2] Note that the conditions imposed here are much stronger than in Parseval's identity.

11.28 Use the Fourier expansion of

$$f(x) = \begin{cases} 0 & \text{for } -\pi < x < 0 \\ x & \text{for } 0 \le x \le \pi \end{cases}$$

to prove that $\sum_{n=1}^{\infty} \frac{1}{(2n-1)^2} = \frac{\pi^2}{8}$.

11.29 Show that

$$\text{(a) } \sum_{n=1}^{\infty} \frac{1}{n^2} = \frac{\pi^2}{6} \qquad \text{(b) } \sum_{n=1}^{\infty} \frac{1}{n^4} = \frac{\pi^4}{90} \qquad \text{(c) } \sum_{n=1}^{\infty} \frac{1}{n^6} = \frac{\pi^6}{945}$$

11.30 Determine whether the given series is the Fourier series of a function in $\mathcal{L}^2(2\pi)$.

$$\text{(a) } \sum_{n=2}^{\infty} \frac{\cos nx}{\sqrt{n} \ln n} \qquad \text{(b) } \sum_{n=1}^{\infty} \frac{\sin nx}{n} \qquad \text{(c) } \sum_{n=1}^{\infty} \frac{\sin nx}{n^2}$$

11.31 Consider the function $f(x) = \cos \alpha x$, $-\pi \le x \le \pi$, where $\alpha \notin \mathbb{Z}$.

(1) Find $\mathcal{F}(f)$.

(2) Show that

$$\cot \pi x = \frac{1}{\pi x} + \frac{2x}{\pi} \sum_{n=1}^{\infty} \frac{1}{x^2 - n^2},$$

$$\operatorname{cosec} \pi x = \frac{1}{\pi x} + \frac{2x}{\pi} \sum_{n=1}^{\infty} \frac{(-1)^n}{x^2 - n^2}.$$

11.32 Show that

$$\frac{\pi^2}{(\sin \pi x)^2} = \lim_{m \to \infty} \sum_{n=-m}^{m} \frac{1}{(x-n)^2},$$

$$\frac{\sin \pi x}{\pi x} = \lim_{m \to \infty} \left[\left(1 - \frac{x^2}{1^2}\right) \left(1 - \frac{x^2}{2^2}\right) \cdots \left(1 - \frac{x^2}{m^2}\right) \right].$$

11.33 Use the previous exercise to show that

$$\sin x = x \left(1 - \frac{x^2}{\pi^2}\right) \left(1 - \frac{x^2}{(2\pi)^2}\right) \left(1 - \frac{x^2}{(3\pi)^2}\right) \cdots$$

and deduce that

$$\frac{\pi}{2} = \frac{2 \cdot 2 \cdot 4 \cdot 4 \cdot 6 \cdot 6 \cdot 8 \cdot 8 \cdots}{1 \cdot 1 \cdot 3 \cdot 3 \cdot 5 \cdot 5 \cdot 7 \cdot 7 \cdots}.$$

11.34 Show that

$$\cos x = \left(1 - \frac{4x^2}{\pi^2}\right)\left(1 - \frac{4x^2}{(3\pi)^2}\right)\left(1 - \frac{4x^2}{(5\pi)^2}\right)\cdots.$$

11.35 Show that for $x \in [-\pi, \pi]$,

$$\frac{\pi \sinh \alpha x}{2 \sinh \alpha \pi} = \frac{\sin x}{\alpha^2 + 1^2} - \frac{2 \sin 2x}{\alpha^2 + 2^2} + \frac{3 \sin 3x}{\alpha^2 + 3^2} + \cdots.$$

11.36 Show that for $x \in [-\pi, \pi]$,

$$\frac{\pi \cosh \alpha x}{2 \sinh \alpha \pi} = \frac{1}{2\alpha} - \frac{\alpha \cos x}{\alpha^2 + 1^2} + \frac{\alpha \sin 2x}{\alpha^2 + 2^2} - \cdots.$$

11.37 Show that

$$\sinh x = x\left(1 + \frac{x^2}{\pi^2}\right)\left(1 + \frac{x^2}{(2\pi)^2}\right)\left(1 + \frac{x^2}{(3\pi)^2}\right)\cdots.$$

11.38 Show that $\frac{\sqrt{2}}{2} = \frac{1\cdot3\cdot5\cdot7\cdot9\cdot11\cdot13\cdot15\cdots}{2\cdot2\cdot6\cdot6\cdot10\cdot10\cdot14\cdot14\cdots}$.

11.39 Let f be a 2π-periodic continuous function. Let $(S_n(f))$ denote the sequence of partial sums of $\mathcal{F}(f)$. Define the Césaro means of $\mathcal{F}(f)$ by

$$\Gamma_n(f) = \frac{1}{n}\left[S_0(f) + S_1(f) + \cdots + S_{n-1}(f)\right].$$

Show that $(\Gamma_n(f))$ converges uniformly to f. Explain why this gives another proof of the Weierstrass theorem.

A
Appendix

A.1 Theorems and Proofs

When one is doing mathematics, one is using reasoning processes to prove results. In order to study mathematics, students must learn to think correctly. It is necessary for students to have a clear understanding of the idea of a mathematical proof. They must know what constitute valid proofs and how to construct them. Here are some basic ideas from Logic that any mathematics student must fully understand.

- A **mathematical statement** is a declarative sentence which is either true or false.

If a sentence is a mathematical statement, then its negation is also a mathematical statement. For example, both the sentence "it is raining" and its negation "it is not raining" are mathematical statements. Of course, if a statement is true, then its negation is false and vice versa.

Two mathematical statements can be combined to give a new statement. Suppose that "P" and "Q" are two statements. Then we have

P	Q	P and Q	P or Q
True	True	True	True
True	False	False	True
False	True	False	True
False	False	False	False

- A **theorem** is a combination of mathematical statements which form one true statement.

Most mathematical Theorems have the logical form "**if** a statement holds, **then** another statement holds". For example: "if f is differentiable, then f is continuous".

Again suppose that "P" and "Q" are two statements. Then we have

P	Q	P implies Q
True	True	True
True	False	False
False	True	True
False	False	True

To prove the statement "if P, then Q", you have to show that when P is true, then Q must be true.

Generally, there are few ways of doing this:

- **direct proof**

    (1) you assume P;

    (2) prove Q.

- **contrapositive**

    (1) you assume the negation of Q;

    (2) prove the negation of P.

- **by contradiction**

    (1) you assume both P and the negation of Q;

    (2) prove a contradiction.

Note

The sentence "P if and only if Q" actually contains two statements: namely

(1) "if P, then Q" and

(2) "if$Q$, then P".

A.2 Set Notations

The concept of **set** plays important role in most fields of modern mathematics. Without going to the axiomatic foundation of set theory, we will recall in this appendix its basic notions.

In set theory, the notion of a set is not defined but rather is described simply by its properties. Other words that are synonymous to, and sometimes used in place of the word "set" are "class" or "collection". Usually, sets are denoted by capital letters.

What is important about a given set is to know: what are its "members", also called "objects" or "elements"? To designate that an object x belongs to a set A, we write "$x \in A$" and read it: "x belongs to A," or "x is an element of A," or "x is a member of A". The negation of "$x \in A$" is "$x \notin A$" meaning the element x does not belong to A.

A pair of braces is often used to describe a set by enumerating its members. Here are some examples:

$$\{a, b, c, d, e, f, g\}\, ;\ \{1, 2, 3, 4\}\, ;\ \{apple, orange, banana\}\, .$$

Sometimes it is tedious or even impossible to list all the elements of a set. In that case, it may be preferable to use a criterion for membership. Example: the set of all prime natural numbers could be described in the following way

$$\{p : p \text{ is prime}\}\, .$$

It is possible that a set may possess no element. For example the set

$$\left\{x : x \text{ rational number and } x^2 = 2\right\}$$

contains no element. A set without any element is called an *empty* set and is denoted by $\varnothing$ or $\{\}$.

Two sets A and B are equal, in symbols $A = B$, if they have exactly the same elements. We say that a set A is included in a set B, (or A is a subset of B, or B contains A), in symbols $A \subset B$, if every elements of A is an element of B. Clearly, we have

$$A = B \text{ if and only if } A \subset B \text{ and } B \subset A.$$

Observe that for any set A, we always have $\varnothing \subset A$.

Let I be a nonempty set. The notation $\{A_i : i \in I\}$, which describes a set whose elements are sets by themselves, is called a **family of sets** . For each element i of the set I, A_i is a set. The set I is called the index set.

Let A and B be arbitrary sets. Then

- the **union** of A and B is defined by

$$A \cup B = \{x : x \in A \text{ or } x \in B\};$$

- the **intersection** of A and B is defined by

$$A \cap B = \{x : x \in A \text{ and } x \in B\};$$

- the **difference** of A and B is defined by

$$A \setminus B = \{x : x \in A \text{ and } x \notin B\};$$

- If A is a subset of X, then its **complement** relative to X is defined by

$$X \setminus A = \{x \in X : x \notin A\}.$$

It is obvious that $X \setminus (X \setminus A) = A$, $A \cap (X \setminus A) = \varnothing$, and $A \cup (X \setminus A) = X$.

More generally, let $\{A_i : i \in I\}$ be a family of sets. Then

- the **union** of the family $\{A_i : i \in I\}$ is defined by

$$\bigcup_{i \in I} A_i = \{x : x \in A_i \text{ for some } i\};$$

- the **intersection** of the family $\{A_i : i \in I\}$ is defined by

$$\bigcap_{i \in I} A_i = \{x : x \in A_i \text{ for all } i\}.$$

- A family $\{A_i : i \in I\}$ is said to be **disjoint** if for each $i \neq j$ in I, $A_i \cap A_j = \varnothing$.

A very useful theorem in set theory is the following:

Theorem A.1 (De Morgan's Law)

Let $\{A_i : i \in I\}$ be a family of subsets of a set X. Then

(1) $X \setminus \left(\bigcup_{i \in I} A_i \right) = \bigcap_{i \in I} (X \setminus A_i)$;

(2) $X \setminus \left(\bigcap_{i \in I} A_i \right) = \bigcup_{i \in I} (X \setminus A_i)$.

A.3 Cantor's Ternary Set

A remarkable and interesting set of points on the real line is Cantor's ternary set, C. It consists of all points in $[0, 1]$ that have ternary (base 3) expansion $\sum_{n=1}^{\infty} a_n 3^{-n}$ with $a_n \in \{0, 2\}$. This representation of points of C is unique for even though many rational numbers have two possible ternary expansion, no number can be written in more than one way without using the digit 1. For example $\frac{1}{3} = 0.1000\ldots$ will be represented by $0.02222\ldots$. Geometrically the Cantor set is constructed as follows.

Consider the interval $C_0 = [0, 1]$. Remove the middle third open interval $(1/3, 2/3)$. Let $C_1 = [0, 1/3] \cup [2/3, 1]$. Remove the middle third open intervals $(1/9, 2/9)$, $(7/9, 8/9)$ of each of the intervals of C_1. Let $C_2 = [0, 1/9] \cup [2/9, 1/3] \cup [2/3, 7/9] \cup [8/9, 1]$.

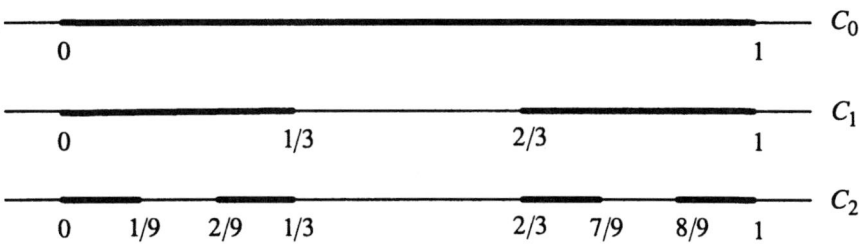

Continuing in this way by removing open middle thirds we obtain

- a set C_n consisting of 2^n closed intervals each of length $1/3^n$;

- $\ell(C_n) = (2/3)^n$;

- for each n, $C_{n+1} \subset C_n$.

The Cantor set is now defined by $C = \bigcap_{n=0}^{\infty} C_n$. We now list important properties of C.

- C is compact. It is clear that C is bounded. It is also closed as intersection of closed sets.

- C is a null set. According to he definition, $C \subset C_n$ which consists of closed intervals of total length $(2/3)^n$ for every n. Therefore since $(2/3)^n \to 0$, $\ell(C) = 0$.

- C has empty interior. C does not contain any interval.

- C is uncountable. The proof is reminiscent of the proof of Example 2.4(36). Let (x_n) be a sequence in C. We write (x_n) as

$$
\begin{array}{ccl}
n & \mapsto & x_n \\
1 & & 0.d_{11}d_{12}d_{13}\ldots \\
2 & & 0.d_{21}d_{22}d_{23}\ldots \\
3 & & 0.d_{31}d_{32}d_{33}\ldots \\
\vdots & & \vdots
\end{array}
$$

  where a_{ij} is either 0 or 2. We consider the number $a = 0.a_1 a_2 a_3 \ldots$ where $a_n = 2$ if $d_{nn} = 0$ and $a_n = 0$ if $d_{nn} = 2$. Then clearly $a \in C$ but $a \neq x_n$ for all n. This shows that the range of any sequence in C cannot cover the whole of C and proves that C is uncountable.

A.4 Bernstein's Approximation Theorem

Definition A.2

Let $f : [0,1] \to \mathbb{R}$ be a continuous function. Then the polynomial

$$
B_n(x;f) = \sum_{k=0}^{n} f\left(\frac{k}{n}\right)\binom{n}{k} x^k (1-x)^{n-k}
$$

is called the n-th **Bernstein polynomial** of f for each $n \in \mathbb{N}$.

Theorem A.3

Suppose that $f \in C([0,1])$. Then the sequence $(B_n(x,f))$ converges uniformly to f.

The proof of the theorem relies on the following fact.

Lemma A.4

For each $x \in [0,1]$, we have

$$
\sum_{k=0}^{n} \left(x - \frac{k}{n}\right)^2 \binom{n}{k} x^k (1-x)^{n-k} \leq \frac{1}{4n}.
$$

Proof

By the binomial theorem we have

$$\sum_{k=0}^{n} \binom{n}{k} x^k b^{n-k} = (x + b)^n. \tag{A.1}$$

Differentiating both sides with respect to x, and then multiplying by x, we obtain

$$\sum_{k=0}^{n} \binom{n}{k} k x^k b^{n-k} = nx (x + b)^{n-1}. \tag{A.2}$$

Again differentiating both sides with respect to x, and then multiplying by x, we obtain

$$\sum_{k=0}^{n} \binom{n}{k} k^2 x^k b^{n-k} = n(n-1) x^2 (x + b)^{n-1} + nx (x + b)^{n-1}. \tag{A.3}$$

Now in each of (A.1), (A.2), and (A.3) we substitute b by $(1 - x)$, we have

$$\sum_{k=0}^{n} \binom{n}{k} x^k (1 - x)^{n-k} = 1. \tag{A.4}$$

$$\sum_{k=0}^{n} k \binom{n}{k} x^k (1 - x)^{n-k} = nx. \tag{A.5}$$

$$\sum_{k=0}^{n} k^2 \binom{n}{k} x^k (1 - x)^{n-k} = n(n-1) x^2 + nx. \tag{A.6}$$

Performing the operations

$$n^2 x^2 \text{ (equation } A.4) + (-2nx) \text{ (equation } A.5) + \text{ (equation } A.6)$$

yield after simplification

$$\sum_{k=0}^{n} \left(x - \frac{k}{n} \right)^2 \binom{n}{k} x^k (1 - x)^{n-k} = \frac{x(1 - x)}{n}.$$

From $0 \leq (2x - 1)^2 = 4x^2 - 4x + 1$, we have $x(1 - x) \leq 1/4$. Hence the desired estimate is established. $\square$

Proof (of the Theorem A.3)

We assume $f \not\equiv 0$ and let $M = \sup \{|f(x)| : x [0, 1]\}$. From equation A.4, it follows that

$$\sum_{k=0}^{n} f(x) \binom{n}{k} x^k (1 - x)^{n-k} = f(x).$$

Therefore we have

$$|f(x) - B_n(x; f)| \leq \sum_{k=0}^{n} \left| f(x) - f\left(\frac{k}{n}\right) \right| \binom{n}{k} x^k (1-x)^{n-k} \qquad (A.7)$$

Thus by uniform continuity, given $\varepsilon > 0$ there exists $\delta > 0$ such that

$$x, y \in [0, 1] \text{ and } |x - y| < \delta \text{ implies } |f(x) - f(y)| < \varepsilon/2.$$

Choose N large enough so that $N \geq \max\left\{\frac{1}{\delta^4}, \frac{M}{\varepsilon^2}\right\}$. For fixed $x \in [0, 1]$, and $n > N$ we let

$$A = \left\{ k \in \{0, 1, 2, \ldots, n\} : \left| x - \frac{k}{n} \right| < \delta \right\}$$

and

$$B = \left\{ k \in \{0, 1, 2, \ldots, n\} : \left| x - \frac{k}{n} \right| \geq \delta \right\}.$$

For $k \in A$ we have $\left| f(x) - f\left(\frac{k}{n}\right) \right| < \varepsilon/2$ and hence

$$\sum_{k \in A} \left| f(x) - f\left(\frac{k}{n}\right) \right| \binom{n}{k} x^k (1-x)^{n-k} \leq \frac{\varepsilon}{2}. \qquad (A.8)$$

For $k \in B$ we have $\left| x - \frac{k}{n} \right|^2 \geq \frac{1}{n^{1/2}}$ and so using the Lemma,

$$\sum_{k \in B} \left| f(x) - f\left(\frac{k}{n}\right) \right| \binom{n}{k} x^k (1-x)^{n-k} \leq \sum_{k \in B} 2M \binom{n}{k} x^k (1-x)^{n-k}$$

$$\leq 2M \sum_{k \in B} \frac{|x - k/n|^2}{|x - k/n|^2} \binom{n}{k} x^k (1-x)^{n-k}$$

$$\leq 2M\sqrt{n} \sum_{k=0}^{n} \left| x - \frac{k}{n} \right|^2 \binom{n}{k} x^k (1-x)^{n-k}$$

$$\leq \frac{M}{2\sqrt{n}} < \frac{\varepsilon}{2} \qquad (A.9)$$

Combining (A.7), (A.8), and (A.9), we obtain for $n > N$

$$|f(x) - B_n(x; f)| < \varepsilon$$

Since x is arbitrary in $[0, 1]$, the proof is complete. □

Note that the above result can easily be extended to continuous functions on any bounded interval $[a, b]$; if f is continuous on $[a, b]$, then the function g defined on $[0, 1]$ by

$$g(t) = f((b - a) t - a)$$

is continuous. Thus the above theorem apply to g. A simple change of variable yields a polynomial approximation of f.

B
Hints for Selected Exercises

Only hints and/or answers to selected problems are given. Students are expected to fill in the precise details of the proofs and/or calculations in order to obtain the complete solution of each selected exercise.

Chapter 1

1.2 Let $x, a \in \mathbb{R}$ be such that $x + a = a$. Add $-a$ to both sides and use A1, A4 and A3 to obtain $0 = a + (-a) = (x + a) + (-a) = x + (a + (-a)) = x + 0 = x$.

1.6 Since $a^2 \geq 0$ and $b^2 \geq 0$, then $a^2 + b^2 = 0$ implies $a^2 = b^2 = 0$.

1.8 (c) Note that $1 + 2^{-1} + 2^{-2} + \cdots + 2^{-n} + 2^{-(n+1)} = 2 - 2^{-n} + 2^{-(n+1)} = 2 - 2^{-(n+1)}$.

(g) Note that $3^{2(n+1)+1} + 2^{(n+1)+2} = 9 \cdot 3^{2n+1} - 2 \cdot 3^{2n+1} + 2 \cdot 3^{2n+1} + 2 \cdot 2^{n+1} = 7 \cdot 3^{2n+1} + 2\left(3^{2n+1} + 2^{n+2}\right)$.

1.9 Note that $x^{n+1} - y^{n+1} = x^{n+1} - x^n y + x^n y - y^{n+1} = x^n (x - y) + y (x^n - y^n)$.

1.15 Note that $\operatorname{int}(n \operatorname{fra} x) \leq n \operatorname{fra} x \leq \operatorname{int}(n \operatorname{fra} x) + 1$. Dividing by n and adding $\operatorname{int} x$, one obtains the desired inequalities.

1.19 Since $\left(\sqrt{|a|} - \sqrt{|b|}\right)^2 \geq 0$, we have $|a| - 2\sqrt{|a| |b|} + |b| \geq 0$ or $\frac{|a| + |b|}{2} \geq \sqrt{|a| |b|}$.

1.27 Suppose that $a > b$. Then $\varepsilon = (a - b)/2 > 0$ and $b + \varepsilon < a$. But since $b + \varepsilon > b$, we must have $b + \varepsilon > a$. Contradiction.

1.29 Since $\inf A \leq a$ and $\inf B \leq b$ for every $a \in A$ and $b \in B$, we have $\inf A + \inf B \leq a + b$. This shows that $A + B$ is bounded below and $\inf(A + B) \geq \inf A + \inf B$. To see the inverse inequality, let $\varepsilon > 0$. There is an element $a \in A$

such that $a < \inf A + \varepsilon/2$ and an element $b \in B$ such that $b < \inf B + \varepsilon/2$. It follows that $\inf (A + B) \leq a + b \leq \inf A + \inf B + \varepsilon$. Since $\varepsilon > 0$ is arbitrary, we have $\inf (A + B) \leq \inf A + \inf B$.

1.36, 1.37 and 1.38 Use the results of Exercise 1.31.

1.39 (a) Note that $f(x) = f(1x) = f(1) + f(x)$ for every x. Hence $f(1) = 0$.
(b) By induction, one obtains $f(a^n) = nf(a)$.

1.41 (3) $x \in f^{-1}(Y \setminus B)$ if and only if $f(x) \in Y \setminus B$ if and only if $f(x) \in Y$ and $f(x) \notin B$ if and only if $x \in f^{-1}(Y)$ and $x \notin f^{-1}(B)$ if and only if $x \in X \setminus f^{-1}(B)$.

1.42 (1) $y \in f\left(\bigcup_{i \in I} A_i\right)$ if and only if there is $x \in \bigcup_{i \in I} A_i$ such that $y = f(x)$ if and only if there is $i \in I$ such that $x \in A_i$ and that $y = f(x)$ if and only if there is $i \in I$ such that $y \in f(A_i)$ if and only if $y \in \bigcup_{i \in I} f(A_i)$.

Chapter 2

2.2 For each $k \in \mathbb{N}$, the set A_k is the range of some sequence $\left(a_n^k\right)_{n \in \mathbb{N}}$. Then $S = \bigcup_{n=1}^{\infty} A_n$ is the range of the sequence defined by

$$\left(a_1^1, a_1^2, a_2^1, a_3^1, a_2^2, a_1^3, a_1^4, a_2^3, a_3^2, a_4^1, a_5^1, a_4^2, a_3^3 \ldots\right).$$

2.3 Suppose that A is countable, i.e. A is the range of some sequence (y_n). Then for each n, y_n is a sequence $(y_{n,k})_{k \in \mathbb{N}}$ consisting of 0's and/or 1's. Consider the sequence defined by $x_k = 1 - y_{k,k}$. Then $(x_k) \in A$ but $(x_k) \neq (y_{n,k})$ for each n.

2.10 (a) Let $A = \{n \in \mathbb{N} : a_{n+1} > a_n\}$. Since $a_1 = 1$ and $a_2 = \sqrt{2}$, $a_1 < a_2$ and hence $1 \in A$. Suppose that $k \in A$. Then $a_{k+1} - a_k > 0$ and hence

$$a_{(k+1)+1} - a_{k+1} = \sqrt{a_{k+1} + 1} - \sqrt{a_{k+1-1} + 1}$$
$$= \frac{a_{k+1} - a_k}{\sqrt{a_{k+1} + 1} + \sqrt{a_{k+1-1} + 1}} > 0.$$

By the principle of mathematical induction we have $A = \mathbb{N}$; that is the sequence (a_n) is increasing.
(b) Let $A = \{n \in \mathbb{N} : a_n < 2\}$. Then clearly $1 \in A$. Suppose that $n \in A$. Then $0 < a_n < 2$. Thus $a_n + 1 < 2 + 1$ and $a_{n+1} = \sqrt{a_n + 1} < \sqrt{3} < 2$; i.e. $n + 1 \in A$. Again by the principle of mathematical induction we have $A = \mathbb{N}$.
(c) Since (a_n) is increasing and bounded, $\lim a_n = l$ exists. It must satisfy $l = \sqrt{l + 1}$. Hence $l = \left(1 + \sqrt{5}\right)/2$.

2.12 The result is trivial if $x = 0$. Suppose that $x \neq 0$ and consider the sequence defined by $x_n = |x|^n$. Then $x_n > 0$ for all n and $x_{n+1} = |x| x_n$. It follows from the assumption $|x| < 1$ that $0 < x_{n+1} < x_n$. Thus since (x_n) is decreasing and bounded below, $l = \lim x_n$ exists and must satisfy $l = |x| l$. Hence $l = 0$.

2.13 Let $n^{1/n} = 1 + x_n$ where $x_n \geq 0$. By the binomial formula we have $n = (1 + x_n)^n = 1 + nx_n + \frac{n(n-1)}{2!}x_n^2 + \cdots + x_n^n$. It follows that $n > \frac{n(n-1)}{2!}x_n^2$ and $0 \leq x_n^2 \leq \frac{2}{n-1}$. Hence $\lim x_n^2 = \lim x_n = 0$. Thus $\lim n^{1/n} = 1$.

2.18 Let $\sigma_n = \frac{a_1 + a_2 + \cdots + a_n}{n}$ for each n and let $a = \lim a_n$. Given $\varepsilon > 0$, there is $N \in \mathbb{N}$ such that $m \geq N$ implies $|a_m - a| < \varepsilon/2$. Also there exists $M > 0$ such that $|a_k| < M$ for all k. For $n > N$, we have

$$\sigma_n - a = \frac{1}{n}\left[(a_1 - a) + (a_2 - a) + \cdots + (a_n - a)\right]$$
$$= \frac{1}{n}\left[(a_1 - a) + (a_2 - a) + \cdots + (a_N - a)\right]$$
$$+ \frac{1}{n}\left[(a_{N+1} - a) + \cdots + (a_n - a)\right].$$

Hence for $n > N$, we have

$$|\sigma_n - a| \leq \frac{N(M + |a|)}{n} + \frac{(n - N)\varepsilon}{2n}.$$

Note that $(n - N)/n < 1$. Choosing n large enough so that $N(M + |a|) < n\varepsilon/2$, we obtain $|\sigma_n - a| < \varepsilon$. Hence $a = \lim \sigma_n$. The sequence defined by $a_n = (-1)^{n+1}$ is divergent C1-summable.

2.23 Note that $\lim \sqrt[n]{n} = 1$.

2.27 (1) Use triangle inequality to show that $\left| |a_n| - |a| \right| \leq |a_n - a|$. (2) Note that $|b_n - b| \leq |b_n - a_n| + |a_n - a|$.

2.36 (a) $-1, 1$; (b) $1, 3$; (c) 0; (d) ∞; (e) $-1, 1$; (f) $-\infty$.

2.38 (1) Let (x_n) and (y_n) be subsequences of (a_n) such that

$$\lim(-x_n) = \limsup(-a_n) \quad \text{and} \quad \lim y_n = \liminf a_n.$$

Then

$$-\liminf a_n = -\lim y_n = \lim(-y_n) \leq \limsup(-a_n)$$
$$= \lim(-x_n) = -\lim x_n \leq -\liminf a_n.$$

Chapter 3

3.1 (a) div. (b) conv. (c) conv. (d) div.

3.3 Note that $\frac{1}{(x+n)(x+n+1)} = \frac{1}{x+n} - \frac{1}{x+n+1}$.

3.6 (a), (b), (d), (i), (k), (m), (o) and (p) are convergent, the rest are divergent.

3.8 Suppose that (a_n) is unconditionally summable and that (3) does not hold. There is a $\delta > 0$ and a sequence (S_m) of finite subsets of $\mathbb{N}$, with $\max S_m < \min S_{m+1}$ and $\left| \sum_{n \in S_m} a_n \right| \geq \delta$ for all m. Let $|S_m|$ be the number of elements in S_n. Consider the permutation σ of $\mathbb{N}$ which maps the set

$$\{\min S_m, \min S_m + 1, \ldots, \min S_m + |S_m|\}$$

into S_m. Then the series $\sum a_{\sigma(n)}$ cannot be Cauchy, contradicting the unconditional summability of (a_n). Conversely, suppose (3) holds and let σ be a permutation of $\mathbb{N}$. Fix $\varepsilon > 0$ and choose $N_\varepsilon > 0$ as in (3). Thus there is an m_ε such that $\{1, \ldots, N_\varepsilon\} \subset \sigma(\{1, \ldots, m_\varepsilon\})$. Thus $\left| \sum_{n=q}^p a_{\sigma(n)} \right| < \varepsilon$ whenever $p > q > m_\varepsilon$; that is $(a_{\sigma(n)})$ is summable.

(c) $\Longrightarrow$(b): Let (a_{n_k}) be a subsequence of (a_n). (Thus $n_k \geq n$ for every n.) Fix $\varepsilon > 0$ and choose $N_\varepsilon > 0$ as in (3). Thus if $p > q > N_\varepsilon$, then $\left| \sum_{n=q}^p a_{n_k} \right| < \varepsilon$. Hence (a_{n_k}) is summable.

(b) $\Longrightarrow$(a): Let (ε_n) be a sequence of ± 1's. Let $A^\pm = \{n \in \mathbb{N} : \varepsilon_n = \pm 1\}$. Then both the series $\sum_{n \in A^+} a_n$ and $\sum_{n \in A^-} a_n$ converge. Let $\varepsilon > 0$. If $q < p$, then $\sum_{n=q}^p \varepsilon_n a_n = \sum_{[q,p] \cap A^+} a_n - \sum_{[q,p] \cap A^-} a_n$. Since the series $\sum_{n \in A^+} a_n$ is convergent, there exists N_+ large enough so that $N_+ < q < p$ implies $\left| \sum_{[q,p] \cap A^+} a_n \right| < \varepsilon/2$. Similarly, there exists $N_- \in \mathbb{N}$ so that $N_- < q < p$ implies $\left| \sum_{[q,p] \cap A^-} a_n \right| < \varepsilon/2$. Thus if $\max \{N_+, N_-\} < q < p$, then $\left| \sum_{n=q}^p \varepsilon_n a_n \right| < \varepsilon$.

(a) $\Longrightarrow$(c): Suppose that (3) does not hold. Then there is a $\delta > 0$ and a sequence (S_m) of finite subsets of $\mathbb{N}$, with $\max S_m < \min S_{m+1}$ and $\left| \sum_{n \in S_m} a_n \right| \geq \delta$ for all m. Let $\varepsilon_n = 1$ if $n \in \bigcup_k S_k$, and -1 otherwise. Then the series $\sum (1 + \varepsilon_n) a_n$ cannot be Cauchy. One of the series $\sum a_n$ and $\sum \varepsilon_n a_n$ must be divergent.

3.9 $\sum_n \left(\sum_m a_{n,m} \right) = 1$; $\sum_m \left(\sum_n a_{n,m} \right) = -1$; $\sum_n s_{nn} = 0$. The double series $\sum a_{n,m}$ diverges.

3.11 Note that $2mn \leq m^2 + n^2$.

3.12 Compare with the divergent series $\sum 1/n$.

3.14 Show that the series $\sum n^p a^n = 0$, for every $p > 0$ provided $|a| < 1$.

3.18 Use integral test. Convergent if $\alpha > 1$. Divergent if $\alpha < 1$. If $\alpha = 1$, convergent if $\beta > 1$, and divergent if $\beta \leq 1$.

3.22 Show that $|a_{n+p} - a_n| \leq M r^n \frac{1 - r^p}{1 - r}$.

3.29 Note that $\sum a_n b_n \leq \left(\sum (a_n)^2 \right)^{1/2} \left(\sum (b_n)^2 \right)^{1/2}$.

3.30 Apply Exercise 3.29 to a_n and $b_n = 1/n$.

3.33 Letting $s_n = \sum_{k=1}^n b_k$ and $\sup s_n = S$, verify that

$$\sum_{k=n}^m a_k b_k = (a_{m+1} s_m - a_n s_{n-1}) + \sum_{k=n}^m (a_k - a_{k+1}) s_k,$$

$$\left| \sum_{k=n}^m a_k b_k \right| \leq \left[(|a_{m+1}| - |a_n|) + \sum_{k=n}^m |a_k - a_{k+1}| \right] S.$$

It follows from this inequality, and conditions (1) and (2) that the series $\sum a_n b_n$ satisfies the Cauchy criterion.

3.34 Both series satisfy Dirichlet's test conditions.

3.35 Use Exercise 3.21 and Exercise 3.33.

Chapter 4

4.1 (a) $\frac{5}{2}$; (b) undefined; (c) $\frac{n}{p}$; (d) 3; (e) undefined; (f) α; (g) $\frac{-\sqrt{(p^2)}+p}{-\sqrt{(q^2)}+q}$; (h) $\frac{1}{2}$; (i) $-\frac{5}{3}$; (j) e^{-1}; (k) e; (l) $\frac{1}{2}$.

4.2 (a) $\frac{1}{9}$; (b) ∞; (c) 0; (d) 3; (e) $\frac{2}{3}$; (f) $a-b$; (g) 1; (h) -1; (i) e^{-1}; (j) $2a$; (k) 0; (l) 1.

4.5 There is $\varepsilon > 0$ such that for every $\delta > 0$ we have $|x - a| < \delta$ and $|f(x) - f(a| \geq \varepsilon$.

4.8 Write $f(x) - f(a) = (x - a)\frac{f(x)-f(a)}{(x-a)}$.

4.9 Use the density theorem.

4.11 Use the result of Exercise 4.10. If f is discontinuous at a point c, then choose a rational r such that $f(c^+) \leq r \leq f(c^-)$.

4.17 Note that $f(0) = 1$ and $f(a + h) - f(a) = f(a) \cdot (f(h) - 1)$.

4.18 Note that $f(0) = 0$ and $f(a + h) - f(a) = hf(1)$.

4.20 If t is rational, consider a sequence of irrationals converging to t, and if t is irrational consider a sequence of rationals converging to t.

4.27 Use intermediate value theorem.

4.28 Apply Exercise 4.27 to $f(x) = x - \cos x$ on the interval $[0, \pi/2]$.

4.38 Note that $|f_A(x) - f_A(y)| \leq |x - y|$.

4.41 A continuous function on a closed bounded interval reaches its maximum M and its minimum m and takes on all the values on the interval $[m, M]$.

Chapter 5

5.3 Let $A = \left\{n \in \mathbb{N} : (fg)^{(n)} = \sum_{k=0}^{n} \binom{n}{k} f^{(n-k)} g^{(k)}\right\}$. Then clearly $1 \in A$. Suppose that $n \in A$. Then

$$
\begin{aligned}
(fg)^{(n+1)} &= \left[\sum_{k=0}^{n} \binom{n}{k} f^{(n-k)} g^{(k)}\right]' \\
&= \sum_{k=0}^{n} \binom{n}{k} \left[f^{(n-k+1)} g^{(k)} + f^{(n-k)} g^{(k+1)}\right] \\
&= \sum_{k=0}^{n} \binom{n}{k} f^{(n+1-k)} g^k + \sum_{k=1}^{n+1} \binom{n}{k-1} f^{(n+1-k)} \\
&= \sum_{k=1}^{n} \binom{n+1}{k} f^{(n+1-k)} + \binom{n}{0} f^{(n+1)} + \binom{n+1}{n} g^{(n+1)} \\
&= \sum_{k=0}^{n+1} \binom{n+1}{k} f^{(n+1-k)} g^k,
\end{aligned}
$$

using the fact that $\binom{n}{k} + \binom{n}{k} = \binom{n+1}{k}$ (show this). Thus by the principle of mathematical induction $A = \mathbb{N}$.

5.7 Set $h = \frac{\alpha}{n}$, $k = \frac{\beta}{n}$. Then

$$\lim_n n\left[f\left(x_0 + \frac{\alpha}{n}\right) - f\left(x_0 - \frac{\beta}{n}\right)\right] = \alpha \lim_{h \to 0} \frac{f(x_0 + h) - f(x_0)}{h}$$
$$+ \beta \lim_{k \to 0} \frac{f(x_0) - f(x_0 - k)}{-k}.$$

5.10 (a) Suppose that we want to show continuity at $x_0 \in (a, b)$. Then the convexity condition is equivalent to

$$f(x_0 + \lambda(x - x_0)) - f(x_0) \leq \lambda[f(x) - f(x_0)]$$

whenever $x \in (a, b)$ and $0 < \lambda < 1$. Consider a sequence (δ_n) converging to 0, $0 < \delta_n < |x - x_0|$. Fix n and let $\lambda = \frac{\delta_n}{|x - x_0|}$. Continuity follows from $|f(x_0 + \delta_n) - f(x_0)| \leq \delta_n |f(x) - f(x_0)|$.

(b) Letting $\lambda = (x - w)/(y - w)$, obtain $\frac{f(x) - f(w)}{x - w} \leq \frac{f(y) - f(w)}{y - w}$.

5.11 (a) $\frac{1}{9}$; (b) $e^{-\frac{1}{2}}$; (c) -1; (d) 1; (e) e^{-1}; (f) 1; (g) 0; (h) e; (i) $\frac{1}{n}$; (j) e^{-1}; (k) -8; (l) 1.

5.14 Apply mean value theorem to $f(x) = \cos x$.

5.15 Note that $0 \leq \frac{|f(x+h) - f(x)|}{|h|} \leq |h|$.

5.18 Apply mean value theorem to $f(x) = x^n$ on $[a, b]$.

5.25 Let $f(x) = e^x - (1 + xe^x)$. Then $f'(x) = -xe^x > 0$ for $x < 0$ and $f'(x) = -xe^x < 0$ for $x > 0$. It follows that $f(0) = 0$ is an absolute minimum; that is $f(x) > 0$ or $e^x < 1 + xe^x$ for $x \neq 0$. One proves the other inequality by applying a similar argument to $g(x) = 1 + x - e^x$.

5.27 If $f(x) = 1 + \frac{x}{2} - \frac{x^2}{8} - \sqrt{1+x}$, then $f'(x) = \frac{1}{2} - \frac{1}{4}x - \frac{1}{2\sqrt{(1+x)}} < 0$ for $x > 0$.
If $f(x) = \sqrt{1+x} - \left(1 + \frac{x}{2}\right)$, then $f'(x) = \frac{1}{2\sqrt{(1+x)}} - \frac{1}{2} < 0$ for $x > 0$.

5.29 Apply mean value theorem to $F(x) = f(x) - g(x)$.

5.32 (1) $\left(f^{-1}\right)''(x) = \frac{-f''(f^{-1}(x))}{[f'(f^{-1}(x))]^3}$.

5.35 Suppose $(a, b) \subset I$ and suppose $f'(a) < \lambda < f'(b)$. Put $g(t) = f(t) - \lambda t$. Then $g'(a) < 0$ and $g'(b) > 0$. Therefore $g(t_1) < g(a)$ for some $t_1 \in (a, b)$ and $g(t_2) < g(b)$ for some $t_2 \in (a, b)$. Hence, g reaches its minimum at some point x in (a, b). Thus $g'(x) = 0$ and hence $f'(x) = \lambda$.

5.37 Integrate $\frac{1}{n!} \int_a^x (x - t)^n f^{(n+1)}(t)\, dt$ by parts. Continue integration by parts until you get the stated formula.

5.38 A polynomial is either increasing or decreasing between two consecutive roots. Note also that a root of odd multiplicity of p' is either a maximum or a minimum point for p.

Chapter 6

6.2 Only (c) and (f) are step functions.

6.3 Note that $-|\varphi| \leq \varphi \leq |\varphi|$.

6.5 Show that if $\sum_{i=1}^{n} \alpha_i \chi_{I_i}$ is a disjoint representation of φ then $\sum_{i=1}^{n} \sqrt{\alpha_i} \chi_{I_i}$ is a disjoint representation of $|\varphi|$.

6.6 $\max \{\varphi, \psi\} = \frac{1}{2} (\varphi + \psi + |\varphi - \psi|)$; $\min \{\varphi, \psi\} = \frac{1}{2} (\varphi + \psi - |\varphi - \psi|)$.

6.8 $\overline{f}^{P_n} = \left(\frac{1}{n}\right)^2 \chi_{\left[0, \frac{1}{n}\right]} + \left(\frac{2}{n}\right)^2 \chi_{\left(\frac{1}{n}, \frac{2}{n}\right]} + \cdots + \left(\frac{n}{n}\right)^2 \chi_{\left[\frac{n-1}{n}, \frac{n}{n}\right]}$ and

$$\underline{f}^{P_n} = (0)^2 \chi_{\left[0, \frac{1}{n}\right)} + \left(\frac{1}{n}\right)^2 \chi_{\left[\frac{1}{n}, \frac{2}{n}\right)} + \cdots + \left(\frac{n-1}{n}\right)^2 \chi_{\left[\frac{n-1}{n}, \frac{n}{n}\right]}.$$

Then $\int \overline{f}^{P_n} - \int \underline{f}^{P_n} = \frac{1}{n} \to 0$. Hence

$$\int_0^1 f(x) \, dx = \lim_n \frac{1}{n} \left(\frac{1}{n^2} + \frac{2^2}{n^2} + \cdots + \frac{n^2}{n^2}\right) = \frac{1}{3}.$$

6.9 Consider the partition $P_n = \{t_0, t_1, \ldots, t_n\}$ where $t_i = a + i \frac{b-a}{n}$ for $0 \leq i \leq n$. Then $\frac{b-a}{n} \sum_{i=1}^{n} f\left(a + i \frac{b-a}{n}\right) = \int \underline{f}^{P_n}$.

6.10 (a) $\lim_{n \to \infty} \frac{1^p + 2^p + \cdots + n^p}{n^{p+1}} = \int_0^1 x^p dx = \frac{1}{p+1}$ for $p \in \mathbb{N}$ and $\lim_{n \to \infty} \frac{1^{1/2} + 2^{1/2} + \cdots + n^{1/2}}{n^{1/2+1}} = \int_0^1 \sqrt{x} dx = \frac{2}{3}$. (b) $\lim_{n \to \infty} \frac{1}{n}(1 + \cos \frac{a}{n} + \ldots + \cos \frac{(n-1)a}{n}) = \int_0^1 \cos ax \, dx = \frac{\sin a}{a}$.

6.15 The function $f - g = 0$ for all but finitely many points.

6.17 Use integration by parts to obtain $\int_0^1 f_n(x) \, dx = \frac{1}{2} + \int_0^1 \frac{x^n}{(1+x)^2} dx$.

6.19 Use integration by parts to obtain $\int_0^1 nx e^{-nx^2} dx = -\frac{1}{2} e^{-n} + \frac{1}{2}$.

6.21 Using Lebesgue dominated convergence theorem we have

$$\int_a^b f(x) \, dx = \int_{[a,b]} f d\ell = \lim \int_{[a,b]} f_n d\ell = \lim \int_a^b f(x) \, dx.$$

6.25–9.30 Use similar arguments as in the case of ordinary Riemann integral.

6.31 Note that the power series converges uniformly on $(-R, R)$.

6.34 (a) $\int_0^1 x d(x^3) = \int_0^1 x (3x^2) \, dx = \frac{3}{4}$. (c) $\int_0^2 x^2 d(\text{int}(x)) = \sum_{j=0}^{2} j^2 = 5$.

6.36 If M is the Lipschitz constant of f, then one has for every $(a_1, b_1), \ldots, (a_n, b_n)$ disjoint subintervals of $[a, b]$

$$\sum_{i=1}^{n} |f(b_i) - f(a_i)| < M \sum_{i=1}^{n} |b_i - a_i|.$$

Chapter 7

7.1 (a) $f_n(0) = 0$ for all n, hence $f(0) = 0$. $f_n(1) = 1$ for all n, hence $f(1) = 1$.
If $x \in (0,1)$, then there exists n large enough so that $x \in [1/n, 1]$. It
follows that $f(x) = 1$ for $x \in (0,1)$.
(b) The convergence cannot be uniform on $[0,1]$ since all the f_n are continuous on $[0,1]$ while the limit f is not.

7.2 (1) $f(x) = \frac{1}{1-x}$; (2) no.

7.3 (1) $f(x) = 0$; (2) Consider the sequence $(1/n)$.

7.4 (1) $f(x) = e^x$; (2) no.

7.6 (1) $f(x) = 0$ if $x \in [0,1)$ and $f(1) = 1$; (2) no.

7.10 Consider $f_n(t) = \begin{cases} \frac{1}{n} & \text{if } t \in (0,1] \cap \mathbb{Q} \\ 0 & \text{if } t \in (0,1] \setminus \mathbb{Q}. \end{cases}$

7.15 Note that f is bounded on $[a,b]$ and that

$$\left| \int_{a_n}^{b_n} f_n(x)\, dx - \int_a^b f(x)\, dx \right|$$

$$= \left| \int_{a_n}^{b_n} f_n(x)\, dx - \left(\int_a^{a_n} f(x)\, dx + \int_{a_n}^{b_n} f(x)\, dx + \int_{b_n}^b f(x)\, dx \right) \right|$$

$$\leq \int_a^b |f_n(x) - f(x)|\, dx + \int_a^{a_n} |f(x)|\, dx + \int_{b_n}^b |f(x)|\, dx.$$

7.16 By induction, show that $|f_n(x)| \leq \sup\{|f(x) : x \in [0,1]|\}/n!$.

7.22 If $\alpha > 0$, the inequality $\left| \frac{\sin nx}{nx} \right| \leq \frac{1}{n\alpha}$ holds for all $x \in [\alpha, \pi]$. Hence $\frac{\sin nx}{nx}$
converges uniformly to 0 on $[\alpha, \pi]$, and thus $\lim_{n\to\infty} \int_a^\pi \frac{\sin nx}{nx}\, dx = 0$.

7.26 (a) $(-2,2)$; (b) $[-1,1]$; (c) $\{0\}$; (d) $(1,3)$; (e) $\mathbb{R}$; (f) $(-e^{-1}, e^{-1})$; (g) $\{0\}$; (h)
$[-1,1)$; (i) $\{0\}$; (j) $(-4,-2)$; (k) $\mathbb{R} \setminus \{0\}$; (l) $[-\sqrt{2}, \sqrt{2}]$.

7.27 (a) For $0 < \alpha < 1$, and for $0 \leq |x|^n \leq \alpha^n \leq \alpha < 1$, we have $\left| \frac{x^n}{1+x^{2n}} \right| \leq$
$\frac{|x|^n}{1-|x|^{2n}} \leq \frac{\alpha^n}{1-\alpha}$. So the series is dominated by the convergent numerical
series $\sum \frac{\alpha^n}{1-\alpha}$. Hence the series converges uniformly and absolutely on
$[-\alpha, \alpha]$. The convergence is not uniform on $(-1,1)$ because for example
$\sum_{k=n}^\infty \frac{x^k}{1+x^{2k}} \geq \frac{1}{2} \sum_{k=n}^\infty x^k = \frac{1}{2} \frac{x^n}{1-x} \to \infty$ as $x \to 1^-$.
(b) The series converges uniformly on any closed interval not containing 0.
(c) For $0 < \alpha < x$, the series converges uniformly on $[\alpha, \infty)$. The convergence
is also uniform on $(-\infty, \beta)$ where $\beta < -1$, and on any closed interval not
containing $\frac{-1}{n^2}$, $n \in \mathbb{N}$.

7.36 (a) $\sum nx^n = x \sum \frac{d}{dx} x^n = x \frac{d}{dx} \sum x^n = \frac{x}{(1-x)^2}$ on $(-1,1)$. (b) $\sum \frac{x^n}{n+1} = \frac{1}{x} \sum \int x^n dx = \frac{1}{x} \int \sum x^n dx = \frac{1}{x} \int \frac{1}{1-x} dx = \frac{1}{x} \ln(1-x)$ on the interval
$(-1,1)$. (c) Let $u = -x$ in (b).

7.40 (a) $\frac{1}{2+x} = \frac{1}{2} - \frac{1}{4}x + \frac{1}{8}x^2 - \frac{1}{16}x^3 + \cdots$; (b) $\frac{1}{1-x^2} = 1 + x^2 + x^4 + \cdots$; (c)
$\ln(1-x^2) = -x^2 - \frac{1}{2}x^4 - \frac{1}{3}x^6 \cdots$.

7.41 (a) $\frac{1}{1+x} = \frac{1}{2} - \frac{1}{4}(x-1) + \frac{1}{8}(x-1)^2 - \frac{1}{16}(x-1)^3 + \cdots$; (b) $\frac{1}{x^2} = 1 - 2(x-1) + 3(x-1)^2 - 4(x-1)^3 + 5(x-1)^4 - \cdots$; (c) $\ln x = (x-1) - \frac{1}{2}(x-1)^2 + \frac{1}{3}(x-1)^3 - \frac{1}{4}(x-1)^4 + \cdots$.

Chapter 8

8.1 Consider $A_n = \left(-\frac{1}{n}, \frac{1}{n}\right)$, $n \in \mathbb{N}$.

8.2 Consider $A_n = \left[-1 + \frac{1}{n}, 1 - \frac{1}{n}\right]$, $n \in \mathbb{N}$.

8.3 Let A be a singleton $\{a\}$. Then if $x R \setminus A$, then letting $\varepsilon = |x - a|$ we notice that $a \notin B(x, \varepsilon/2)$. Thus $B(x, \varepsilon/2) \subset \mathbb{R} \setminus A$, i.e. $\mathbb{R} \setminus A$ is closed and A is open. Now a finite set is the finite union of singletons and therefore it is closed.

8.7 Since $A \subset A \cup B$ and $B \subset A \cup B$, $A^\circ \subset (A \cup B)^\circ$ and $B^\circ \subset (A \cup B)^\circ$. Hence $A^\circ \cup B^\circ \subset (A \cup B)^\circ$. If $A = [0, 1]$ and $B = (1, 2)$, then $A^\circ \cup B^\circ = (0, 1) \cup (1, 2)$ and $(A \cup B)^\circ = (0, 2)$.

8.9 Let $x \in A^- \cap B$. Then $x \in A^-$ and $x \in B$. Since B is open there is $\delta_1 > 0$ such that $B(x, \delta_1) \subset B$. Since $x \in A^-$, $B(x, \delta_1) \cap A \ni x_1$. For each n, let $\delta_n = \delta_1/n$ and pick $x_n \in B(x, \delta_n) \cap A$. Hence we have constructed a sequence (x_n) in $A \cap B$ converging to x.

8.16 $\partial [a, b] = \partial (a, b] = \partial [a, b) = \partial (a, b) = \{a, b\}$; $\partial \varnothing = \partial \mathbb{R} = \varnothing$; $\partial \mathbb{Q} = \mathbb{R}$.

8.18 A is open if and only if $A = A^\circ$ if and only if $\partial A = A^- \setminus A^\circ = A^- \setminus A$ if and only if $A \cap \partial A = \varnothing$. A is closed if and only if $A = A^-$ if and only if $\partial A = A^- \setminus A^\circ = A \setminus A^\circ$ if and only if $\partial A \subset A$.

8.30 (a) Since $\mathbb{N}^- = \mathbb{N}$ and $\mathbb{N}^\circ = \varnothing$, $\mathbb{N}$ is nowhere dense in $\mathbb{R}$. (b) $\mathbb{N}$ and $\mathbb{Q}$ can be written as countable unions of singletons which are nowhere dense. (c) If A is a nonempty open set, then $A \subset \left(A^-\right)^\circ$. (d) Let $A = \bigcup_{n=1}^\infty A_n$ where the A_n are of the first category. Then for each n we have $A_n = \bigcup_{m=1}^\infty B_{n,m}$ where the $B_{n,m}$ are nowhere dense sets. We now see that $A = \bigcup_{n=1}^\infty \bigcup_{m=1}^\infty B_{n,m}$ is of the first category.

8.34 Let $(U_i)_{i \in I}$ be an open cover of $K = \{x_n : n \in \mathbb{N}\} \cup \{x\}$. Fix $i_0 \in I$ such that $x \in U_{i_0}$. Since $x_n \to x$, there exists $N \in \mathbb{N}$ such that $n > N$ implies $x_n \in U_{i_0}$. For $n \le N$, we choose $i_n \in I$ so that $x_n \in U_{i_n}$. Then it is clear that $K \subset \bigcup_{n=0}^N U_{i_n}$. This completes the proof.

8.37 $A \cap B$ is closed and bounded.

8.39 If A is compact and $x \notin A$ in $\mathbb{R}$. Then for every $a \in A$, there exist two disjoint open sets U_a and V_a such that $a \in U_a$ and $x \in V_a$. $(U_a)_{a \in A}$ is easily seen as an open cover of A. Since A is compact there exists a finite subset of A, say $\{a_1, a_2, \ldots, a_n\}$ such that $A \subset U = \bigcup_{i=1}^n U_{a_i}$. We let $V = \bigcup_{i=1}^n V_{a_i}$. We notice that both U and V are open, $U \cap V = \varnothing$, $A \subset U$ and $x \in V$.

Chapter 9

9.1 Note that $A = f^{-1}((\alpha, \infty))$ and $B = f^{-1}([\alpha, \infty))$.

9.3 Suppose that f is continuous. Let $A \subset \mathbb{R}$ and $a \in A^-$. Let V be a neighborhood of $f(a)$. Then there is a neighborhood U of a such that $f(U \cap A) \subset V \cap f(A)$. Since $U \cap A \neq \varnothing$, we have $V \cap f(A) \neq \varnothing$. Hence $f(a) \in [f(A)]^-$. Conversely, suppose $f(A^-) \subset [f(A)]^-$ for every $A \subset \mathbb{R}$. Let V be open in $\mathbb{R}$. Let $A = f^{-1}(\mathbb{R} \setminus V)$. If $x \in A^-$, then $f(x) \in f(A^-) \subset [f(A)]^- \subset R \setminus V$. Thus $x \in A$ and A is closed. Hence $f^{-1}(V) = \mathbb{R} \setminus A$ is open.

9.5 Note that the function $f_x(y) = |x - y|$ is continuous on A.

9.7 Suppose that f is continuous on every compact subset of R. Let $x_0 \in R$, and let $x_n \to x_0$. Then $A = \{x_n : n \in \mathbb{N}\} \cup \{x_0\}$ is compact. Therefore $\lim_n f(x_n) = f(x_0)$. This proves f is continuous.

9.8 A constant function.

9.10 It suffices to notice that the closure of the set A is compact.

9.12 $A = (f - g)^{-1}(\{0\})$.

9.14 If (f_n) does not converge uniformly to f on A. Then there is $\varepsilon > 0$, a subsequence (g_n) of (f_n), and a sequence (x_n) in A such that $|g_n(x_n) - f(x_n)| \geq \varepsilon$ for each n. Since A is compact, there is a subsequence (x_{k_n}) of (x_n) converging to some x in A. By continuity, $f(x_{k_n}) \to f(x)$. By our assumption, $g_{k_n}(x_{k_n}) \to f(x)$. Hence $|g_{k_n}(x_{k_n}) - f(x_{k_n})| \to 0$. Contradiction.

9.15 Use Exercise 8.7.

9.17 $f(x) = x^2$ is continuous on $[0, 1]$ and separates the points of $[0, 1]$. Apply Stone–Weierstrass to the algebra generated by $\{1, x^2\}$. -1 and 1 cannot be separated by the closure of the algebra generated by $\{1, x^2\}$ in $C([-1, 1])$.

9.19, 9.20, 9.21 Show that the collection of functions of the required form satisfies the hypotheses of the Stone–Weierstrass theorem.

9.22 Note that f can be uniformly approximated by a sequence of polynomials, say (p_n). Then $p_n f \to f^2$ uniformly and by our assumption $\int_0^1 p_n(x) f(x)\, dx = 0$. Hence $\int_0^1 f^2(x)\, dx = 0$ and thus $f(x) = 0$ for all $x \in [0, 1]$.

9.26 (a) If $f(x) = x, f(y) = y$ and $x \neq y$, then $|x - y| < |f(x) - y| \leq |x - y|$. Impossible. (b) Let $g(x) = |f(x) - x|$. Then $|g(x) - g(y)| \leq |f(x) - f(y)| + |x - y| \leq 2|x - y|$ shows that g is continuous. Since A is compact, there is $a \in A$ a minimum point for g. Now $g(f(a)) = |f(f(a)) - f(a)| \leq |f(a) - a| = g(a)$.

9.37 Let $\varepsilon > 0$. Fix $a \in A$. By equicontinuity, there is a neighborhood V_a of a such that $|f_n(x) - f_n(a)| < \varepsilon/5$, for all n and for all $x \in V_a$; by continuity of f, there is a neighborhood W_a of a such that $|f(x) - f(a)| < \varepsilon/5$, for all $x \in W_a$. We let $U_a = V_a \cup W_a$. By compactness, there are $a_1, a_2, \ldots, a_n$ in A such that $A \subset \bigcup_{i=1}^n U_{a_i}$. For every $a \in A \cap U_{a_i}$, there is $N_i \in N$ such that $|f_n(a) - f(a)| < \varepsilon/5$ for $n > N_i$. Let $N = \max\{N_i\}$. Now for every $x \in A$, there exists i such that $x \in U_{a_i}$ and

$$|f_n(x) - f(x)| \leq |f_n(x) - f_n(a_i)| + |f_n(a_i) - f_n(a)|$$
$$+ |f_n(a) - f(a)| + |f(a) - f(x)|.$$

Hence $|f_n(x) - f(x)| < \varepsilon$ for all $x \in A$ and for all $n > N$.

9.38 Let $x \in A$ and fix $\varepsilon > 0$. By equicontinuity, there is a neighborhood U of x such that $|f_n(x) - f_n(y)| < \varepsilon/3$ for all n and for all $y \in U$. Fix $y \in U$, and choose N so that $|f_n(x) - f(x)| < \varepsilon/3$ and $|f_n(x) - f(x)| < \varepsilon/3$ whenever $n \geq N$. Hence

$$|f(x) - f(y)| \leq |f(x) - f_n(x)| + |f_n(x) - f_n(y)| + |f_n(y) - f(y)| < \varepsilon.$$

Chapter 10

10.1 (1) If $(I_k) \in \mathcal{I}(A)$, then $(I_k^-) \in \mathcal{I}(A)$ and $\ell(I_k^-) = \ell(I_k)$. (2) If $(I_k) \in \mathcal{I}(A)$, consider $(J_k) \in \mathcal{I}(A)$ such that $I_k \subset J_k$ and $\ell(J_k) = \ell(I_k) + \varepsilon/2^k$. (4) If $(I_k) \in \mathcal{I}(A)$, divide each I_k into disjoint subintervals of length less than δ.

10.4 Note that if $(I_k) \in \mathcal{I}(A)$, then $(I_k + t) \in I(A + t)$ and $\ell(I_k + t) = \ell(I_k)$.

10.5 (1) is straightforward. (2) If $\varphi > 0$, then $\int \varphi \geq \int \varphi \chi_A \geq \int k \chi_A = k\ell^*(A)$.

10.10 By our assumption, $\ell^*(\{x \in A : f(x \leq 0)\}) = 0$. For each n let

$$A_n = \left\{ x \in A : f(s) \geq \frac{1}{n} \right\}.$$

Then $A = \left(\bigcup_{n=1}^{\infty} A_n \right) \cup B$ and $f\chi_{A_n} \leq f\chi_A$ a.e. for each n. Thus $0 \leq \frac{1}{n}\ell^*(A_n) \leq \int f\chi_{A_n} d\ell \leq \int f\chi_A d\ell = 0$. Hence $\ell^*(A_n) = 0$ for each n and therefore $\ell^*(A) = 0$.

10.15 Suppose that $\int f_n d\ell \to 0$, and $f_n \to f$ a.e. Then $\int f d\ell = 0$ and thus $f \sim 0$.

10.17 Use monotone convergence theorem on the sequence $g_n = \sum_{k=1}^{n} f_k$.

10.18 It is enough to establish the result for $f = \chi_I$ where I is any bounded interval, say $I = [a, b]$. The inequality $|\chi_I(x)\cos nx| \leq |\chi_I(x)|$ implies that $\chi_I(x)\cos nx \in \mathcal{L}$ for each n, and $\left| \int \chi_I(x) \cos nx d\ell(x) \right| = \left| \int_a^b \cos nx dx \right| = \frac{1}{n}|\sin nb - \sin na| \leq \frac{2}{n} \to 0$.

10.20 (1) is straightforward. (2) It is enough to establish the result for $f = \chi_{[a,b]}$ where $[a, b] \subset [0, 1]$. Clearly, we have $\int_{[0,1]} f(x) r_n(x) d\ell(x) = \int_a^b r_n(x) dx$. Fix $\varepsilon > 0$. Choose n_0 such that $2^{-n_0} < \min\left\{ \frac{\varepsilon}{3}, \frac{b-a}{4} \right\}$. Let $P_n = \{t_0, t_1, \ldots, t_n\}$ be a partition of $[0, 1]$ where $t_i = \frac{i}{2^n}$ and $n \geq n_0$. Note that $\int_{t_{i-1}}^{t_i} r_n(x) dx = 0$, $a \in [t_{k-1}, t_k]$ and $b \in [t_{k+m-1}, t_{k+m}]$ for some k and m such that $k+m \leq n$. It follows that

$$\left| \int_a^b r_n(x) dx \right| \leq \int_a^{t_k} |r_n(x)| dx + \int_c^b |r_n(x)| dx$$
$$= (t_k - a) + (b - c) < \varepsilon,$$

where c is either t_{k+m-1} or t_{k+m-2}.

10.27 Fix $0 < a < b$. Let $u = x^2$. Then

$$\left| \int_a^b \sin\left(x^2\right) dx \right| = \left| \int_{a^2}^{b^2} \frac{\sin\left(u\right)}{2\sqrt{u}} du \right|$$

$$= \frac{1}{2} \left(\left[\frac{\cos u}{\sqrt{u}} \right]_{a^2}^{b^2} - \int_{a^2}^{b^2} \cos u d \left(\frac{1}{\sqrt{u}} \right) \right)$$

$$\leq \frac{1}{2} \left(\frac{1}{a} + \frac{1}{b} + \int_{a^2}^{b^2} d \left(\frac{1}{\sqrt{u}} \right) \right) = \frac{1}{b}.$$

This shows that $g \in \mathcal{R}\left([0,\infty)\right)$. To see that $g \notin \mathcal{L}\left([0,\infty)\right)$, show that

$$\int_0^{\sqrt{n\pi}} \left| \sin\left(x^2\right) \right| dx = \sum_{k=1}^n \int_{\sqrt{k\pi-\pi}}^{\sqrt{k\pi}} \left| \sin\left(x^2\right) \right| dx \geq \frac{1}{\sqrt{\pi}} \sum_{k=1}^n \frac{1}{\sqrt{k}}.$$

10.37 Let $r = \frac{p}{q}$ and $\frac{1}{r} + \frac{1}{s} = 1$. Then

$$\int |f|^p \, d\ell = \int |f|^p \cdot 1 d\ell \leq \left(\int |f|^{pr} \, d\ell \right)^{1/r} \left(\int 1^s d\ell \right)^{1/s}$$

$$= \left(\int |f|^{pr} \, d\ell \right)^{1/r} (b-a)^{1/s}.$$

10.38 Make use of the inequalities $1 - \frac{1}{t} \leq \ln t \leq t - 1$, for $t > 0$.

10.39 Consider $A = \{x : |f(x)| \geq 1\}$ and define $g = |f|^2 \chi_A + |f| \chi_A$. Then $g \in \mathcal{L}^1$. Let $1 \leq p \leq 2$. Then $|f|^p \leq g$. Hence $f \in \mathcal{L}^p$. Let (p_n) be any sequence with $1 \leq p_n \leq 2$ and $p_n \to 2$. Then $|f|^{p_n} \leq g$ holds for each n. Since $|f|^{p_n} \to |f|$ a.e., the Lebesgue dominated convergence theorem implies $\lim_{p \to 1+} \left(\int |f|^p \, d\ell \right)^{1/p} = \lim_{n \to \infty} \left(\int |f|^{p_n} \, d\ell \right)^{1/p_n} = \int |f| \, d\ell$.

Chapter 11

11.3 (a) $x = \pi - 2 \sum \frac{\sin nx}{n}$.

11.4 (a) $x \neq (2k+1)\pi$; (b) for all x; (c) $x \neq 2k\pi/(2k+3)$.

11.6 Put $x = lt/\pi$ and consider the 2π-periodic function $\varphi(t) = f(lt/\pi)$.

11.8 (1) $|x| = \frac{l}{2} - \frac{4l}{\pi^2} \sum_{p=0}^{\infty} \frac{1}{(2p+1)^2} \cos \frac{(2p+1)\pi}{l} x$;

(2) $e^x = \frac{e^l - e^{-l}}{2l} + l \left(e^l - e^{-l} \right) \sum_{n=1}^{\infty} \frac{(-1)^n \sin \frac{n\pi x}{l}}{l^2 + n^2 \pi^2}$.

11.11 (b) $\frac{2}{\pi} + \frac{4}{\pi} \sum_{k=1}^{\infty} \frac{\cos 2kx}{1-(2k)^2}$; (c) $\frac{1}{2} + \frac{2}{\pi} \sum_{k=1}^{\infty} \frac{\cos(2k-1)x}{(2k-1)}$; (d) $\frac{\pi^2}{6} - \sum_{k=1}^{\infty} \frac{\cos 2kx}{k^2}$.

11.13 (b) $\frac{8}{\pi} \sum_{k=1}^{\infty} \frac{n \sin 2nx}{4n^2 - 1}$; (d) $\pi - x = \pi - 2 \sum_{k=1}^{\infty} (-1)^{k+1} \frac{\sin kx}{k}$.

11.14 Note that the series $\sum_{n=1}^{\infty} (|a_n| + |b_n|)$ dominates the series

$$\frac{a_0}{2} + \sum_{n=1}^{\infty} (a_n \cos nx + b_n \sin nx).$$

Use Weierstrass' M-test.

11.17 For (b) note that $S_n (f) (x) = \dfrac{1}{\pi} \displaystyle\int_{-\pi}^{\pi} f (x+t) \dfrac{\sin \left(n + \frac{1}{2}\right) t}{2 \sin (t/2)} dt$ and hence

$$\sigma_n (f) (x) = \frac{1}{\pi n} \int_{-\pi}^{\pi} \frac{f (x+t)}{2 \sin (t/2)} \sum_{k=0}^{n-1} \sin \left(k + \frac{1}{2}\right) t\, dt.$$

11.20 Use the substitution $x = t - \pi$.

11.29 Use Parseval's identity.

11.31 (a) $\cos \alpha x = \frac{\sin \alpha \pi}{\pi} \left[\frac{1}{\alpha} + 2\alpha \sum_{n=1}^{\infty} \frac{(-1)^n \cos nx}{\alpha^2 - n^2}\right]$. (b) Setting $x = \pi$ in part (a) we have $\cot \pi \alpha = \frac{1}{\pi} \left(\frac{1}{\alpha} + 2\alpha \sum_{n=1}^{\infty} \frac{1}{\alpha^2 - n^2}\right)$.

11.32 Differentiate (resp. integrate) term-by-term the first series obtained in Exercise 11.31(2).

11.33 Replace x with x/π in the second equation of Exercise 11.32. Finally let $x = 1/2$ in the obtained equation.

11.39 Let

$$K_n (t) = \frac{1}{n} [D_0 (t) + D_1 (t) + \cdots + D_{n-1} (t)]$$

$$= \begin{cases} \frac{1}{2n} \left(\frac{\sin \frac{1}{2} nt}{\sin \frac{1}{2} t}\right)^2 & \text{if } 0 < |t| \le \pi \\ \frac{1}{2} n & \text{if } t = 0. \end{cases}$$

Since $\frac{1}{\pi} \int_{-\pi}^{\pi} D_k (t)\, dt = 1$ for every k, it follows that $\frac{1}{\pi} \int_{-\pi}^{\pi} K_n (t)\, dt = 1$. Hence $|\Gamma_n (f) - f (x)| \le \frac{1}{\pi} \int_{-\pi}^{\pi} |f (x+t) - f (x)| K_n (t)\, dt$. By the uniform continuity of f on $[-\pi, \pi]$ for every $\varepsilon > 0$, there is δ with $0 < \delta < \pi$ such that

$$\frac{1}{\pi} \int_{-\delta}^{\delta} |f (x+t) - f (x)| K_n (t)\, dt \le \frac{1}{\pi} \int_{-\delta}^{\delta} \varepsilon K_n (t)\, dt = \frac{\varepsilon}{2}.$$

On the other hand, if $M = \sup \{f (x) : x \in [-\pi, \pi]\}$, then

$$\frac{1}{\pi} \int_{\delta}^{\pi} |f (x+t) - f (x)| K_n (t)\, dt \le \frac{1}{n} \frac{\pi^2 M}{4 \delta}.$$

The same estimate holds for the interval $[-\pi, -\delta]$. Choosing n large enough so that $\frac{1}{n} \frac{\pi^2 M}{4 \delta} < \frac{\varepsilon}{2}$, we obtain $|\Gamma_n (f) - f (x)| < \varepsilon$ for all $x \in [-\pi, \pi]$. Since $\Gamma_n (f)$ is seen to be a trigonometric polynomial, we have given another proof of the Weierstrass theorem.

Bibliography

[1] C.D. Aliprantis and O. Burkinshaw, *Principles of Real Analysis*, Academic Press, London, 1990

[2] T.M. Apostol, *Mathematical Analysis*, Addison-Wesley, London (2nd edition) 1974

[3] R.G. Bartle, *The Elements of Real Analysis,* John Wiley, NewYork, 1976

[4] D.S. Bridges, *Foundations of Real Analysis*, Springer Verlag, New York, 1997

[5] A. Brown and C. Pearcy, *An Introduction to Analysis*, Springer Verlag, New York, 1994

[6] N. Piskunov, *Differential and Integral Calculus,* Vol I & II, Mir Publishers, 1974

[7] M.H. Protter and C.B. Morrey, *A First Course in Real Analysis*, Springer Verlag, New York, 1991

[8] K.A. Ross, *Elementary Analysis: The Theory of Calculus*, Springer Verlag, New York, 1980

[9] W. Rudin, *Principles of Mathematical Analysis*, McGraw-Hill, Singapore, 1976

Index